FIRE AND EMERGENCY SERVICES INSTRUCTOR
NINTH EDITION

Walter G. M. Schneider III, Ph.D., P.E., CFO, MCP, CBO
Technical Writer

Brad McLelland and Cindy Brakhage
Lead Senior Editors

Lynn Hughes
Lead Instructional Developer

Ben Brock
Graphic Designer

Validated by the International Fire Service Training Association
Published by Fire Protection Publications • Oklahoma State University

The International Fire Service Training Association (IFSTA) was established in 1934 as a *nonprofit educational association of fire fighting personnel dedicated to the advancement of fire fighting techniques and firefighter safety through training*. To carry out the IFSTA mission, Fire Protection Publications (FPP) was established at Oklahoma State University in the College of Engineering, Architecture and Technology. The primary purpose of FPP is to publish and distribute training materials as proposed, developed, and validated by IFSTA. FPP also develops state of the art teaching and learning materials to support the adoption and use of the manuals. The print products are supported by a full array of eBooks, online learning and apps.

IFSTA holds two meetings each year, one in January and one in July. During these meetings, selected committees of subject matter experts, who are acknowledged leaders in their fields, review draft materials and ensure that the content is accurate and meets the National Fire Protection Association professional qualification standards. These assemblies bring together individuals from several related and allied occupations:

- Fire department executives, training officers, and line personnel
- Educators from colleges and universities
- Representatives from governmental agencies
- Delegates of firefighter associations and industrial organizations

Validation committee members receive no fees for their work and their travel expenses are only partially funded by FPP. Members participate because of a personal commitment to the fire service and its future through training. Serving on a committee is prestigious in the fire and emergency response community. This unique process to validate the content of training textbooks creates a close relationship between IFSTA, the publications and the fire service community.

The manuals and associated instructional materials are the primary source of training and education for most fire departments and training agencies in North America, the U.S. Department of Defense, fire service related higher education programs, and other emergency response organizations worldwide. Some of the manuals are available in other languages including French, Arabic, Japanese and Chinese.

Copyright © 2019 by the Board of Regents, Oklahoma State University

All rights reserved. No part of this publication may be reproduced in any form without prior written permission from the publisher.

ISBN 978-0-87939-696-1   *Library of Congress Control Number: 2018956841*

Ninth Edition, First Printing, January 2019   *Printed in the United States of America*

10   9   8   7   6   5   4   3   2   1

> If you need additional information concerning the International Fire Service Training Association (IFSTA) or Fire Protection Publications, contact:
>
> **Customer Service**, Fire Protection Publications, Oklahoma State University
> 930 N. Willis St., Stillwater, OK. 74078-8045
> 800-654-4055   Fax: 405-744-8204   customer.service@osufpp.org
>
> For assistance with training materials, to recommend material for inclusion in an IFSTA manual, or to ask questions or comment on manual content, contact:
> **Editorial Department**, Fire Protection Publications, Oklahoma State University
> 930 N. Willis St., Stillwater, OK. 74078-8045
> 405-744-4111   Fax: 405-744-4112   E-mail: editors@osufpp.org

Oklahoma State University, in compliance with Title VI of the Civil Rights Act of 1964 and Title IX of the Educational Amendments of 1972 (Higher Education Act), does not discriminate on the basis of race, color, national origin, or sex in any of its policies, practices, or procedures. This provision includes but is not limited to admissions, employment, financial aid, and educational services.

Chapter Summary

Chapters

Section A: Instructor I
1. The Instructor as a Professional 9
2. Principles of Learning 23
3. Instructional Planning 45
4. Instructional Materials and Equipment 61
5. Learning Environment 87
6. Classroom Instruction 105
7. Student Interaction 139
8. Skills-Based Training Beyond the Classroom 159
9. Testing and Evaluation 183
10. Records, Reports, and Scheduling 203

Section B: Instructor II
11. Lesson Plan Development 213
12. Training Evolution Supervision 247
13. Test Item Construction 263
14. Supervisory and Administrative Duties 291
15. Instructor and Class Evaluations 319

Section C: Instructor III
16. Course and Curriculum Development 331
17. Training Program Evaluation 351
18. Training Program Administration 369

Appendices 397
A. Chapter and Page Correlation to NFPA 1041 Requirements 399
B. Americans with Disabilities Act (ADA) and World Wide Web Consortium (W3C) 401
C. NFPA Standards Applicable to Training 402
D. Short Answer Evaluation Rubric 403
E. Risk Management Formulas 404
F. Level I Instructor Evaluation Rubric 406

Index 409

Table of Contents

List of Tables x
List of Key Terms xi
Preface xiii
Introduction 1
Purpose and Scope 1
Book Organization 1
Terminology 2
Key Information 3

Section A. Instructor I

1 The Instructor as a Professional 9
Characteristics of Effective Instructors 11
    Desire to Teach 12
    Motivation 12
    Subject and Teaching Competencies 12
    Leadership Abilities 12
    Strong Interpersonal Skills 13
    Preparation and Organization 14
    Ingenuity, Creativity, and Flexibility 14
    Empathy 14
    Conflict-Resolution Skills 14
    Fairness 15
    Personal Integrity 15
    Honesty 15
    Sincerity 15
Instructor Responsibilities 16
    Obligations 16
    Challenges 17
Laws, Regulations and Standards Applicable to the Instructor 19
    Regulations 20
    Codes and Standards 20
Chapter Review 21
Discussion Questions 22
Key Terms 22

2 Principles of Learning 23
Foundations of Learning 25
    Sensory-Stimulus Theory 25
    Knowles' Assumptions of Adult Learners 26
    Thorndike's Laws of Learning 27
    Maslow's Hierarchy of Needs 29
Domains of Learning 29
    Cognitive (Knowledge) 30
    Psychomotor (Skills) 30
    Affective (Attitude) 30
Student Diversity 31
    Age 32
    Gender 34
    Cultural and Ethnic Background 34
    Learning Styles and Learner Characteristics 34
    Self-Regulated Learning 35
Interpersonal Communication 35
    Verbal Component 38
    Nonverbal Component 39
Listening Skills 40
Factors That Affect Learning 41
    Learning Obstacles 41
    Learning Plateaus 42
Motivation 43
Chapter Review 43
Discussion Questions 44
Key Terms 44

3 Instructional Planning 45
Planning to Teach 47
    Organization 47
    Session Preparation 48
    Session Logistics 49
Training Aid Selection 51
    Learning Objectives and Lesson Content 52
    Class Size and Interaction 52
    Learning Pace 53
    Practice Factors 53
Learning Environment Continuity 54
    Instructor Changes 54
    Weather Variations 55
    Instructional Resource Variations 56
    Differences in Learner Characteristics 56
    Differences in Knowledge Level 57
        Remedial Knowledge Level 57
        Advanced Knowledge Level 57
Course Consistency 57
    Safety Factors 57
    Training and Resource Materials 58
Chapter Review 58
Discussion Questions 59
Key Terms 59

4 Instructional Materials and Equipment 61
Purposes and Benefits of Lesson Plans 63
    Lesson Plan Components 64
    Lesson Outlines and Learning Objectives 64
        Basics of Learning Objectives and Lesson Outlines 66
        Lesson Outline Resources 68
Copyright Laws and Permissions 69
Teaching Aids 71
    Nonprojected Teaching Aids 72
        Marker Boards and Easel Pads 72

Table of Contents

    Illustration or Diagram Displays 73
    Duplicated Materials 73
    Models 74
    Recordings 75
  Projected Teaching Aids 75
    Multimedia Projectors/Large-Screen Images 76
    Interactive Display Systems 77
    Visual Presenters 78
    Video Presentations 78
    Video Conferencing 79
  Simulators 79
    Computer Simulations 79
    Virtual Reality Simulations 79
    Casualty Simulations 80
    Anatomical/Physiological Manikins 80
  Training Props 80
Cleaning, Care, and Maintenance
  of Teaching Aids 81
Benefits of Teaching Aids and Props 82
Chapter Review 83
Discussion Questions 83
Key Terms 84
Skill Sheet 85

5 Learning Environment 87
Classroom Environment 89
  Seating Arrangements 89
  Lighting 92
  Temperature and Ventilation 92
  Noise Level 93
  Audiovisual Equipment 93
  Other Learning Environment Considerations 94
    Power Outlet Access 94
    Internet, Phone, and Cable Television Access 95
    Visual Distractions 95
    Comfort Facilities 95
    Safety Hazards and Emergency Exits 95
Training Ground Environment 96
  Remote Sites 96
    Weather Conditions 99
    Site Environment Considerations 99
    Vehicle Traffic 99
    Training Ground Noise 100
    Lighting 100
    Access/Egress 100
  Permanent Training Facilities 100
Chapter Review 101
Discussion Questions 101
Key Terms 102
Skill Sheet 103

6 Classroom Instruction 105
Presentation Techniques 107
  Characteristics of Effective Speakers 107
  Presentation Planning 108
  Organizing the Presentation 109
  Methods of Sequencing 110
  Transitions 110
    Verbal Transitions 111
    Nonverbal Transitions 112
Four-Step Method of Instruction 112
  Preparation 112
  Presentation 114
  Application 114
  Evaluation 114
Instructional Methods 115
  Giving an Illustrated Lecture 115
  Providing Demonstration 116
  Leading Class Discussions 118
    Whole Group Discussions 118
    Small Group Discussions 119
    Leading Discussions 120
    Discussion Techniques 120
  Asking Effective Questions 121
    Question Types 122
    Responding to Students' Answers 123
    Answering Students' Questions 124
  Psychomotor Skill Instruction 124
Structured Exercises 125
  Case Studies 125
  Role Playing 126
  Simulations 126
  Field and Laboratory Experiences 127
Competency-Based Learning in the Fire and
  Emergency Services 127
Teaching Strategies 128
  Traditional Instructor-Led Training (ILT) 128
    Multiple Instructors 129
    Generating and Maintaining Student Interest 129
    Reinforcing Learning 130
  Blended or Hybrid Learning 131
  Student-Centered Learning 131
  Flipping the Classroom 131
  Computer Simulation 131
  Self-Directed Learning 133
  Individualized Instruction 134
Chapter Review 135
Discussion Questions 135
Key Terms 135
Skill Sheets 137

Table of Contents

7 Student Interaction 139
Individual Student Needs..141
    Diversity.. 141
    Low Literacy Levels 142
    Special Needs.. 142
    High Learning Abilities................................... 146
Students' Rights..146
Student Behavior Management............................147
    Reviewing Policies... 147
    Counseling.. 148
    Coaching.. 149
    Motivating and Encouraging Students ... 150
    Providing Peer Assistance............................. 150
    Mentoring... 151
Strategies for Dealing with Classroom Issues 151
    Nondisruptive, Nonparticipating Students 152
        Shy or Timid *153*
        Bored or Uninterested.............................. *153*
    Disruptive, Nonparticipating Students 153
        Talkative Students *153*
        Attention-Seeking Students.................. *154*
    Reasons for Disruptive Behavior 154
        Instructor-Caused Disruptive Behavior......... *154*
        Student-Caused Disruptive Behavior *155*
Taking Formal Disciplinary Action 156
Chapter Review 158
Discussion Questions 158
Chapter 7 End Notes 158
Key Terms 158

8 Skills-Based Training Beyond the Classroom 159
Resources: Safety Guidelines, Regulations, and Information...................................161
    Federal Government Agencies 161
        State/Provincial and Local Safety and Health Agencies....................................... 161
        Standards-Writing Organizations............ 163
        Professional and Accrediting Organizations......... 164
Instructor as Safety Role Model 165
Planning for Safe Training 166
    Verifying Instructor Skill Level................. 166
    Inspecting and Repairing Facilities and Props 166
    Identifying Training Hazards..................... 167
Psychomotor Skills Demonstrations 167
Evolution Control...168
Simple Training Evolutions...................................168
Increased Hazard Exposure Training 170
    Live-Fire Training.. 171
        Safety during Live-Fire Training Evolutions........... *171*
        Acquired Structures *172*
        Purpose-Built Structures......................... *176*
        Exterior and Wildland Fires *176*
    Increased Hazard Exposure Training...... 177
Legal Liability...178
    Vicarious Liability.. 179
    Foreseeability.. 179
    Liability Reduction 180
Chapter Review 181
Discussion Questions 181
Key Terms 182

9 Testing and Evaluation............................... 183
Approaches to Student Assessment 185
    Norm-Referenced Assessments 186
    Criterion-Referenced Assessments 186
Test Classifications 187
    Purpose Classification 187
        Prescriptive (Pretest).................................. *187*
        Formative (Progress) *187*
        Summative (Comprehensive)................ *187*
    Administration Classification 188
        Oral Tests.. *188*
        Written Tests *188*
        Performance Tests...................................... *189*
Test Bias 189
Test Administration 190
    Administering Written Tests 190
    Administering Performance Tests 191
Test Scoring 192
    Scoring Written Tests 192
    Scoring Oral Tests 194
    Scoring Performance Tests 194
    Grading Fire and Emergency Services Tests......... 194
    Grading Bias 194
Grade Reporting..195
Test Security 195
Evaluation Feedback..196
Chapter Review 196
Discussion Questions 197
Key Terms 197
Skill Sheets 197

10 Records, Reports, and Scheduling............... 203
Difference Between Records and Reports........205
Types of Training Records and Reports............205
    Training Records.. 206
    Training Reports... 207
Report Writing 208
Scheduling Training Sessions 209
Chapter Review 211

Table of Contents

Discussion Questions 211
Key Terms 211
Skill Sheet 212

Section B. Instructor II

11 Lesson Plan Development 213
Laws of Learning Applicable to Lesson Plan Development 215
Lesson Plan Creation 216
   Eliminating Bias in Instructional Materials 218
   Learning Objective Development 219
      Cognitive Levels of Learning 220
      Psychomotor Levels of Learning 222
      Affective Domain 223
   Conducting Basic Research 223
      Data Collection 224
      Information Sources 224
      Reference Material Citations 226
   Lesson Outline Development 227
   Instructional Method Selection 227
   Lesson Activity Development 227
      Whole Group Discussion Development 227
      Small Group Discussion Development 228
      Case Study Development 228
      Role Play Development 229
   Ancillary Components 229
      Handout 230
      Skill Sheet 230
      Worksheet 230
      Study Sheet 231
      Assignment Sheet 231
   Technology Tools 231
Lesson Plan First Use 232
Lesson Plan Evaluation and Revision 233
   Lesson Plan Evaluation 233
   Lesson Plan Revision 234
Chapter Review 234
Discussion Questions 234
Key Terms 234
Skill Sheets 235

12 Training Evolution Supervision 247
The Safety Challenge 249
   Organizational and Administrative Support 249
   Unsafe Behavior 251
   Hazard and Risk Analysis 253
Using IMS to Supervise Training 254
   Incident Management System Duties and Functions 254
   Training Plan or Incident Action Plans (IAP) 254

   Training Evolution Evaluation 255
Environmental Issues at Training Evolutions 257
   Water 257
   Atmosphere 258
   Soil 258
Accident Investigation 259
Chapter Review 259
Discussion Questions 259
Key Terms 259
Skill Sheet 261

13 Test Item Construction 263
Common Considerations for all Tests 265
   Test Formatting and Item Arrangement 267
   Test Item Level of Cognition and Difficulty 268
   Test Instructions and Time Requirements 268
   Testing Bias 269
Student Evaluation Instruments 269
   Written Tests 270
      Multiple-Choice 270
      True-False 272
      Matching 273
      Short-Answer/Completion 275
      Essay 275
      Interpretive Exercises 277
   Oral Tests 277
   Performance (Skills) Tests 277
Test Planning 281
   Determining Test Purpose and Classification 281
   Identifying Learning Objectives 282
   Constructing Appropriate Test Items 283
      Selecting Proper Level of Test Item Difficulty ... 283
      Determining the Appropriate Number of Test Items 283
      Eliminating Language and Comprehension Barriers 284
      Avoiding Giving Clues to Test Answers 284
      Ensuring Test Usability 284
      Ensuring Validity and Reliability 285
Test Scoring Method Selection 285
Chapter Review 286
Discussion Questions 286
Key Terms 287
Skill Sheets 288

14 Supervisory and Administrative Duties 291
Supervising Other Instructors 293
   Establishing and Communicating Goals and Objectives 294
   Promoting Professional Development 294
   Empowering Instructors 295

Table of Contents

    Celebrating Instructor Accomplishments................295
    Offering Recognition for Quality Performance 295
    Resolving Conflicts..295
    Maintaining Positive Examples/Attitudes 297
Scheduling Resources and Instructional Delivery 297
  Assessing Factors That Affect Scheduling 297
  Determining Scheduling Needs.............................298
  Determining Requirements....................................300
  Determining Availability 300
  Coordinating Training..300
  Creating a Schedule 301
  Publishing the Schedule..301
  Revising the Schedule 302
Recommending Budget Needs.....................................302
  Funding Needs Determination 302
  AHJ Budget Policies 304
  Sources of Funds 304
    Charitable Contributions 305
    Grants..305
  Budget Request Justification.................................306
Purchasing Process...307
  Determining Funding Sources...............................308
  Determining Purchasing Needs.............................309
  Contacting Vendors 309
  Purchase Orders..309
Managing Training Records..310
  Training Information 310
  Record Management Systems 311
  Record Auditing Procedures.................................312
  Legal Requirements for Training Records............312
    Retention Length of Records and Reports 312
    Privacy of Records and Reports 312
    Public Access to Records and Reports 313
    Open Records Act Exemptions..............................313
Chapter Review 313
Discussion Questions 314
Key Terms 314
Skill Sheets 315

15 Instructor and Class Evaluations 319
Supervisory Instructor Evaluations............................321
  Supervisory Evaluation Tools................................321
  Performance Evaluation Processes.......................322
  Course Evaluations 324
Findings from Evaluations..328
  Instructor Strengths...328
  Instructor Weaknesses..328
Chapter Review 328
Discussion Questions 328
Skill Sheets 329

Section C. Instructor III

16 Course and Curriculum Development........... 331
Four-Step Development Model 334
Identifying Training Needs..334
  Needs Analysis 335
  Job Performance Requirements 336
  Gap Analysis 337
  Cost/Benefit Analysis 338
Designing a Program or Curriculum 338
  Curriculum Design...338
    Identify a Curriculum Outcome 338
    Identify Courses for the Curriculum......................338
    Sequence Courses into a Curriculum....................338
  Course Design...339
Implementing the Curriculum 339
Evaluating the Curriculum 341
Evaluating Testing Instruments..................................342
  Test Validity and Reliability....................................342
  Test-Result Analysis 342
  Test-Item Analysis 343
Chapter Review 345
Discussion Questions 345
Key Terms 345
Skill Sheets 346

17 Training Program Evaluation...................... 351
Methodologies..353
Evaluation Types and Categories 354
Evaluation Plans..356
  Observation of Instructors 357
  Supervisor Surveys 357
  Organizational Evaluations 357
  Performance Measurement 358
  Course and Instructional Design Evaluations 359
Evaluation Results and Recommendations 360
Chapter Review 362
Discussion Questions 363
Key Terms 363
Skill Sheets 364

18 Training Program Administration................ 369
Record-Keeping Systems..371
Disclosure of Information...374
AHJ-Level Needs Analysis..374
  Committee Meetings...374
  Data Organization and Interpretation 375
Development of Training Policies, Procedures, and
  Guidelines..375

Table of Contents

    Policies 375
    Procedures and Guidelines................................... 376
    Identifying a Need for a New Policy, Procedure, or Guideline 377
    Developing a Draft Document............................. 378
    Submitting a Draft for Review 378
    Adopting a Policy, Procedure, or Guideline.......... 379
    Publishing a Document 379
    Implementing a Document's Contents 379
    Evaluating Effectiveness...................................... 379
Standards that Influence Training 380
Formulating Budget Needs 381
    Equipment 381
    Sources of Instructional Materials......................... 382
    Maintenance and Repair of a Training Center 382
Equipment Purchasing Policies 383
    Developing a Request for Proposal (RFP)............. 383
    Creating Purchasing Specifications........................ 385
    Bid Evaluation and Awarding of the Contract...... 387
Human Resource Management.............................388
    Instructional Staff Selection 388
    Instructor Qualifications 388
    Position Advertising.. 389
Personnel Evaluations 389
Chapter Review 389
Discussion Questions 390
Key Terms 390
Skill Sheets..391

Appendicies 397

Appendix A Chapter and Page Correlation to NFPA 1041 Requirements......................................399
Appendix B Americans with Disabilities Act (ADA) and World Wide Web Consortium (W3C) 401
Appendix C NFPA Standards Applicable to Training..402
Appendix D Short Answer Evaluation Rubric 403
Appendix E Risk Management Formulas 404
Appendix F Level I Instructor Evaluation Rubric 406

Index 409

List of Tables

Table 2.1	Obsolete Terms and Their Replacements	35
Table 2.2	Areas of Student Frustration	42
Table 4.1	Common Action Verbs for Learning Objectives	66
Table 6.1	Skills Demonstration	117
Table 7.1	Categories of Disability under IDEA Law	144-145
Table 7.2	Unfavorable Student Reactions to Instructor Actions	155
Table 10.1	Report Writing Responsibilities	208
Table 11.1	Useful Words for Expressing Objectives	222
Table 11.2	Possible Action Verbs for Psychomotor Objectives	223
Table 11.3	Instructional Method Characteristics	227
Table 17.1	Course Performance Ratings Examples	361

List of Key Terms

A
Academic Misconduct ... 197
Accreditation ... 182
Acquired Structure ... 182
Administration Classification ... 197
Agenda-Based Process ... 135
Alternatives ... 287
Ancillary Components ... 234
Andragogy ... 44
Associative Phase ... 135
Authority Having Jurisdiction (AHJ) ... 22
Autonomous Phase ... 135

B
Block Grant ... 314
Blog ... 135
Burn Building ... 102

C
Capital Budget ... 314
Case Study ... 136
Coaching ... 158
Code ... 22
Code of Ethics ... 22
Cognition ... 44
Cognitive Phase ... 136
Competency-Based Learning (CBL) ... 136
Computer-Based Training (CBT) ... 136
Computer-Generated Slide Presentations ... 84
Cone of Learning ... 44
Controlled Burning ... 102
Cost/Benefit Analysis ... 345
Counseling ... 158
Course ... 345
Course Consistency ... 59
Criteria ... 197
Criterion ... 197
Criterion-Referenced Assessment ... 197
Critical Criteria ... 234
Curriculum ... 345
Customer Approval Rating (CAR) ... 363

D
Disclosure ... 390
Distance Learning ... 136
Distractors ... 287

E
Education ... 22
Evaluation ... 197

F
Fair Use ... 84
Family Educational Rights and Privacy Act (FERPA) ... 390
File Sharing ... 136
Flammable/Combustible Liquid Pit ... 102
Foreseeability ... 182
Formative Evaluation ... 363
Formative Test ... 197
Four-Step Method of Instruction ... 136

G
Gap Analysis ... 345
Goals-Based Evaluation ... 363
Grant ... 314
Guideline ... 390

H
Harassment ... 22
Hazard and Risk Analysis ... 259

I
Incident Action Plan (IAP) ... 259
Individualized Instruction ... 136
Instructional Materials ... 84
Instructor ... 22
Invasion of Privacy ... 84

J
Job Performance Requirement (JPR) ... 84

K
Key Point ... 234

L
Learner Characteristics ... 44
Learning Disability ... 22
Learning Domain ... 44
Learning Environment Continuity ... 59
Learning Management System (LMS) ... 136
Learning Objective ... 84
Learning Outcome ... 84
Learning Plateau ... 44
Learning Style ... 44
Legal Precedent ... 22
Liability ... 182
Live-Fire Exercises ... 102
Logistics ... 59

M
Mastery Learning ... 136
Mixed Methods Research ... 363
Mnemonic Device ... 44
Model Release ... 84
Moulage Kit ... 84

N
Needs Analysis ... 345
Negligence ... 22
Nominal Group Process ... 136
Norm-Referenced Assessment ... 197
Normalization of Deviance ... 260

List of Key Terms

O
Occupational Safety and
Health Administration (OSHA) 22
Operational Budget 314
Operational Step 234
Ordinance 22
Outcomes-Based Evaluation 363

P
Pilot Course 345
Policy 390
Postincident Analysis 260
Prescriptive Test 197
Procedure 390
Process-Based Evaluation 363
Program 345
Public Domain 84
Purpose-Built Structure 182
Purpose Classification 197

Q
Qualitative Methodology 363
Quantitative Methodology 363

R
Reasonable Accommodation 22
Records 211
Regulations 22
Reliability 345
Reports 211
Request for Proposal (RFP) 390
Resource Analysis 390

Resource Management 390
Restrictive Bid (Sole-Sourced) 390
Risk-Management Plan 260

S
Scoring Rubric 197
Self-Directed Learning 136
Sensory Memory 44
Sexual Harassment 22
Smokehouse 102
Social Networking 136
Standard 22
Stem 287
Summative Evaluation 363
Summative Test 197

T
Teleconferencing 136
Test Item 287
Testing Instrument 287
Training 22
Training Aids 59

V
Validity 345
Vehicle Driving Course 102
Vicarious Liability 182

W
Web Conferencing 136
Wiki 136
Wildland Fire 182
Wildland/Urban Interface 182

Preface

Acknowledgments

This ninth edition of **Fire and Emergency Services Instructor** is designed to meet the requirements of NFPA 1041, *Standard for Fire Service Instructor Professional Qualifications (2019)*.

Acknowledgement and special thanks are extended to the members of the IFSTA validating committee who contributed their time, wisdom, and knowledge to the development of this manual.

IFSTA Fire and Emergency Services Instructor, Ninth Edition Validation Committee

Chair
Gary Wilson
Training Chief
Overland Park Fire Department
Overland Park, KS

Vice-Chair
Randal E. Novak
Bureau Chief, Retired
Iowa Fire Service Training Bureau
Ames, IA

Secretary
Lynelle Vetsch
Senior Instructor
Nova Scotia Firefighters School
Waverly, Nova Scotia, Canada

Committee Members

Shane Alexander
Battalion Chief
Ocala Fire Rescue
Ocala, FL

Shawn Bayouth
Department Chair
Arkansas State University
Jonesboro, AR

Bryn Crandell
Instructional Systems Specialist
Department of Defense Fire Academy
Goodfellow AFB, TX

Dr. Jason Decremer, Ph.D
Director of Certification
Connecticut Commission on
Fire Prevention and Control
Windsor Locks, CT

Rodney Foster
Chief of Training
Midwest City Fire Department
Midwest City, OK

Joseph Guarnera
Assistant Chief
Berlin Fire-EMS
Berlin, MA

Donald E. Hansen, Ph.D
Branch Chief, Retired
Virginia Department of Fire Programs
Glen Allen, VA

Brett Lacey
Fire Marshal
Colorado Springs Fire Department
Colorado Springs, CO

Michael Luna
Lieutenant
McAllen Fire Department
McAllen, TX

Richard Merrell
Captain
Fairfax County Fire & Rescue Department
Dumfries, VA

Committee Members (continued)

Richard K. Murray, JD/Ph.D
Chief
Wilderness Ranch Fire Protection District
Boise, ID

Demond Simmons
Battalion Chief
Oakland Fire Department &
International Association of
Black Professional Firefighters
Oakland, CA

Mark Wilson
Program Director
The Broward Fire Academy
Davie, FL

Lynn Wojcik
Department Head of Public Safety
Oklahoma State University-OKC
Yukon, OK

Sincere appreciation is given to the following individuals and organizations for contributing information, text, photos, photo shoots, or other contributions instrumental in the development of this manual:

Joe Elam, Chief of Training/Assistant Chief, Edmond, OK, and the Edmond Fire Department

Brett Noakes, Fire Protection Engineer, Los Alamos National Laboratory, Los Alamos, NM

Chris Mickal, District Chief, New Orleans (LA) FD Photo Unit

Rod Smith, Assistant Chief, Lane County Fire District No. 1, Veneta, OR

Richard Valenta, eLearning Program Director, University of Illinois Fire Service Institute

California State Fire Training and the Oakland Fire Department

Fire and Rescue Training Institute, University of Missouri

Iowa Fire Service Training Bureau

Midwest City Fire Department, Midwest City, OK

Oklahoma State Firefighters Association (OSFA) State Fire School

Owasso Fire Department, Owasso, OK

Overland Park Fire Department, Overland Park, KS

National Fire Protection Association

National Institute for Literacy

Spokane Valley Fire Department, Spokane, WA

Stillwater Fire Department, Stillwater, OK

Tulsa Regional Fire Academy and the Tulsa Fire Department Training Center

U.S. Air Force

Special thanks go to the directors, instructors, and staff of Oklahoma State University Fire Service Training in Stillwater, OK. Their guidance on technical information and dedication to fire and emergency services training go above and beyond.

Last, but certainly not least, gratitude is extended to the following members of the Fire Protection Publications staff whose contributions made the final publication of this manual possible.

Fire and Emergency Services Instructor, 9th Edition Project Team

Lead Senior Editors
Brad McLelland, Senior Editor
Cynthia Brakhage, Senior Editor

Lead Instructional Developer
Lynn Hughes, Instructional Developer

Contract Writer
Walter G. M. Schneider III, Ph.D., P.E., CFO,
MCP, CBO
　Agency Director
　Centre Region Code Administration
　State College, PA
　Chief of Department
　Bellefonte Fire Department
　Bellefonte, PA

Director of Fire Protection Publications
Craig Hannan

Associate Director of Fire Protection Publications
Mike Wieder

Managing Editor
Colby Cagle

Editorial Manager
Clint Clausing

Curriculum Managers
Colby Cagle
Leslie Miller

Production Coordinators
Missy Hannan
Ann Moffat

eProducts Manager
Justin Smola

Editorial Staff
Beckie Bigler, Senior Editor
Kimberly Edwards, Senior Editor
Jeff Forney, Senior Editor
Mike Fox, Senior Editor
Tara Gladden, Editorial Assistant
Leslie Miller, Senior Editor
Lynne Murnane, Senior Editor
Libby Snyder, Senior Editor

Illustrators and Layout Designers
Ben Brock, Senior Graphic Designer
Missy Hannan, Production Coordinator

Curriculum Staff
Colby Cagle, Curriculum Manager/Managing Editor
Lindsey Dugan, Instructional Developer
Beth Ann Fulgenzi, Instructional Developer
Matthew Keith, Instructional Developer
Angel Musik, Instructional Developer
Simone Rowe, Instructional Developer
Reinaldo Sanchez, Instructional Developer
David Schaap, Instructional Developer

IFSTA/FPP Photographers
Colby Cagle
Clint Clausing
Kim Edwards
Jeff Fortney
Brad McLelland
Leslie Miller
Angel Musik
Brett Noakes
Clint Parker
Veronica Smith
Fred Stowell
Mike Sturzenbecker
Mike Wieder

eProducts Staff
Janice French
Kelly Naas
Ryan Souders

Student Assistants
Aimee Cagle, Curriculum
Samantha Sanchez, Curriculum

Indexer
Nancy Kopper

The IFSTA Executive Board at the time of validation of **Fire and Emergency Services Instructor, 9th Edition**, was as follows:

IFSTA Executive Board

Chair
Bradd Clark
Fire Chief
Owasso Fire Department, Retired
Program Manager/Instructor
Florida State Fire College
Ocala, FL

Vice-Chair
Mary Cameli
Fire Chief
City of Mesa Fire Department
Mesa, AZ

Executive Director
Mike Wieder
Associate Director
Fire Protection Publications
Oklahoma State University
Stillwater, OK

Board Members

Steve Ashbrock
Fire Chief
Madeira & Indian Hill Fire Department
Cincinnati, OH

Steve Austin
Project Manager
Cumberland Valley Volunteer Firemen's Association
Newark, DE

Dr. Larry Collins
Associate Dean
Eastern Kentucky University
Safety, Security, & Emergency Department
Richmond, KY

Dennis Compton
Fire Chief
Mesa & Phoenix Fire Departments, Retired
Chairman, Board of Directors
National Fallen Firefighters
Mesa, AZ

Elizabeth Hendel
Deputy Chief
Phoenix Fire Department
Phoenix, AZ

John Hoglund
Director Emeritus
Maryland Fire & Rescue Institute
New Carrollton, MD

Tonya Hoover
Superintendent
National Fire Academy
Emmitsburg, MD

Tom Jenkins
Fire Chief
Rogers Fire Department
Rogers, AR

Dr. Scott Kerwood
Fire Chief
Hutto Fire Rescue
Hutto, TX

Wes Kitchel
Assistant Chief, Retired
Sonoma County Fire & Emergency Services Department
Santa Rosa, CA

Brett Lacey
Fire Marshal
Colorado Springs Fire Department
Colorado Springs, CO

Board Members (continued)

Dr. Lori Moore-Merrell
Assistant to the General President
International Association of Fire Fighters
Washington, DC

Jeff Morrissette
State Fire Administrator
State of Connecticut
Commission on Fire Prevention and Control
Windsor Locks, CT

Dan Ripley
Fire Captain
Lincoln Fire and Rescue
Lincoln, NE

Josh Stefancic
Fire Chief
Safety Harbor Fire Department
Safety Harbor, FL

DEDICATION

This manual is dedicated to the men and women who hold devotion to duty above personal risk, who count on sincerity of service above personal comfort and convenience, who strive unceasingly to find better and safer ways of protecting people, homes and property from the ravages of fire, medical emergencies, and other disasters...

THE FIREFIGHTERS OF ALL NATIONS

Introduction

Fire and emergency services instructors train personnel on the skills they need to perform their jobs safely and effectively. Training subjects include basic skills taught to entry-level personnel as well as specialist-level and in-service training for current personnel. The authority having jurisdiction (AHJ), which may include local, state/provincial, or national legislation or guidelines, usually mandates the level and type of skills-based training that is required. In addition, instructors today provide both vocational skills training and adult education for their organizations.

NFPA 1041, *Standard for Fire Service Instructor Professional Qualifications* (2019), establishes widely accepted job performance requirements (JPRs) for fire and emergency instructors. NFPA 1041 also supports the company-level training requirements of Fire Officer Level I in NFPA 1021, *Standard for Fire Officer Professional Qualifications* (2014). The learning objectives included in this manual are written to reinforce NFPA 1041. The JPRs for NFPA 1041 are correlated to the text and are presented in **Appendix A**.

Purpose and Scope

This ninth edition of **Fire and Emergency Services Instructor** is written for personnel intending to satisfy Instructor Levels I, II, and III professional qualifications. The **Purpose** is to provide personnel with the basic information necessary to meet the job performance requirements (JPRs) of NFPA 1041, *Standard for Fire Service Instructor Professional Qualifications*, 2019 Edition for Instructor Levels I, II, and III. In addition, company officer candidates who wish to meet the Fire Officer Level I JPRs of NFPA 1021, *Standard for Fire Officer Professional Qualifications* must also certify to Level I Instructor requirements in NFPA 1041.

The **Scope** of this manual is to provide current fire and emergency services instructors and instructor candidates with basic instructional knowledge. This knowledge is necessary to develop skills for preparing and presenting training for personnel of fire and emergency services organizations using a variety of instructional methods.

Book Organization

This book is organized according to the certification levels presented in NFPA 1041. Chapters 1-10 present information for Instructor Level I. Chapters 11-15 present information for Instructor Level II. Finally, Chapters 16-18 present information for Instructor Level III. Each level has distinct duties and responsibilities as defined by NFPA 1041, including the following:

- **Level I Instructor** — Delivers instruction from a prepared lesson plan. *Other responsibilities:*
  — Assembles course materials
  — Uses instructional aids and evaluation tools
  — Reviews and adapts lesson plans to meet the needs of individual students, groups, and the AHJ
  — Organizes the teaching environment to maximize the learning experience and provides a safe learning environment

- Presents a lesson from a prepared lesson plan, adjusting the presentation as required to ensure that objectives are attained

- Prepares and maintains training records in accordance with the requirements of the jurisdiction

- **Level II Instructor** — Satisfies the Instructor I professional qualifications and has the knowledge and ability to develop individual lesson plans for a specific topic, learning objectives, instructional aids, and evaluation instruments. *Other responsibilities:*

  - Manages instructional resources, including facilities, personnel, time, funds, and records
  - Schedules training sessions based on overall training requirements of the AHJ
  - Supervises and coordinates the activities of other instructors
  - Evaluates Level I instructors
  - Develops instructional materials, including the creation of new lesson plans and modification of existing lesson plans
  - Develops student, course, and instructor evaluation instruments
  - Analyzes the results of student evaluations to determine test validity

- **Level III Instructor** — Satisfies the Instructor II professional qualifications and has demonstrated the knowledge and ability to develop comprehensive training programs, curricula, and courses for use by single or multiple organizations. *Other responsibilities:*

  - Administers organizational/agency policy and procedures
  - Administers training records system
  - Selects training staff
  - Creates instructor evaluation plan
  - Conducts organization needs analyses
  - Develops training goals and implementation strategies
  - Creates or modifies programs, curricula, and courses required to fulfill the organization's training needs
  - Creates a program evaluation plan

Lastly, each Instructor level section in this text is organized around the Four-Step Method of Instruction presented in Chapter 6 as follows:

- **Preparation** — Planning instruction
- **Presentation** — Providing lecture, demonstrating psychomotor skills, imparting knowledge
- **Application** — Working with students to develop and practice skills
- **Evaluation** — Testing students, evaluating instructors, courses, and programs

NOTE: With the exception of the *application* step in the Instructor III section, the chapters have been arranged to move the reader through these concepts in order.

Terminology

This manual is written with a global, international audience in mind. For this reason, it often uses general descriptive language in place of regional- or agency-specific terminology (often referred to as *jargon*). Additionally, in order to keep sentences uncluttered and easy to read, the word *state* is used to represent both state and provincial level governments (or their equivalent). This usage is applied to this manual for the purposes of brevity and is not intended to address or show preference for only one nation's method of identifying regional governments within its borders.

Key Information

Various types of information in this book are given in shaded boxes marked by symbols or icons. See the following definitions:

Case Study
A case study analyzes an event. It can describe its development, action taken, investigation results, and lessons learned.

Safety Alert
Safety alert boxes are used to highlight information that is important for safety reasons. (In the text, the title of safety alerts will change to reflect the content.)

Information
Information boxes give facts that are complete in themselves but belong with the text discussion. It is information that needs more emphasis or separation. (In the text, the title of information boxes will change to reflect the content.)

Review Box
Review boxes summarize information from lower certification levels that should be revisited when studying for higher certification levels. Review boxes are primarily used when information in NFPA standards is repeated from one certification level to the next.

What This Means to You
These boxes take information presented in the text and synthesize it into an example of how the information is relevant to (or will be applied by) the intended audience, essentially answering the question, "What does this mean to you?"

A **key term** is designed to emphasize key concepts, technical terms, or ideas that Level I, II, and III Instructors need to know. They are highlighted in the text. Definitions are listed in alphabetical order at the end of the chapter. An example of a key term is:

Learning Domain — Distinct sphere or area of knowledge, such as the affective, cognitive, and psychomotor domains.

Three key signal words are found in the book: **WARNING, CAUTION,** and **NOTE.** Definitions and examples of each are as follows:

- **WARNING** indicates information that could result in death, harm, or serious injury to Level I, II, and III Instructors. See the following example:

WARNING: An instructor can be held liable for injuries and fatalities on a fire ground.

- **CAUTION** indicates important information or data that Level I, II, and III Instructors need to be aware of in order to perform their duties safely. See the following example:

CAUTION: Altering, deleting, or editing course evaluations in any way is unethical and may be a violation of the law.

- **NOTE** indicates important operational information that helps explain why a particular recommendation is given or describes optional methods for certain procedures. See the following example:

  NOTE: Instructors who regularly plan small group discussions should coach students in the leadership skills they will need to serve as group facilitators.

Chapter End Notes refer the reader to sources referenced within the chapter. Not all chapters have Chapter End Notes as they are only included when appropriate to do so.

Taymans, J. M., National Institute for Literacy, "Learning to Achieve: A Professional's Guide to Educating Adults with Learning Disabilities," Washington, DC 20006.

To find curriculum or study materials associated with this manual and its contents, please go to ifsta.org and use the search tool in the shop to find accompanying products. You can also search for IFSTA apps using your smartphone or tablet device.

Metric Conversions

Throughout this manual, U.S. units of measure are converted to metric units for the convenience of our international readers. Be advised that we use the Canadian metric system. It is very similar to the Standard International system, but may have some variation.

We adhere to the following guidelines for metric conversions in this manual:

- Metric conversions are approximated unless the number is used in mathematical equations.
- Centimeters are not used because they are not part of the Canadian metric standard.
- Exact conversions are used when an exact number is necessary such as in construction measurements or hydraulic calculations.
- Set values such as hose diameter, ladder length, and nozzle size use their Canadian counterpart naming conventions and are not mathematically calculated. For example, 1 1/2 inch hose is referred to as 38 mm hose.

The following two tables provide detailed information on IFSTA's conversion conventions. The first table includes examples of our conversion factors for a number of measurements used in the fire service. The second shows examples of exact conversions beside the approximated measurements you will see in this manual.

U.S. to Canadian Measurement Conversion

Measurements	Customary (U.S.)	Metric (Canada)	Conversion Factor
Length/Distance	Inch (in) Foot (ft) [3 or less feet] Foot (ft) [3 or more feet] Mile (mi)	Millimeter (mm) Millimeter (mm) Meter (m) Kilometer (km)	1 in = 25 mm 1 ft = 300 mm 1 ft = 0.3 m 1 mi = 1.6 km
Area	Square Foot (ft^2) Square Mile (mi^2)	Square Meter (m^2) Square Kilometer (km^2)	1 ft^2 = 0.09 m^2 1 mi^2 = 2.6 km^2
Mass/Weight	Dry Ounce (oz) Pound (lb) Ton (T)	gram Kilogram (kg) Ton (T)	1 oz = 28 g 1 lb = 0.5 kg 1 T = 0.9 T
Volume	Cubic Foot (ft^3) Fluid Ounce (fl oz) Quart (qt) Gallon (gal)	Cubic Meter (m^3) Milliliter (mL) Liter (L) Liter (L)	1 ft^3 = 0.03 m^3 1 fl oz = 30 mL 1 qt = 1 L 1 gal = 4 L
Flow	Gallons per Minute (gpm) Cubic Foot per Minute (ft^3/min)	Liters per Minute (L/min) Cubic Meter per Minute (m^3/min)	1 gpm = 4 L/min 1 ft^3/min = 0.03 m^3/min
Flow per Area	Gallons per Minute per Square Foot (gpm/ft^2)	Liters per Square Meters Minute (L/[m^2.min])	1 gpm/ft^2 = 40 L/(m^2.min)
Pressure	Pounds per Square Inch (psi) Pounds per Square Foot (psf) Inches of Mercury (in Hg)	Kilopascal (kPa) Kilopascal (kPa) Kilopascal (kPa)	1 psi = 7 kPa 1 psf = .05 kPa 1 in Hg = 3.4 kPa
Speed/Velocity	Miles per Hour (mph) Feet per Second (ft/sec)	Kilometers per Hour (km/h) Meter per Second (m/s)	1 mph = 1.6 km/h 1 ft/sec = 0.3 m/s
Heat	British Thermal Unit (Btu)	Kilojoule (kJ)	1 Btu = 1 kJ
Heat Flow	British Thermal Unit per Minute (BTU/min)	watt (W)	1 Btu/min = 18 W
Density	Pound per Cubic Foot (lb/ft^3)	Kilogram per Cubic Meter (kg/m^3)	1 lb/ft^3 = 16 kg/m^3
Force	Pound-Force (lbf)	Newton (N)	1 lbf = 0.5 N
Torque	Pound-Force Foot (lbf ft)	Newton Meter (N.m)	1 lbf ft = 1.4 N.m
Dynamic Viscosity	Pound per Foot-Second (lb/ft.s)	Pascal Second (Pa.s)	1 lb/ft.s = 1.5 Pa.s
Surface Tension	Pound per Foot (lb/ft)	Newton per Meter (N/m)	1 lb/ft = 15 N/m

Conversion and Approximation Examples

Measurement	U.S. Unit	Conversion Factor	Exact S.I. Unit	Rounded S.I. Unit
Length/Distance	10 in	1 in = 25 mm	250 mm	250 mm
	25 in	1 in = 25 mm	625 mm	625 mm
	2 ft	1 in = 25 mm	600 mm	600 mm
	17 ft	1 ft = 0.3 m	5.1 m	5 m
	3 mi	1 mi = 1.6 km	4.8 km	5 km
	10 mi	1 mi = 1.6 km	16 km	16 km
Area	36 ft^2	1 ft^2 = 0.09 m^2	3.24 m^2	3 m^2
	300 ft^2	1 ft^2 = 0.09 m^2	27 m^2	30 m^2
	5 mi^2	1 mi^2 = 2.6 km^2	13 km^2	13 km^2
	14 mi^2	1 mi^2 = 2.6 km^2	36.4 km^2	35 km^2
Mass/Weight	16 oz	1 oz = 28 g	448 g	450 g
	20 oz	1 oz = 28 g	560 g	560 g
	3.75 lb	1 lb = 0.5 kg	1.875 kg	2 kg
	2,000 lb	1 lb = 0.5 kg	1 000 kg	1 000 kg
	1 T	1 T = 0.9 T	900 kg	900 kg
	2.5 T	1 T = 0.9 T	2.25 T	2 T
Volume	55 ft^3	1 ft^3 = 0.03 m^3	1.65 m^3	1.5 m^3
	2,000 ft^3	1 ft^3 = 0.03 m^3	60 m^3	60 m^3
	8 fl oz	1 fl oz = 30 mL	240 mL	240 mL
	20 fl oz	1 fl oz = 30 mL	600 mL	600 mL
	10 qt	1 qt = 1 L	10 L	10 L
	22 gal	1 gal = 4 L	88 L	90 L
	500 gal	1 gal = 4 L	2 000 L	2 000 L
Flow	100 gpm	1 gpm = 4 L/min	400 L/min	400 L/min
	500 gpm	1 gpm = 4 L/min	2 000 L/min	2 000 L/min
	16 ft^3/min	1 ft^3/min = 0.03 m^3/min	0.48 m^3/min	0.5 m^3/min
	200 ft^3/min	1 ft^3/min = 0.03 m^3/min	6 m^3/min	6 m^3/min
Flow per Area	50 gpm/ft^2	1 gpm/ft^2 = 40 L/(m^2.min)	2 000 L/(m^2.min)	2 000 L/(m^2.min)
	326 gpm/ft^2	1 gpm/ft^2 = 40 L/(m^2.min)	13 040 L/(m^2.min)	13 000 L/(m^2.min)

Continued on next page

Conversion and Approximation Examples (Cont.)

Measurement	U.S. Unit	Conversion Factor	Exact S.I. Unit	Rounded S.I. Unit
Pressure	100 psi	1 psi = 7 kPa	700 kPa	700 kPa
	175 psi	1 psi = 7 kPa	1225 kPa	1 200 kPa
	526 psf	1 psf = 0.05 kPa	26.3 kPa	25 kPa
	12,000 psf	1 psf = 0.05 kPa	600 kPa	600 kPa
	5 psi in Hg	1 psi = 3.4 kPa	17 kPa	17 kPa
	20 psi in Hg	1 psi = 3.4 kPa	68 kPa	70 kPa
Speed/Velocity	20 mph	1 mph = 1.6 km/h	32 km/h	30 km/h
	35 mph	1 mph = 1.6 km/h	56 km/h	55 km/h
	10 ft/sec	1 ft/sec = 0.3 m/s	3 m/s	3 m/s
	50 ft/sec	1 ft/sec = 0.3 m/s	15 m/s	15 m/s
Heat	1200 Btu	1 Btu = 1 kJ	1 200 kJ	1 200 kJ
Heat Flow	5 BTU/min	1 Btu/min = 18 W	90 W	90 W
	400 BTU/min	1 Btu/min = 18 W	7 200 W	7 200 W
Density	5 lb/ft^3	1 lb/ft^3 = 16 kg/m^3	80 kg/m^3	80 kg/m^3
	48 lb/ft^3	1 lb/ft^3 = 16 kg/m^3	768 kg/m^3	770 kg/m^3
Force	10 lbf	1 lbf = 0.5 N	5 N	5 N
	1,500 lbf	1 lbf = 0.5 N	750 N	750 N
Torque	100	1 lbf ft = 1.4 N.m	140 N.m	140 N.m
	500	1 lbf ft = 1.4 N.m	700 N.m	700 N.m
Dynamic Viscosity	20 lb/ft.s	1 lb/ft.s = 1.5 Pa.s	30 Pa.s	30 Pa.s
	35 lb/ft.s	1 lb/ft.s = 1.5 Pa.s	52.5 Pa.s	50 Pa.s
Surface Tension	6.5 lb/ft	1 lb/ft = 15 N/m	97.5 N/m	100 N/m
	10 lb/ft	1 lb/ft = 15 N/m	150 N/m	150 N/m

Chapter 1

The Instructor as a Professional

SECTION A INSTRUCTOR I

Chapter Contents

- **Characteristics of Effective Instructors....11**
  - Desire to Teach 12
  - Motivation 12
  - Subject and Teaching Competencies...... 12
  - Leadership Abilities...... 12
  - Strong Interpersonal Skills 13
  - Preparation and Organization 14
  - Ingenuity, Creativity, and Flexibility 14
  - Empathy...... 14
  - Conflict-Resolution Skills 14
  - Fairness...... 15
  - Personal Integrity 15
  - Honesty...... 15
  - Sincerity 15
- **Instructor Responsibilities...... 16**
  - Obligations 16
  - Challenges...... 17
- **Laws, Regulations, and Standards Applicable to the Instructor 19**
  - Regulations 20
  - Codes and Standards 20
- **Chapter Review 21**
- **Discussion Questions 22**
- **Key Terms 22**

JPRs addressed in this chapter

This chapter provides information that addresses the following job performance requirements of NFPA 1041, *Standard for Fire Service Instructor Professional Qualifications*, 2019 Edition.

4.2.1
4.2.2
4.5.2

Learning Objectives

1. Describe the characteristics of an effective instructor.

2. Describe the duties and responsibilities of an Instructor I. [4.2.1, 4.2.2, 4.5.2]

3. Describe laws, regulations, and standards applicable to the fire and emergency services instructor. [4.5.2]

Chapter 1
The Instructor as a Professional

Section A. Instructor I

Training is vital to all fire and emergency services organizations. It ensures that all organization personnel conduct emergency and nonemergency operations in a safe, effective, efficient, and consistent manner. Ensuring that personnel attain the proper proficiency level requires that they receive both education and training.

Characteristics of Effective Instructors

While on the surface some may find little difference between the terms *teacher* and *instructor*, there can actually be large distinctions between the two. This manual will apply the term **instructor** to discuss the responsibilities of the individual charged with training personnel in the fire and emergency services.

Similarly, the terms **education** and **training** are often used interchangeably in fire and emergency services organizations; however, they have different meanings. Education is generally understood as the acquisition of knowledge, usually through academic means such as college or university courses. Training is primarily the transfer of knowledge regarding vocational or technical skills. Fire and emergency service training and education are usually provided by one of the following entities:

- Fire and emergency services organizations' training divisions
- State/provincial, regional, or national training academies
- Vocational/technical schools, community colleges, and universities
- Professional organizations
- Private training providers

Effective instructors possess some of the following characteristics:

- Desire to teach
- Motivation
- Subject and teaching competencies
- Leadership abilities
- Strong interpersonal skills (**Figure 1.1**)
- Preparation and organization
- Ingenuity, creativity, and flexibility
- Empathy
- Conflict-resolution skills
- Fairness
- Personal integrity
- Honesty
- Sincerity

Figure 1.1 Instructors must be able to interact well with students.

Desire to Teach

Instructors often telegraph their desire to teach as enthusiasm. Enthusiasm is contagious! Effective instructors strive to instill their vision and inspiration within the organization. When they show interest in a subject, instructors can generate enthusiasm in both students and administrators alike. The following expressions foster and maintain student interest and engagement:

- Lively and varied vocal delivery (tone, pitch, and volume)
- High energy levels **(Figure 1.2)**
- Obvious love of teaching and the subject matter

When the educational experience becomes fun and exciting, students' willingness to participate increases and learning outcomes improve. As a result, administrators are more likely to support instructor and curriculum needs.

Figure 1.2 An instructor sets an example in the classroom by exhibiting a high energy level while instructing.

Motivation

Instructors must have the motivation to achieve goals and to encourage similar motivation in their students. Instructors must show students their willingness to help each individual learn by giving each of them every opportunity to do so. Instructors can motivate others by:

- Clearly communicating what action must be performed and how it must be performed
- Showing students the importance of the presented information
- Making the knowledge and skills easy to understand and learn
- Allowing for mistakes as students practice and improve
- Encouraging students during the learning process
- Using positive reinforcement
- Using constructive criticism as necessary

Subject and Teaching Competencies

Instructors must have the background knowledge and experience to teach a subject and its skills, and they must have the ability to transfer that knowledge and experience to others. Learning is a lifelong process. Instructors must continually seek to increase their knowledge and skills in technical subject matter and educational methodology. They must be open-minded and attempt to learn and understand alternative methods and ideas. Instructors should maintain and improve their teaching skills and mastery of their subject matter through experience, study, and professional development.

Leadership Abilities

Effective instructors must be effective leaders. Evidence of good leadership includes:

- Guiding students through the requirements, knowledge, and skills of a class
- Ensuring each student's needs are met
- Providing opportunities for students to think, discuss, and develop conclusions
- Providing opportunities for students to foster their own leadership skills

Effective leaders are effective followers. Instructors set an example for their students by following the rules and policies of their hosting organization, and applying them fairly.

Instructors should study various leadership models and determine the most appropriate model and methods for their own personal strengths, the instructional situation, and the students. Effective leaders know when to exhibit characteristics of different leadership models, recognizing times when they need to give instructions that must be followed closely or when they can allow students to make their own decisions.

Instructors lead by example; therefore, instructors never ask students to do anything that they themselves have not done or would not do. Decisions must be based on accurate information and able to withstand the application and scrutiny of logic.

An effective instructor must also possess the following leadership qualities:

- **Self-confidence** — Progresses through self-examination and clear self-appraisal.
- **Trustworthiness** — Can be earned with accurate and fair performance over time and positive experiences shared among groups of people, including the instructor and his or her supervisor, and the instructor and his or her students.
- **Consistency** — Routinely makes and maintains appropriate decisions, actions, and relationships.
- **Responsibility** — Accepts consequences for both good and bad results from decisions.
- **Acceptance** — Recognizes that not all problems can be resolved to everyone's satisfaction.
- **Expertise** — Teaches only from developed skills and abilities based upon knowledge and experience.

Strong Interpersonal Skills

Instructors must have strong interpersonal skills that include clarity, sensitivity, and fairness. Clarity involves the ability to precisely and clearly explain concepts and processes through a systematic presentation of material. When students do not understand the material, the instructor must be able to restate the concept in a style that students will understand **(Figure 1.3)**.

NOTE: Standard methods for presenting information to adult learners are presented in Chapter 6, Classroom Instruction.

Instructors must also be able to apply interpersonal skills when dealing with other instructors, staff members, supervisors, and the public. These skills enable instructors to work well with other people. They must be able to develop relationships that are built on mutual respect, rapport, and confidence.

Figure 1.3 An instructor who demonstrates interest in students' performance will benefit the entire class.

Preparation and Organization

Preparation and organization are accomplished through detailed course outlines, established course objectives, defined evaluation procedures, and class preparation. Immediately before a class session, instructors prepare in the following ways:

- Practice presentations to determine whether sufficient time and materials have been allocated for the topic (**Figure 1.4**).
- Arrive early to the class session area.
- Assemble and arrange all materials, handouts, audio or visual materials, props, and equipment in the classroom or training area.
- Test equipment for proper operation, ensure that replacement parts (as needed) are available, and prepare alternative plans.
- Eliminate learning barriers such as audible and visual distractions, uncomfortable environmental temperatures, and poor lighting.

A well-organized and prepared instructor can manage time efficiently. This requirement is particularly important when it is necessary to develop lesson plans for a new course or topic.

NOTE: Development of class documents and selection of teaching aids and equipment will be discussed in Chapter 4, Instructional Materials and Equipment.

Figure 1.4 Instructors must be organized and practice their lessons in order to use class time wisely.

Ingenuity, Creativity, and Flexibility

An effective instructor understands that a teaching or learning technique suitable for one student or group of students may not be suitable for another. Instructors can demonstrate ingenuity and creativity by developing or using various training aids and supplemental materials. They can also discover and use innovative means of presenting information to meet the needs of every student. Instructors must also be flexible and able to alter the training process quickly when there are changes in the environment, props, equipment, or class size.

Empathy

Empathy is the ability to understand another person's feelings and attitudes. Instructors must exhibit sensitivity to be able to understand the students' points of view, opinions, problems, challenges, or barriers to learning and communication. Empathetic instructors have a sincere desire and personal interest to help individuals learn, without patronizing or acting superior or threatening. Having and demonstrating empathy is especially important when working with students who have learning difficulties.

Conflict-Resolution Skills

At times an instructor must act to resolve conflict between students and the organization, and between students themselves. Instructors may have to resolve disputes in the following situations:

- In the training environment
- On evaluations and tests
- With many types of personalities and responsibilities
- On a variety of other issues that may arise during the course

In these situations, an instructor listens to both sides and suggests solutions, and may have to assist both sides in formulating a solution without showing preference for a particular side of the issue **(Figure 1.5)**. An instructor must work to create mutually beneficial situations and maintain positive relationships between all parties, while ensuring compliance with legal mandates and safe operating procedures.

Fairness

Instructors must treat all students equally and provide them with the same learning opportunities. Instructors must evaluate student performance against an established objective standard and not against a subjective set of expectations. If a student perceives that an instructor is biased against him or her, or favors another student, that student or others may assume the instructor will not maintain the same rules across the classroom.

Figure 1.5 Instructors should handle disagreements between students professionally and courteously.

Instructors must be fair and impartial to all students, open-minded, and willing to hear, consider, and discuss ideas with them. In particular, instructors must be able to listen to and understand students' needs.

Personal Integrity

Personal integrity is based on the individual's values and morals. It can be stated as a personal **code of ethics** that provides the instructor with specific guidelines for action and decisions. Personal integrity must also be consistently applied to all situations and people. Students will respect instructors who consistently follow their personal ethical codes. Student respect is easier to maintain when an instructor adheres to personal integrity but difficult to gain when that integrity is compromised or questioned.

Honesty

Instructors must always be truthful and honest. Students realize that an instructor may not know all the answers. Students want and expect honesty and prefer instructors who are willing to admit that they do not know but are willing to find the answers to questions. Instructors do not need to be embarrassed when they cannot answer a question during class. Instructors should be prepared to say, "I don't know the answer to that question, but I'll find out for you," or "Does someone in the class know the answer?" Instructors should not attempt to bluff their way through a question quickly because that can cost them their credibility with students.

Sincerity

Sincerity is the personal quality of being open and truthful. Sincere attitudes and responses that show an interest in helping students to learn are important traits for instructors to possess. Students react, respond, and cooperate more positively and willingly with instructors who demonstrate a concern for them.

Most instruction is based upon the instructor's ability to communicate information, both verbally and nonverbally, to students. Communications that are not sincere, such as sarcastic remarks or offensive jokes, undermine an instructor's educational message and either distract students from their learning or put them on the defensive. The emotional reactions of the class may hinder effective communication.

Instructor Responsibilities

Fire and emergency services instructors are professionals who meet a standard that is based on a high level of personal performance. Instructors are providers of adult education and adult training. As members of the fire and emergency services profession, they must give their students the same respect that they give emergency responders at an incident.

Like their students, instructors develop their own knowledge and skill from practical experience with a variety of subjects related to fire and emergency services operations as well as keeping current on subjects through studying periodicals and journals and attending courses themselves **(Figure 1.6)**. As a result, instructors are both teacher and practitioner. This arrangement is a benefit to instructors because it increases the base knowledge that they teach from and their credibility with their students.

As both teacher and practitioner, instructors need to possess characteristics that are associated with both of these roles. They must understand that they, as fire and emergency services instructors, meet the criteria for members of a *profession*: calling or vocation that requires specialized knowledge and lengthy, intense preparation that includes (1) learning scientific, historical, or scholarly principles that apply to specific skills, processes, and methods; (2) maintaining high standards of personal achievement and conduct; and (3) committing to continued study and educational advancement — all with the prime purpose of providing a public service.

Figure 1.6 A good instructor strives to remain current on instructional topics and professional skills.

Obligations

As indicated earlier in this chapter, instructor obligations outside of preparation and delivery of classroom sessions include considerations for:

- **The student** — Effective training ensures that students will perform their duties safely and skillfully in the fire and emergency services.
- **The organization** — An instructor provides effective training that supports the mission, policies, and procedures of the organization. The training should also meet all applicable federal, state/provincial, and local regulations and codes.
- **The fire and emergency services profession** — An instructor provides an important link between the student and the fire and emergency services profession by providing a positive role model and effective leadership. Instructors are also role models for safe behavior in the fire and emergency services **(Figure 1.7)**.
- **Themselves** — An instructor has an obligation to continue professional development through the acquisition of knowledge and improvement of skills. Because the field of fire and emergency services constantly changes, instructors must always be aware of new improvements or developments.

Figure 1.7 Donning the same PPE that students wear during training is an example of how instructors can model appropriate and safe behavior.

Ultimately, the instructor's role is to provide the most efficient and safest training opportunities possible for the student. By meeting these obligations effectively, instructors ensure that the public is served by the best trained fire and emergency services personnel.

> ### Observe Other Instructors
>
> Instructors should take every opportunity to watch other instructors as they plan, develop, teach, and work with others. This observation will show some indications of which organization and teaching methods work well under what conditions. When instructors compare experiences and share knowledge with each other, they can adapt others' best practices to improve their own effectiveness.

Challenges

Some of the following responsibilities may pose additional challenges to the instructor on a day-to-day basis:

- **Familiarization with standards** — Instructors must be familiar with standards and regulations that may apply to the scheduled training. Sources for these standards and regulations may include the National Fire Protection Association (NFPA), the **Occupational Safety and Health Administration (OSHA)**, applicable EMS regulations (U.S. Department of Transportation [DOT], state/provincial, and jurisdictional).

- **Instructor priorities** — Instructors need to balance multiple responsibilities within the organization. It is critical that the instructor devote an adequate amount of time to prepare and provide quality training. Instructors must understand that they are responsible for estimating the time required for course tasks and assignments, and for making suitable schedules and arrangements to fulfill each goal (**Figure 1.8**).

Figure 1.8 Instructors may perform many additional duties, such as completing administrative tasks, developing additional curriculum, and providing demonstrations.

- **Student priorities** — Instructors should be aware of time constraints and other outside influences on students' ability to train.
- **Student diversity** — Instructors must be prepared to teach students and not discriminate on the basis of race, color, national origin, genetic information, sex, age, sexual orientation, gender identity, religion, disability, or status as a veteran, treating all individuals in a way that accommodates all without showing preference.
- **Students with disabilities** — Instructors should be prepared to make reasonable accommodation for these individuals' needs should those students seek modifications. Chapter 7, Student Interaction, will discuss students with disabilities and individual student needs in greater detail.
- **Organizational apathy** — Organizations sometimes may suffer an apathetic attitude toward training, especially if overwhelmed by strict budget restrictions or lack of additional funding. Whatever the budgetary constraints, instructors should strive to be instructional champions and gain their organization's respect by providing thorough, safe, and effective training.

- **Collaborative relationships** — Instructors must learn to collaborate with officials from other agencies and government levels along with leaders in the private sector in both business and education (**Figure 1.9**). Doing so provides additional resources to instructors and their organizations.

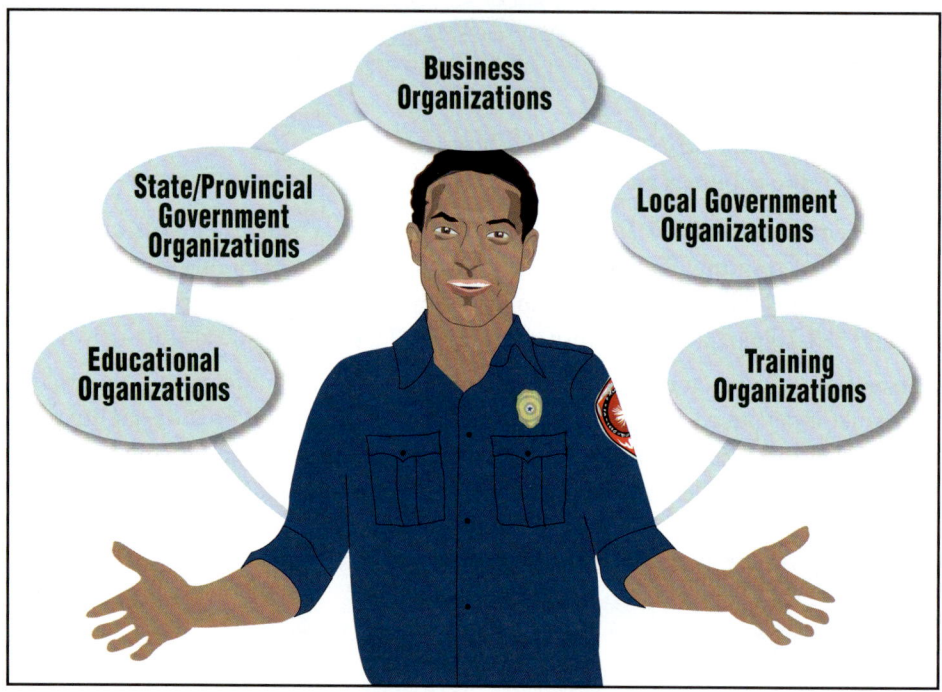

Figure 1.9 Developing relationships with other organizations will bring additional resources to the instructional process.

- **Organizational promotion** — Instructors must promote the organization's benefits, assets, and training programs to the public and other organizations and agencies that may send students as external customers.
- **Management directives** — As subordinates, instructors must adhere to the management directives and mandates of the organization and its leaders. They should also be advocates for change when directives need to be updated, and work together with superiors to affect change within organizations.
- **Knowledge of instructional environment** — Instructors should have a working knowledge of instructional environments available to them for training. If these locations are not applicable to a planned lesson or training session, the instructor must seek out an appropriate environment.
- **Safe training environments** — Instructors must be familiar with local, state/provincial, and federal safety regulations as they apply to all fire and emergency services training. Students must be assured that safety is the primary concern of the organization.
- **Course schedules** — Instructors may be asked to take a role in planning and scheduling appropriate and required training sessions. They should be prepared to seek out any training formats that help to meet necessary training requirements for their organizations.
- **Funds and resources** — Instructors are obligated to use funds and resources effectively and efficiently. When organizational funding decreases, training programs may be the first to be affected. Lack of funding is not an excuse for lack of training. Instructors must seek out training opportunities that are in keeping with available funding and resources. Remember that a proven level of quality training might ensure more success with future funding requests.
- **Documentation systems** — Instructors may find that systems used to document training and student records can pose challenges or difficulties in their day-to-day workload. It will be imperative to find and utilize the most efficient organizational methods for records and documents.

> **Ethics and the Emergency Services Instructor**
>
> Society views the field of fire and emergency services as one of the most trustworthy professions. Ways they demonstrate this trust is to provide emergency responders with keyed access to private residences and businesses, and to indicate to responders the locations of children and beloved pets. This trust is partially based on the moral and ethical obligation that emergency responders hold for the profession and to the people they serve. This moral and ethical obligation acts as a compass to remind and guide emergency responders in all that they do. This is captured in writing in documents such as the *Fire Fighter Code of Ethics* published by the National Fire Academy and the National Society of Fire Executive Officers. In addition, most emergency responders are sworn in using an oath of office that includes a pledge to uphold many of the ideals that are included in this code.

Laws, Regulations, and Standards Applicable to the Instructor

Instructors must know and understand which laws apply to the fire and emergency services and be aware of alterations in those laws and the creation of new laws. Federal, state/provincial, and local laws exist that apply to fire and emergency instruction. Information on copyright laws and infringement is included in Chapter 4, Instructional Materials and Equipment. Information on instructor liability is included in Chapter 8, Skills-Based Training Beyond the Classroom.

Some examples of federal laws that apply to instructors are as follows:

- **Title VII of the Civil Rights Act** — Prohibits employment practices that discriminate based on race, color, religion, sex, or national origin. The law also protects employees from physical, verbal, and **sexual harassment**. The Canadian equivalent of this law is the Canadian Human Rights Act.

- **Americans with Disabilities Act (ADA)** — Prohibits discrimination against persons with an identified disability. Instructors and training organizations must provide **reasonable accommodation** for students with documented disabilities (**Figure 1.10**). For example, an individual with a documented **learning disability** might be given additional time on written exams. (See **Appendix B** for the formal, legal definition of ADA.)

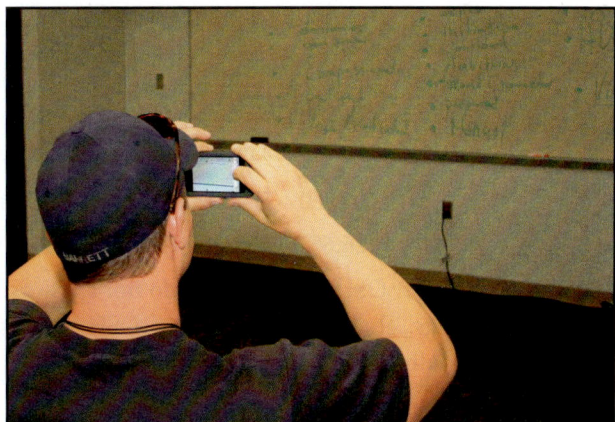

Figure 1.10 Students who require additional assistance must receive reasonable accommodations. For example, another student may take photos of notes on the board for a student who has difficulty taking notes.

- **Privacy Act** — Establishes a code of fair information practices that governs the collection, maintenance, use, and dissemination of information about individuals that is maintained in federal records systems. The act prohibits the disclosure of a record about an individual from a system of records absent the written consent of the individual, unless the disclosure is pursuant to one of twelve statutory exceptions. The act also provides individuals with a means by which to seek access to and amendment of their records, and sets forth various agency record-keeping requirements.

- **Family Educational Rights and Privacy Act (FERPA)** — Is a federal law that protects the privacy of student education reports. These provisions are applicable to all schools that receive funds under applicable programs of the U.S. Department of Education. According to FERPA, a student is defined as an individual who is enrolled in and actually attends an educational institution, including by correspondence. FERPA prohibits disclosure of a student's protected information to a third party whether it is verbally, in writing, or by electronic transmission. FERPA gives parents certain rights with respect to their children's education records. These rights transfer to the students when they reach the age of 18 or attend a school beyond the high school level. Students to whom the rights have transferred are considered "eligible students." In terms of a Canadian equivalent to FERPA, each province and territory in Canada holds its own legislation regarding freedom of information and the protection of privacy.

Instructors should seek a legal interpretation of the application of federal and state/provincial acts on their organizations. The restrictions imposed by these acts may or may not apply to fire and emergency services training divisions based on that interpretation. It is recommended that the instructor confer with his or her supervisor as to the specific requirements and policies of the organization regarding these laws and regulations.

> **Privacy as it Applies to Medical Records**
>
> In the event that a student is injured or an EMS trainee is working with patients, students and instructors may have to maintain the privacy of medical records. Instructors should be knowledgeable of laws in their jurisdictions that apply to the release or privacy of medical records. One source for this information is the Health Insurance Portability Accountability Act (HIPAA). HIPAA outlines what medical information is public knowledge and what information is not. The HIPAA privacy rule is complicated and beyond the scope of this manual; however, instructors who may be required to access or handle medical records as part of their training should become familiar with HIPAA and how it is applied in their jurisdictions. HIPAA only applies as federal law in the United States.

States/provinces also create laws that may affect the instructor. These laws could include training requirements, licensure, and they may also authorize state/provincial agencies to develop and adopt regulations. State/provincial laws vary dramatically among jurisdictions and are beyond the scope of this manual to explain in detail.

Municipal corporations, such as those formed by counties/parishes, cities, or townships, often have their own local needs and create laws called **ordinances** that address matters beyond federal or state/provincial laws. Regardless of the adoption method, an instructor must take the time to learn which laws, regulations, and standards have been adopted and apply to their job functions.

Regulations

A legislative body has the legal sanction to establish agencies that can establish rules regarding a particular activity or action. An example is OSHA, established by the U.S. Government. These rules are known as **regulations** and carry the force of law. Some examples of regulations include:

- 29 *CFR* 1910.120 – Hazardous Waste Operations and Emergency Response
- 29 *CFR* 1926.651 – Specific Excavation Requirements (trench operations)

NOTE: OSHA regulations do not apply as law to all states in the U.S., but may still be cited in criminal proceedings. That said, even in non-OSHA states, many local agencies follow OSHA regulations because they are nationally recognized safety standards.

Codes and Standards

Codes and **standards** are not laws unless adopted by the **Authority Having Jurisdiction (AHJ)**. They are, however, recognized and developed by subject matter experts. Instructors should recognize that just because the AHJ has not adopted a code or standard as law does not mean that an instructor may not be held accountable under that standard in a court of law **(Figure 1.11)**. **Legal precedent** has been established that, because these codes and standards are developed by an instructor's peers, the code or standard should be taken under consideration even when it does not rise to the level of law. The instructor should be familiar with the following standards:

- NFPA 1001, *Standard for Fire Fighter Professional Qualifications*
- NFPA 1403, *Standard on Live Fire Training Evolutions*
- NFPA 1500™, *Standard on Fire Department Occupational Safety and Health Program*

NOTE: A list of NFPA consensus standards that apply to instructors can be found in **Appendix C**.

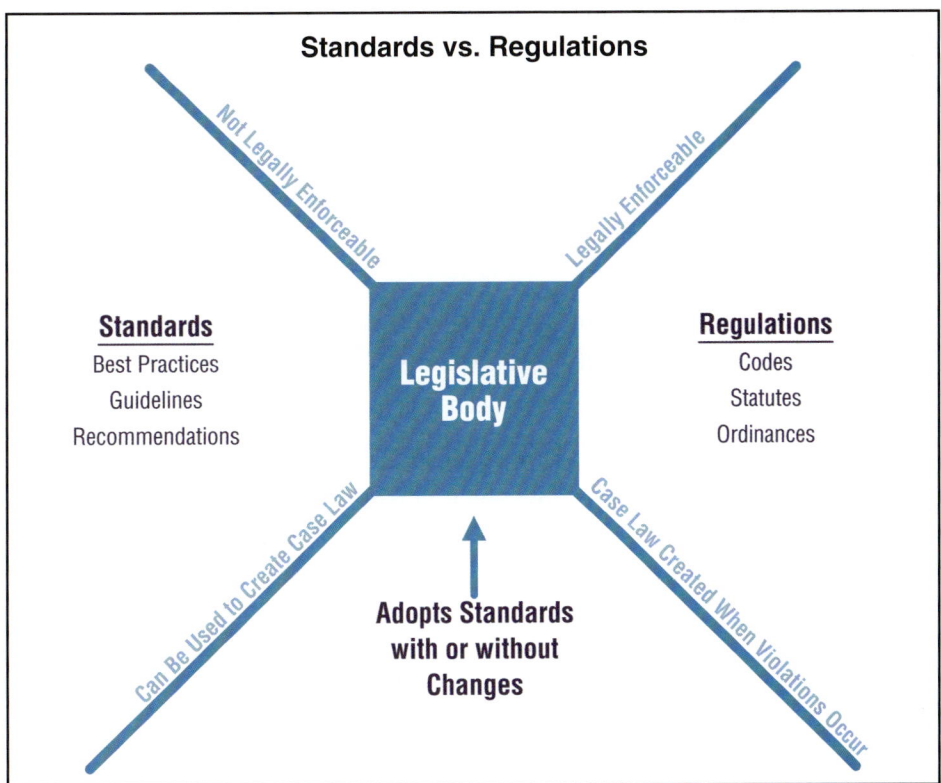

Figure 1.11 Regulations carry the force of law, while standards must be adopted by a legislative body before they are enforceable as law.

What This Means to You

As an instructor in the fire and emergency services, you will routinely interact with students. You will be partly responsible for their classroom success and fully responsible for their safety during skills training. Knowing the legal parameters for these interactions is essential to maintaining a good relationship with students while keeping them safe.

If you fail to stay current on the legal issues that apply to your position, you may find yourself in court for **negligence**, **harassment**, or civil rights violations. A good general rule is to treat everyone fairly. Fair treatment can be summed up as allowing all students access to all possible accommodations, though some accommodations may be based on preset rules, such as requiring students to ask for those modifications. In social interactions with students, the instructor should be friendly, but maintain a professional level of decorum.

If you have any concern that you may be facing a legal issue, try to resolve the issue with the accusing party and begin documenting your interactions with that person. You may find that a little understanding and communication can prevent issues from escalating to legal conflicts. Regardless of whether you believe you have achieved resolution of the issue, be sure to share all documentation and communication in a timely manner with your superior(s) and/or supervisor(s).

Chapter Review

Answer the following questions to review the information provided in this chapter.

1. What are some characteristics of an effective fire and emergency services Instructor I?
2. What are some common duties and responsibilities of an Instructor I?
3. What federal, state/provincial, and organizational/AHJ laws apply to the fire and emergency services instructor?

Discussion Questions

The following questions are intended to generate discussion, expand your understanding of the chapter text, and allow you to think critically about what you have learned. Answers to these questions may vary.

1. What are some of the characteristics of favorite instructors you have had in the past?
2. What duties and responsibilities do you believe would be the most challenging for an Instructor I?

Key Terms

Authority Having Jurisdiction (AHJ) — An organization, office, or individual responsible for enforcing the requirements of a code or standard, or approving equipment, materials, an installation, or a procedure.

Code — A collection of rules and regulations that has been enacted by law in a particular jurisdiction. Codes typically address a single subject area; examples include a mechanical, electrical, building, or fire code.

Code of Ethics — Statement of behavior that is right and proper conduct for an individual functioning within an organization or society as a whole.

Education — The acquisition of knowledge, usually through academic means such as college or university courses.

Harassment — Course of conduct directed at a specific person that causes substantial emotional distress in said person and serves no legitimate purpose.

Instructor — Individual deemed qualified by the authority having jurisdiction to deliver instruction and training in fire and emergency services; charged with the responsibility to conduct the class, direct the instructional process, teach skills, impart new information, lead discussions, and cause learning to take place.

Learning Disability — Cognitive disorder that diminishes a person's capacity to interpret what he or she sees and hears, and/or to link information from different parts of the brain.

Legal Precedent — History of rulings made in courts of law that can be referenced and used to make court decisions in future cases or influence laws outside of the court system.

Negligence — Breach of duty in which a person or organization fails to perform at the standard required by law, or that would be expected by a reasonable person under similar circumstances.

Occupational Safety and Health Administration (OSHA) — U.S. federal agency that develops and enforces standards and regulations for occupational safety in the workplace.

Ordinance — Local or municipal law that applies to persons and things of the local jurisdiction; a local agency act that has the force of a statute; different from law that is enacted by federal or state/provincial legislatures.

Reasonable Accommodation — Changes or adjustments in a work or school site, program, or job that makes it possible for an otherwise qualified employee or student with a disability to perform the duties or tasks required.

Regulations — Rules or directives of administrative agencies that have authorization to issue and enforce them.

Sexual Harassment — Superior offering advancement or special treatment in return for sexual favors from a subordinate; also may refer to any situation in which an employee, regardless of gender, believes that the workplace is a hostile environment because of sexually offensive or sexist behavior.

Standard — A set of principles, protocols, or procedures that explain how to do something or provide a set of minimum standards to be followed. Adhering to a standard is not required by law, although standards may be incorporated in codes, which are legally enforceable.

Training — The transfer of knowledge regarding vocational or technical skills.

Principles of Learning

Chapter 2

SECTION A INSTRUCTOR I

Chapter Contents

Foundations of Learning.....................25
  Sensory-Stimulus Theory 25
  Knowles' Assumptions of Adult Learners .. 26
  Thorndike's Laws of Learning 27
  Maslow's Hierarchy of Needs........... 29

Domains of Learning 29
  Cognitive (Knowledge)................ 30
  Psychomotor (Skills) 30
  Affective (Attitude) 30

Student Diversity 31
  Age 32
  Gender 34
  Cultural and Ethnic Background.......... 34

  Learning Styles and Learner Characteristics 34
  Self-Regulated Learning................ 35

Interpersonal Communication..............35
  Verbal Component 38
  Nonverbal Component 39

Listening Skills...............................40

Factors That Affect Learning 41
  Learning Obstacles.................... 41
  Learning Plateaus..................... 42

Motivation.....................................43

Chapter Review 43

Discussion Questions 44

Key Terms 44

JPRs addressed in this chapter

This chapter provides information that addresses the following job performance requirements of NFPA 1041, *Standard for Fire Service Instructor Professional Qualifications*, 2019 Edition.

4.3.2   4.4.3
4.3.3   4.4.4
4.4.2

Learning Objectives

1. Describe the foundations of learning. [4.4.3]
2. Differentiate among the domains of learning. [4.4.3]
3. Explain student diversity and how it affects instruction in the learning environment. [4.3.2, 4.4.4]
4. Summarize interpersonal communication purposes, elements, and components. [4.4.3, 4.4.4]
5. Identify aspects of the listening process. [4.4.3]
6. Describe learning obstacles and learning plateaus. [4.3.2, 4.3.3, 4.4.2, 4.4.4]
7. Discuss motivation as it relates to student success. [4.4.4]

Chapter 2
Principles of Learning

In order to become successful teachers, instructor candidates must also have an understanding of teaching methodology and the basic principles of learning. These principles and concepts are explained in this chapter and include the following:

- Foundations of learning
- Domains of learning
- Student diversity
- Interpersonal communication
- Listening skills
- Factors that affect learning
- Motivation

Instructor candidates should supplement the information in this manual with additional reading. Educational theories and methodologies continue to evolve, and new information frequently becomes available.

In addition, instructors may wish to participate in higher education courses in adult education and teaching methods. Other resources are available through the U.S. Fire Administration's National Fire Academy (NFA), state/provincial training agencies, libraries, and the Internet.

Foundations of Learning

Many educational psychologists have done extensive research on how humans learn and remember information. This section covers a few research areas that explain how instruction methods and techniques can affect or influence learning.

The instructor should be aware that the field of education and learning is not static. The foundations of learning presented here are considered the classic learning theories. If the instructor is interested, additional learning theories currently being researched and validated may be useful.

Sensory-Stimulus Theory

Sensory-stimulus theory states that there is a lifelong reliance on the five senses as the primary tool set for learning. According to this theory:

- People can only change their behavior or knowledge base by engaging the five senses.
- The sense of sight takes in the most information, with hearing next.
- People learn very little through the remaining three senses, although those senses often stimulate memories **(Figure 2.1 p. 26)**.
- Students pay more attention to sensory experiences than to mental processes or emotional involvement.

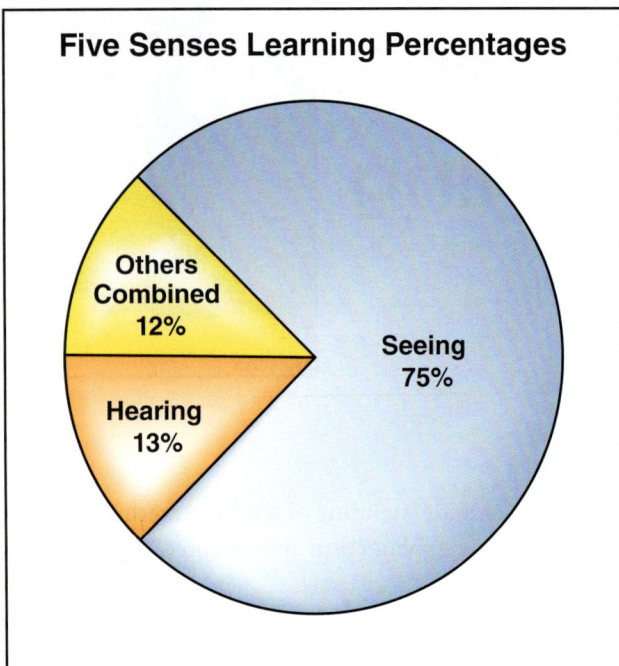

Figure 2.1 Students gather most of their information through sight, but other senses also play a part in learning.

A sensory stimulus is either important enough to remember, so commonplace it is disregarded, or unimportant enough that it is forgotten. The mental storage system for attention-getting sensory stimuli or input (such as odors, sights, sounds, and sensations) is **sensory memory**. It is difficult to attend to more than one stimulus at a time and remember it well. In order to remember information, students must give an appropriate amount of attention time to the sensory stimuli they are receiving. They should do this before they can attend to other stimuli.

In the fire and emergency services, this sensory-stimulus approach has evolved and been more accurately defined using the **Cone of Learning** (Figure 2.2). This cone illustrates the approximate amount of information retained:

- Ten percent of what is read
- Twenty percent of what is heard
- Thirty percent of what is seen
- Fifty percent of what is seen and heard together
- Seventy percent of what is said and repeated
- Ninety percent of what is said and done at the same time

Because people learn more when they actively participate, versus when they passively listen, the most effective mode of learning is the one that includes receiving or learning a new idea by a combination of methods. In other words, the blending of approaches causes individuals to engage more successfully with the learning process.

As illustrated by the Cone of Learning, performing a task while describing it results in the highest retention. On the other hand, the cone illustrates that individuals recall very little from passive methods such as reading an assignment or listening to a lecture. The sensory-stimulus theory, and the associated Cone of Learning, concludes that as more senses are used in the learning process, the more information is remembered.

Knowles' Assumptions of Adult Learners

Dr. Malcolm Knowles was among the first American theorists to use the term **andragogy** (an-druh-go-je), which refers to the art of teaching adults. It describes the characteristics of adult students and provides a set of assumptions for most effectively teaching adults. The theory of andragogy is now widely accepted and includes the following assumptions:

- **Self-concept** — Adults need to be self-directed while still relying on an instructor or training course to provide the knowledge they desire.
- **Experience** — Adults have accumulated extensive and varied quantities of experiences that serve as resources for them and

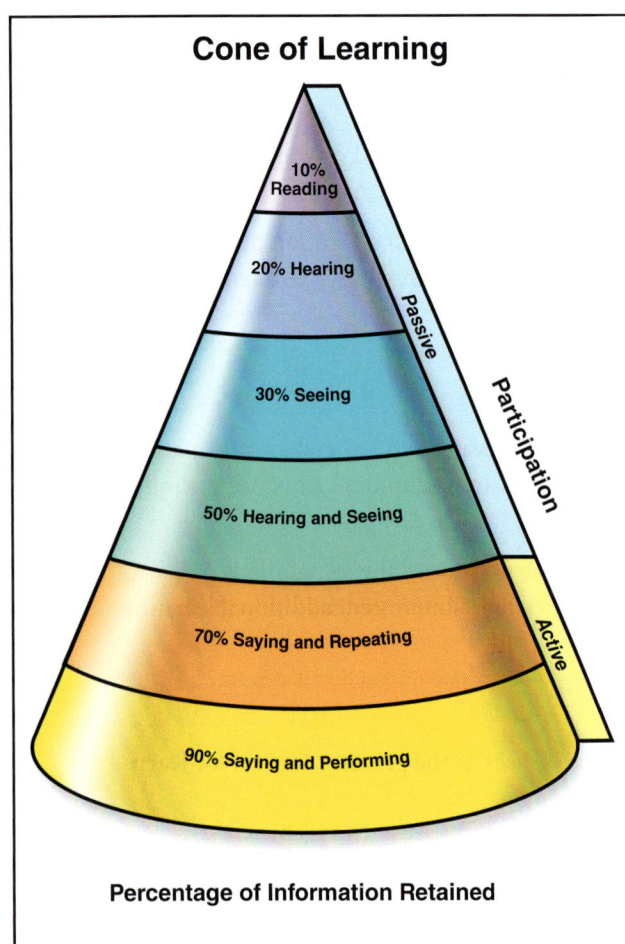

Figure 2.2 The Cone of Learning illustrates the sensory-stimulus approach to learning.

to which they can relate new information. They also have more personal experiences to contribute to the learning process than children **(Figure 2.3)**.

- **Readiness to learn** — Adults are ready to learn whatever they need to know or do in order to meet job requirements or social roles.
- **Learning orientation** — Adults' learning orientation is problem-centered because they have specific purposes for learning and want skills or knowledge that can be applied to real-life problems or situations.
- **Motivation** — Adults have internal incentives or motivators. They are motivated to learn by such factors as increased self-esteem resulting from the successful completion of the learning process, and the desire to attain a goal.

Figure 2.3 Experienced instructors have a great deal of information to share with students. Similarly, students can benefit from being grouped with other students of different experience levels.

Thorndike's Laws of Learning

The Laws of Learning, as theorized by Professor Edward L. Thorndike, suggest that there are certain laws or traits of adult learners that can be used to effectively instruct adult students. Instructors need to understand these laws to lead adult education. The laws of adult learning are:

- **Readiness** — Adult students must be prepared to learn and must place themselves in a state where they are mentally and physically able to learn new knowledge or skills. In the fire and emergency services, adult students recognize that the information they will receive is necessary and important to their success and safety. There may be barriers to readiness such as attending a class after an all-night shift; however, adult students still have to recognize the importance of learning.
- **Exercise** — Adults learn best when they are allowed to exercise skills; the more an act is practiced, the faster and surer the learning becomes **(Figure 2.4)**. Fire and emergency services instructors should, therefore, include as much time as possible for adult learners to practice concepts and skills. The amount of exercise required will vary among students.
- **Effect** — Adult learners need to see the positive effect of what they are learning. This effect could be the satisfaction of learning a new skill or mastering new information. A good instructor continually reinforces why adult learners need to master the information or skills presented.

Figure 2.4 Students should practice the skills they learn in order to gain proficiency.

- **Disuse** — Among adult learners, it can be assumed that habits and memories used repeatedly are strengthened, and habits not reinforced are weakened. One way that instructors address the disuse of previous training topics is by reviewing necessary precursor information at the beginning of a lesson. The effects of disuse form the foundation of the argument for continued training and education. Follow-up practice at the company level reinforces skills learned during instruction.
- **Association** — Adult learners tend to try to associate new information with information they have already learned. Creating association during instruction is very simple, such as expanding on prerequisite knowledge and adding another layer of information. Sometimes, information is entirely new to students. Instructors should still attempt to find associations for the information even though the connection to what students already know may take more advanced thought to discover.

- **Recency** — Skills and information practiced or learned most recently are also the best remembered. For example, adult learners may not remember the information on a certification test years after passing it if they do not use the information in the intervening years. Instructors need to recognize when skills or information have fallen into disuse so that they know when it is appropriate to review information that adult learners already know. Instructors need to include enough time to summarize the day's lesson in order for the information to be most recent in the students' minds when they leave.
- **Primacy** — Primacy is similar to recency. Primacy assumes that the first of a series of learned acts is remembered better than others. With this effect in mind, instructors should begin each lesson with a strong overview of content, including the session's learning objectives, major concepts, and the way the lesson will connect to overall training.
- **Intensity** — If a stimulus (experience) is vivid and real, it will more likely change or have an effect on the behavior (learning). For example, an instructor who involves students in a rescue demonstration and then demonstrates how to use a certain tool or extrication technique provides an experience more likely to be remembered than a lecture or video recording on how to perform the skill (**Figure 2.5**).

Figure 2.5 Hands-on experience is an effective training tool.

Certification vs. Competence

The difference between certification and competence may be a fine line, but an important one. An instructor who is certified possesses proof that he or she completed course work and passed a written and/or practical exam. In contrast, competence means that the individual has practiced and maintained mastery of a skill or knowledge. Most training situations will require instructors to provide certifications to indicate that they are competent to meet basic benchmarks. In some cases, the AHJ may encourage the instructor to seek out opportunities to maintain his or her own competencies.

Being competent to teach knowledge or skills, especially in a fire and emergency services environment, requires continually maintaining mastery over the skills taught during training. As mastery is achieved, instructors should also maintain their certifications. Although certifications may not prove competence, they do show a timeline of an instructor's willingness to take classes and stay current on documentation.

Maslow's Hierarchy of Needs

Maslow's Hierarchy of Needs presents a five-stage hierarchy to explain human motivation. Psychologist Abraham Maslow published the concept in 1943 in an academic paper titled, "A Theory of Human Motivation." The five basic levels Maslow identified are **(Figure 2.6)**:

- **Level 1 — Physiological or biological.** Need for air, water, food shelter, warmth, sleep, sex, etc. Until a person is reasonably satisfied with these needs, the focus will always be on satisfying these and will not progress.
- **Level 2 — Safety.** Need for security, order, stability, law, and freedom from fear.
- **Level 3 — Social.** Need to belong to a societal group, be accepted, be loved, and valued by others.
- **Level 4 — Esteem.** Need to achieve and master, need for self-respect, prestige, respect of others, status, and dominance.
- **Level 5 — Self-actualization.** Need to seek self-fulfillment, personal growth, and peak or culminating experiences.

Maslow postulated that all people are capable of reaching the higher levels, but life circumstances or other considerations may disrupt that progress. In a training environment, this theory explains why a student may not perform well in a classroom setting when he or she is unable to meet the lowest-level requirements.

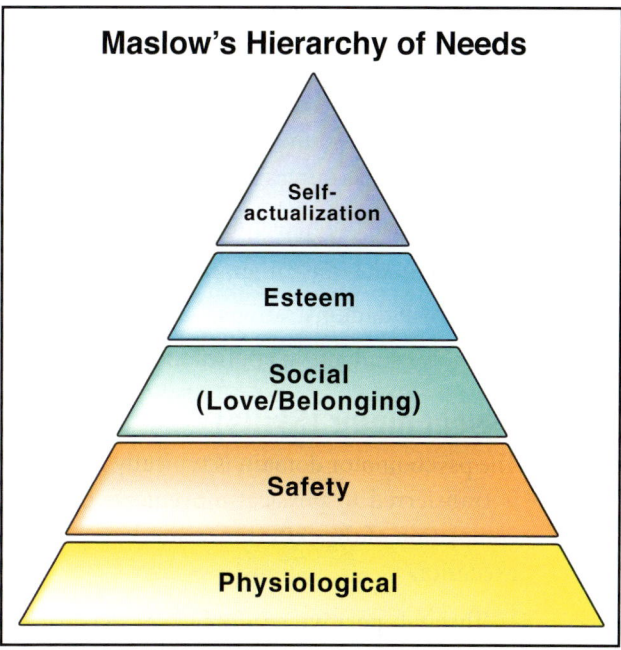

Figure 2.6 A five-stage pyramid most commonly illustrates Maslow's Hierarchy of Needs. More basic human requirements, such as the need for air and water, are represented at the bottom.

Domains of Learning

Domains of learning refer to interrelated areas in which learning occurs. When the domains are used together, a student is encouraged to understand a concept, perform a task, and alter a behavior. The three **learning domains** are:

- **Cognitive (Knowledge)** — Domain that encompasses "what" information a student should learn
- **Psychomotor (Skills)** — Domain that encompasses "how" a student should apply knowledge
- **Affective (Attitude)** — Domain that encompasses "why" the information is useful

Having an understanding of these domains and how they interact will assist the instructor in presenting effective instruction. Through the cognitive domain, students gain understanding about a concept or topic. Through the psychomotor domain, students perform the skills associated with that concept or topic. Through the affective domain, students develop a willingness to perform the behavior correctly and safely. The cognitive, psychomotor, and affective domains are the what, how, and why of the learning process **(Figure 2.7)**.

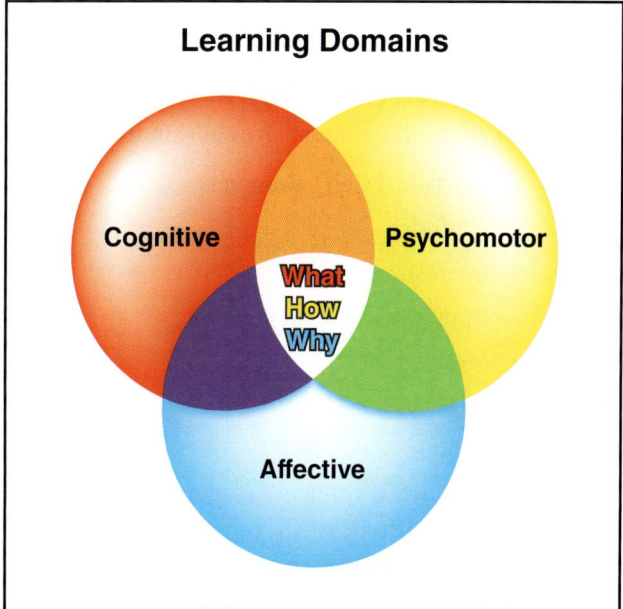

Figure 2.7 The three domains of learning interact and create the what, how, and why of the learning process.

Cognitive (Knowledge)

Cognition is a general concept that refers to all forms of knowing, including perceiving, imagining, reasoning, and judging. It is the foundation for the other two domains.

Cognitive information is usually presented in a technical or factual presentation, usually in lecture and discussion form. To describe and illustrate cognitive material and make it interesting and dynamic, instructors may use the following resources and techniques:

- Use instructional technology tools and training aids.
- Show models and other displays.
- Perform demonstrations.
- Involve students in application activities.

Psychomotor (Skills)

The psychomotor domain is typically referred to as *hands-on training*. Knowledge gained in the cognitive domain is transferred to physical movement. Learning is developed through repeated practice of the skill. Successful completion of the skill is measured in speed, precision, distance, adherence to known techniques, or sequence of execution.

Through positive reinforcement (feedback) and continued practice, students develop correct techniques and become proficient so the skill becomes a habit that is performed automatically. With practice and experience, students modify actions to fit other situations as needed.

Instructors must be aware that students learn at different rates of speed and levels of ability. Students must be comfortable in one psychomotor level before advancing to the next. Some students may want to observe longer than others before they begin to practice. Others may want more guidance and coaching before feeling confident to work on their own.

Instructors must watch for and understand student abilities at each level and provide appropriate time and opportunity for learning. Studies of vocational training have indicated that most students will master each level if they are provided enough time.

Affective (Attitude)

Affective learning involves how individuals deal with issues emotionally and includes the following traits:

- Individual (self) awareness
- Attitudes
- Interests
- Appreciation
- Motivations
- Enthusiasm
- Values

Instructors influence students' perception of the merits of a class, institution, or environment when they demonstrate attitudes toward authority, respect, responsibility, and safety (among other values). When an instructor's attitude is positive, the students will place higher value on these qualities and resources. On the other hand, negative attitudes can result in students having a low regard for these values.

Learning outcomes of the affective domain take time to achieve and are not as readily observable as the results of the cognitive and psychomotor domains. While learning new cognitive information and performing new psychomotor skills, students may alter old attitudes, values, and beliefs. For example, a trainee may begin to value safety and demonstrate this by maintaining a safe environment, wearing safety equipment, or following safety rules without being reminded. Instructors should reinforce correct affective behaviors with positive feedback and note the indicators of behavioral change **(Figure 2.8)**.

Figure 2.8 An instructor can teach students important safety concepts by modeling the appropriate behaviors, such as wearing a safety belt when driving or riding in an apparatus.

Student Diversity

Student diversity is one factor that will contribute to a learning environment. Examples of diversity found within a classroom include life experiences, motivation, time demands or responsibilities, confidence, and differences in learner characteristics.

What This Means to You

Student diversity basically means acknowledging that we are all human beings with unique backgrounds. Each of us brings something different to the learning environment. Instructors are no different. You come to the learning environment with your own experiences and biases as well. While you use your experience to help students, you have an obligation not to let your biases influence how you treat students. Always remember that all students should be treated fairly.

Above all, learn to embrace diversity in training. When you allow each individual to offer his or her particular experience to the training environment, all students benefit and the instructional environment is enhanced. The instructor should be very careful not to treat a student differently in any way based on his or her race, color, religion, sex, or national origin. A violation of this nature is very serious, no matter how innocent it may seem at the time. On a fundamental basis, it will negatively affect the learning environment for both the student directly involved and the other students in the class or program, possibly making the instructor ineffective. In addition, a violation of this nature is a violation of the student's civil rights and can be prosecuted in federal court.

Instructors have a challenge in meeting student needs and assisting diverse groups to find the most effective learning method. When an instructor can tailor a lesson plan to his or her students, the students will be more engaged. Instructors should consider the following characteristics of adult learners:

- **Life experiences and responsibilities** — Adult students possess a variety of life experiences that are gained through work, leisure, and family responsibilities **(Figure 2.9, p. 32)**. Instructors should use discussion

Figure 2.9 Adult students bring their own unique life experiences to the classroom. These experiences influence how they learn.

techniques to help students establish their own connections between their past experiences and the new materials being taught. These discussions help to involve students in the learning process and allow the instructor to determine the gap between what students already know and what they need to know to fulfill the needs of their jobs or learn new skills.

- **Motivation** — Adults take classes for a variety of reasons. Sometimes they are motivated by an internal desire to gain knowledge or skills they believe will help them to be successful. At other times, they are in class because an employer or supervisor requires them to attend. Whether the motivation comes from the student (internal) or from the employer/supervisor (external), instructors should use these motivations to the advantage of effective instruction, even when the student may perceive some aspect of the class negatively.

NOTE: Motivation will be explained further as part of student behavior management concepts introduced in Chapter 7, Student Interaction.

- **Self-confidence** — Some students enter a course very confident about their ability to learn. Other students may be ready to learn but unfamiliar with the subject matter. In still other cases, students may have been away from the school environment for years or have little confidence in their abilities to be successful based upon prior, negative learning experiences. Regardless of how a student gained his or her level of confidence, instructors should be aware of how all self-confidence levels affect students.

In addition to the diversity of students' experiences, adult learners may have different ages, gender, cultural and ethnic origins, educational backgrounds, physical capabilities, and sexual orientations. Students should be respected for their abilities, experiences, individuality, and willingness to learn and perform. Increased student diversity brings a diversity of thought to the instructional environment which in turn enhances learning. The sections that follow discuss some of the types of diversity instructors will encounter.

NOTE: Generalities based upon aspects of diversity may be inaccurate. Instructors are cautioned to avoid broad generalizations based upon any one aspect of diversity.

Age

Adults of different generations may have different experiences and skill sets that they bring to the learning environment. One generation may have a more hands-on understanding of machinery, tools, and construction. Another generation may be more inclined toward emerging technologies and computers. Instructors should take these differences into account and ask students with a particular strength to assist students who have less experience.

Generational Characteristics

The following are broadly worded generalizations about the generations that instructors will encounter in their learning environments:

Baby Boomers:
- Born between 1946 and 1964
- Are typically idealistic
- Place a high value on fairness, equality, hard work, and competition
- Have a history of questioning authority and wanting to know why something is important
- Tend to place a high value on education, family, and personal leisure time
- Tend to be workaholics and work efficiently
- Tend to be team players and desire quality
- Respond positively to achievement titles
- Want to be valued and needed

Gen-X:
- Born between 1965 and 1980
- Require personal flexibility and thrive on feedback from instructors and supervisors
- Prefer to work independently with minimal supervision
- Seek a balance between work and leisure time in their lives
- Are accustomed to change
- Self-reliance is important, more entrepreneurial
- May exhibit some qualities of the Baby Boomers (their parents) and some qualities of the Millennials (their children)
- Referred to as the "baby bust" because their population is much smaller than that of their parents

Gen-Y, Dot Comers, Millennials, Nexters:
- Born between 1980 and 1995
- Use technology as part of their daily lives
- Use online social networks as a way of creating relationships and sharing information
- Are generally optimistic
- Appreciate diversity
- Demand instant gratification in the form of tangible results for their efforts
- Take a broad worldview
- Have high expectations for educational outcomes
- Often require detailed explanations of theories and tasks
- Embrace multitasking
- Goal-oriented
- Work is a means to an end
- Embrace email and voicemail over in-person communications

Generation Z, Post-Millennials:
- Born between 1995 and 2012
- Always have known the Internet and use it at a very young age
- Use online social networks as a way of creating relationships and sharing information
- Move from PC to mobile computing resources

Figure 2.10 Instructors must allow differing genders to bring their own experiences to the classroom, but prevent gender biases from causing distractions or disruptions.

Gender

Differing genders bring their own unique experiences to the classroom and the training ground **(Figure 2.10)**. In addition, students may also bring gender biases to the learning environment which may affect their relationship with the instructor, other students' perceptions of the instructor, and the learning process. Allowing gender biases to enter a learning environment is disruptive to effective instruction. These differences may lead to class discussion but should not derail the entire topic of the class.

NOTE: For the purposes of this discussion, the term *gender* has been used to refer to the differences between men and women.

Cultural and Ethnic Background

Individuals from different cultural and ethnic backgrounds bring unique customs, behaviors, attitudes, and values to the learning environment. Instructors need to recognize and understand situations in which ethnic and cultural differences may have an effect on classroom instruction or student interaction. Because of the diverse group of individuals who participate in training courses, it is not always possible for instructors to be familiar with the customs of every culture and ethnic group. Instructors should regularly seek out and attend diversity training courses and opportunities.

> ### Cultural Concept of Words
>
> The meaning or symbolism that people place on words depends on their cultural backgrounds. Generally, the meanings of words used in North American English are based on a Eurocentric culture (European-based worldview). Therefore, words have been used to compare other people with this traditionally dominant group. The result has been the common use of terms that place others at subordinate positions in society by stereotyping or generalizing certain characteristics or traits of a group of people. For example, the obsolete terms in **Table 2.1** are gender biased.
>
> A person can use language to stereotype people by gender, ethnicity, age, religion, political association, education, and regional background. Avoid words that draw attention to these classifications in a negative context. To demean, put down, or degrade people based on the words they use only builds barriers to real communication. It is more productive to attempt to understand other people and show respect for their cultural backgrounds.

Learning Styles and Learner Characteristics

A **learning style** is the consistent way a person gathers and processes information. Students in the learning environment use sight, hearing, and touch to gather information. Students will favor one or a combination of these senses individually. Students may not be aware that they use any particular style to participate in learning nor that they may use different learning styles for different tasks or circumstances. They develop their individual learning styles based upon their personal **learner characteristics**, which are comprised of academic, social/emotional,

Table 2.1
Obsolete Terms and Their Replacements

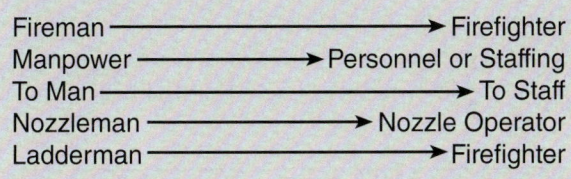

Fireman	Firefighter
Manpower	Personnel or Staffing
To Man	To Staff
Nozzleman	Nozzle Operator
Ladderman	Firefighter

and cognitive skills. These learner characteristics determine a student's learning style. Instructors can recognize different learning styles in the way individual students:

- Perceive, remember, and think about information and solve problems (**Figure 2.11**)
- See and make sense of their world and attend to their environment
- Attend to instruction and participate in activities

These differences are all representative of differing learning styles based on which sense or combination of senses provide students with the most accurate amount of information that is acceptable to them. To meet these different learning styles, instructors should plan a variety of teaching methods and learning activities in their lessons. Using a variety of methods helps instructors reach the many individual learning styles so that all students can participate in a style that enables them to learn.

In some situations, students will have developed learning styles that are not productive and can even contribute to their own failure. They are often not aware that their studying and learning habits are not working. Through the use of a variety of teaching strategies, an instructor can guide students into developing more effective learning styles.

Figure 2.11 Many students will review notes during class lectures.

Self-Regulated Learning

Some students will indicate a preference to regulate their learning environment to take advantage of their skills, focus, and direction. When an instructor is able to accommodate this self-regulation, a student can set his or her own goals, learn content, and provide feedback to the instructor independent of the rest of the class.

The instructor should be aware that the self-regulated learner is likely working toward a goal for personal satisfaction. To that end, the self-regulated learner may ask more probing explorative questions, and seek additional information beyond the other students. This may at times present a challenge for the instructor and the rest of the class since the self-regulated student may require additional attention, time, and resources compared with the traditional student. The instructor is encouraged to support the self-regulated learning activity, but must balance this with the needs of the entire class.

Interpersonal Communication

Interpersonal communication takes place between individuals every day and can reveal both the similarities and polar differences in the way people interact and disperse information. Some characteristics of interpersonal communication include:

- Informal language
- Informal nonverbal clues
- Frequent changes of the speaker and listener roles

- Spontaneity
- Formality (command vs. discussion)
- Intensity (tone, pitch, and volume)

The tone of the conversation can change based on the perceptions of the two parties. Therefore, it is important that all individuals understand and master the skills involved in interpersonal communication. **Figure 2.12** provides a model of communication that illustrates possible barriers between messages that the instructor may attempt to relay to the students.

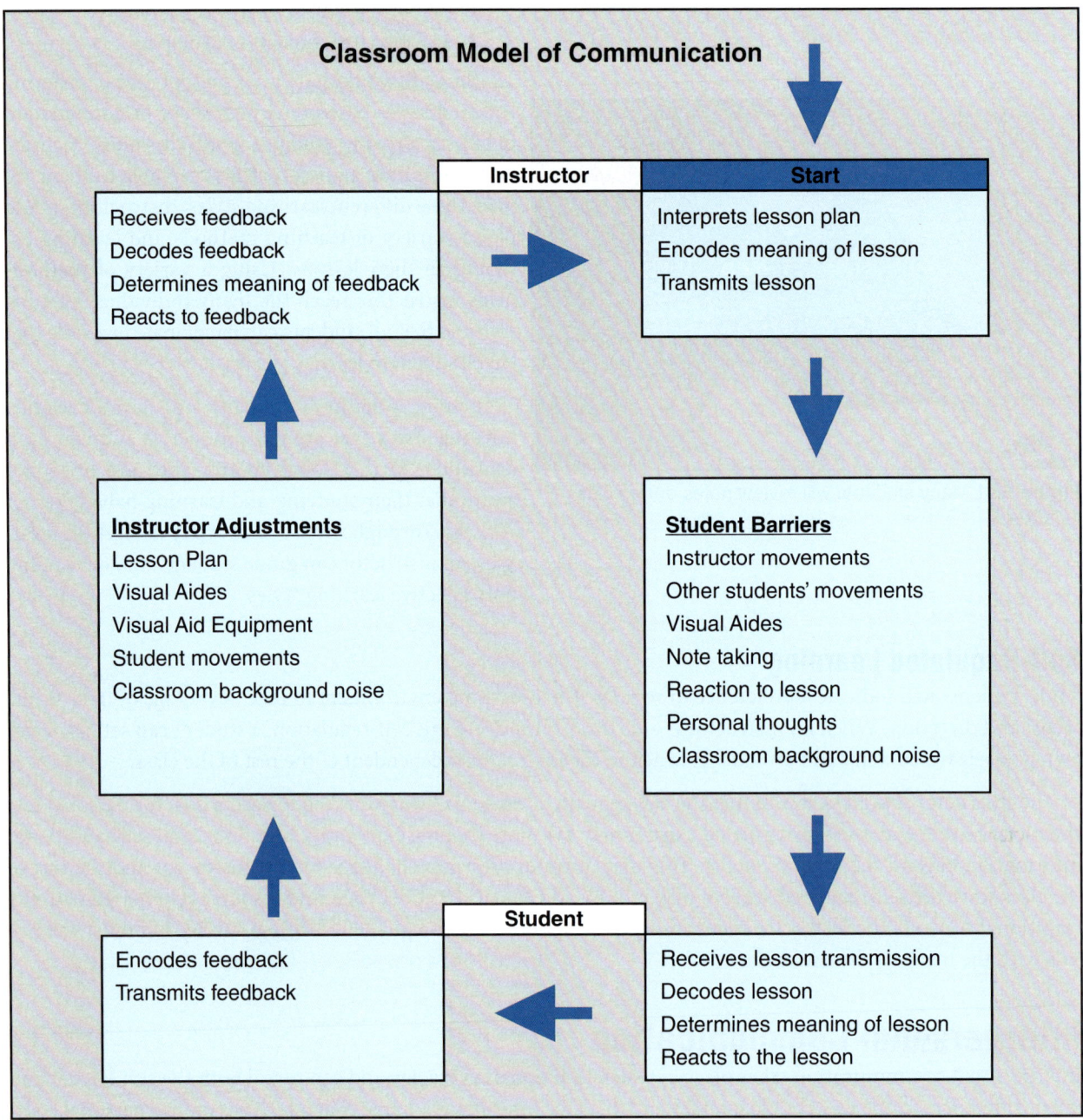

Figure 2.12 Students and instructors may face a number of communication barriers in the classroom.

The following list describes five general purposes for interpersonal communication:

- **Learning** — Acquire knowledge or skills.
- **Relating** — Establish a new relationship or maintain an existing one.

- **Influencing** — Control, direct, or manipulate behavior.
- **Playing** — Create a diversion and gain pleasure or gratification, as with positive humor.
- **Helping** — Attend to another person's needs or console someone in a time of tragedy or loss **(Figure 2.13)**.

In addition to understanding the purposes for these interactions, the instructor must also understand the following five basic elements of interpersonal communication during a lecture:

- **Sender (instructor)** — Verbal and nonverbal communication.
- **Message (the lesson)** — Content that the sender is trying to communicate. The message may consist of information intended for multiple human senses (sight, hearing, taste, smell, touch).
- **Receiver (student)** — Individual or individuals to whom the sender is attempting to communicate.
- **Feedback to the sender** — Reaction of the receiver to the message and its tone. If this feedback is verbal, the receiver becomes the sender and relates a new message to the original sender, who becomes the receiver. Receiving feedback allows the original sender to confirm reception of the message and to assess the receiver's level of understanding.
- **Interference** — Anything that may prevent the receiver from completely understanding the message. This can include information overload, when a student tries to absorb too much information at a time.

In a classroom model of instruction, it is the instructor's responsibility to interpret and encode the information in a lesson plan so that students will understand it. All communication takes place within a larger frame of reference. Senders encode their messages based on their education level, position of authority, personal or ethnic background, and other characteristics. The receiver will then decode the message based upon similar characteristics of his or her own. When the sender takes these characteristics into account, the receiver is more likely to understand the message clearly. Instructors must learn to encode information at the student's level, and not attempt to include more information than necessary **(Figure 2.14)**.

As the receivers of information, students will attempt to decode their instructor's message. After decoding the message, students will relate it to what they already know and determine what its meaning is for them. If instruction is well-planned and communicated, this meaning will closely match what the instructor intended.

Students respond to an instructor through feedback. Feedback may be verbal, such as a student asking a question, or nonverbal, as when students appear bored and unmotivated. When instructors pay attention to student feedback, they can modify the lesson to better serve students.

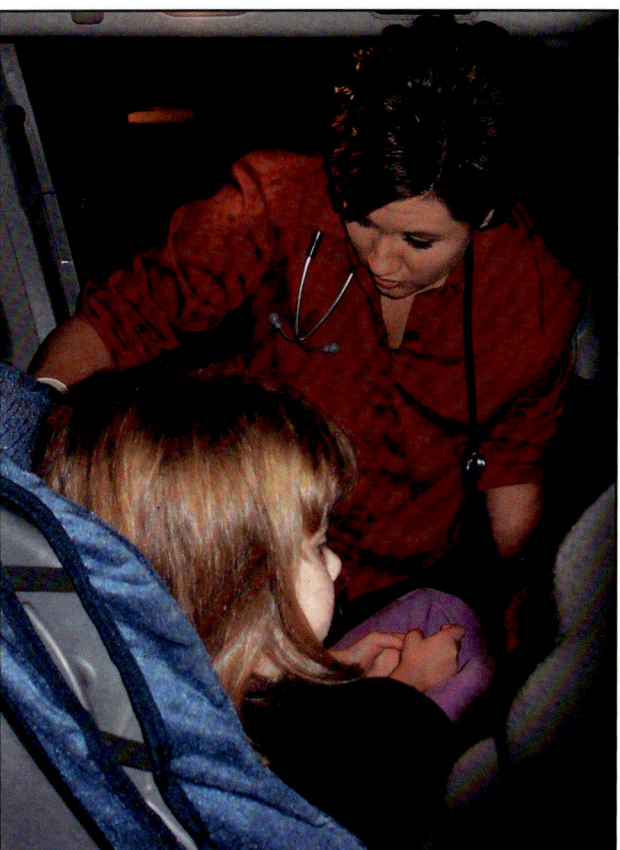

Figure 2.13 Good interpersonal communication is especially important when helping others cope with hardships.

Figure 2.14 A student's silence or lack of feedback may indicate that he or she doesn't understand. Improving ways to encode information may increase the student's comprehension.

The delivery or reception of a message may be distorted or blocked through interference. Types of interference include:

- An instructor not recognizing or misinterpreting a student's feedback cues
- A student not listening to an instructor
- Inappropriate use of electronic devices

For communication to be effective, both the sender and receiver must agree on its purpose. A shared situation usually creates this agreement. For instance, when an instructor and student meet in a learning environment, both agree that the purpose of their relationship is for the student to learn information that the instructor knows. In some cases, the purpose must be explicitly stated to ensure that both parties understand it fully.

The following sections focus on verbal and nonverbal components of communication. Instructors must know the strengths and weaknesses of both components, and must be able to minimize the inherent interference as much as possible.

Verbal Component

Instructors must understand both the power and weakness of words as part of a message. The words in a message typically account for only a very small percent of the overall total information communicated; however, the words in a message carry most of the abstract meaning in communication.

For an instructor to communicate effectively, he or she must have some idea of the students' familiarity with the technical aspects of the class content. This is particularly important when the students do not have a shared experience with the speaker. When a subject is completely new to most students in a class, the instructor should limit any industry-specific and technical terms and abbreviations.

Avoid any technical language and fire service jargon when speaking with the public, elected officials, media, and others from outside the profession. Also avoid language that might be considered offensive, gender biased, racist, or otherwise stereotyped.

By practicing the following guidelines, instructors can hone their verbal communication skills:

- **Engage in dual perspective** — Be aware of the receiver's frame of reference. Recognize the listener as having a different perspective and attempt to relate to it by asking questions and listening openly.
- **Take responsibility for personal feelings and thoughts** — When conveying a personal opinion, or offering a corrective course of action, use language that is *I-based* such as "I believe" or "I think." Avoid phrases such as "You disappoint me" or "You are wrong." Focus instead on language that owns one's feelings and offers a way for the student to correct his or her action such as "I hope you will consider."
- **Show respect for the feelings and thoughts of the other person** — Avoid trying to apply personal feelings to another person such as saying, "I know how you feel." Instead, understand and respect others' positions and build upon those concepts to create strong relationships. A better way of responding in this type of situation is to say, "I'm sorry you have to go through this."
- **Try to gain accuracy and clarity in speaking** — Avoid abstract language that can cause misunderstandings. Avoid generalizations that result in stereotypes. Generalizations are, in themselves, inaccurate. Be clear and accurate in all types of communication.
- **Be aware of any special needs of the receiver** — When communicating with a person who is deaf or hard of hearing, speak normally and face him or her directly to aid in lip reading. Do not exaggerate lip or mouth movements.
- **Avoid speaking or addressing a problem while angry or otherwise experiencing strong emotions** — Pause and place the conversation on hold until emotions are under control. While attending to any emotions, stay aware of all safety concerns, especially in a training environment.

Nonverbal Component

Nonverbal communication transmits the vast majority of any message—a large percentage belonging to body language, another significant amount belonging to vocal tone and inflection. Only a small percent of the transmitted message is actually verbal communication **(Figure 2.15)**.

Nonverbal communication consists of the following elements:

- Body language
- Vocal tone and volume
- Personal appearance

An understanding of the importance of each of the elements of nonverbal communication assists the instructor in recognizing and interpreting those signals, thereby improving nonverbal communication. Nonverbal communication can project a person's self-perception, emotional state, approachability, or cultural background. Brief descriptions of nonverbal communication and recommendations are as follows:

- **Vocal characteristics** — Practice speaking slowly, using variation in pitch to provide emphasis; use volume appropriate to the situation, and proper diction to ensure that words are clearly understood.
- **Vocal interferences** — Eliminate filler words and empty phrases. Instructors with a regional or cultural accent may also need to closely monitor their speech to make sure their audience can understand them.
- **Eye contact** — Maintain eye contact while speaking and modify the amount or duration of eye contact when appropriate. In Eurocentric cultures, good eye contact can convey self-confidence, honesty, trust, and credibility. Averting one's gaze can indicate deceit, dishonesty, insecurity, or anxiety. However, eye contact is also a function of cultural background, so it must be appropriate to the situation, the relationship, and the culture. Examples:

    — In the wrong context, too much eye contact can be as damaging as too little. Staring into the eyes of a member of the opposite sex can be considered too personal or intimidating.

    — Many Native American and Asian societies believe that it is disrespectful to make direct eye contact with a person who is not of the same status.

- **Facial expression** — Match facial expressions to the message. The face can show the basic emotions: happiness, sadness, surprise, fear, anger, and disgust. To effectively communicate the correct message in a relationship, the facial expression must match the verbal message.
- **Gestures** — Learn to use gestures to emphasize and illustrate the message **(Figure 2.16)**. In situations where noise prevents

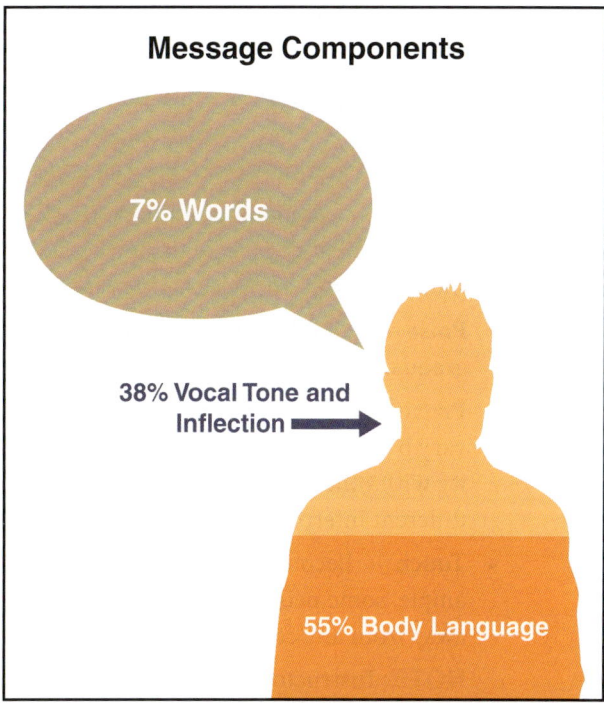

Figure 2.15 Instructors should recognize that their words only account for a small percentage of interpersonal communication.

Figure 2.16 Nonverbal gestures help to accentuate the message.

Chapter 2 • Principles of Learning 39

verbal communication, gestures are effective for sending messages such as *come here* and *stop*. Identify and control gestures that are annoying or distracting to others.

- **Personal appearance** — Maintain a professional appearance at all times. Set an example for students, and other members of the organization. When determining personal attire and grooming, consider the lesson being taught and the experience and expectations of the students/audience.
- **Posture** — Maintain good posture when standing or sitting in front of a classroom or assembly. Sitting or standing erect can create the impression of self-confidence and authority. Slouching or standing with stooped shoulders gives an appearance of insecurity, disinterest, or intimidation.
- **Poise** — In addition to good posture, poise includes a sense of calmness and control. Instructors develop poise gradually as they become more confident in their role and subject matter, and in public speaking.
- **Proximity** — The physical distance between the instructor as he or she talks to an individual or a group may vary based on a number of factors including the class assignment, ambient noise, and the individuals' familiarity with others and the environment. In addition, the instructor should be aware that different cultures have different interpretations of how close people should be when they interact.
- **Touch** — Become conscious of the effect, both positive and negative, that touch can have on others. For example, some people may perceive a light pat on the shoulder as an affirmation of good work. Others may shy away from any contact they do not initiate.

NOTE: Instructors are encouraged to seek out other sources of information on the complex topic of nonverbal communication.

Listening Skills

According to speech communication professionals, listening constitutes from 45 to 53 percent of a person's average day. In the learning environment, it is estimated that students spend 50 to 75 percent of class time listening to the instructor, other students, or audiovisual presentations. Therefore, improving listening skills is essential to effective communication. As role models for their students, instructors must practice good listening skills.

Listening, unlike simply hearing, is an active process. Instructors can become better listeners by paying close attention to the following unique aspects of the listening process (**Figure 2.17**):

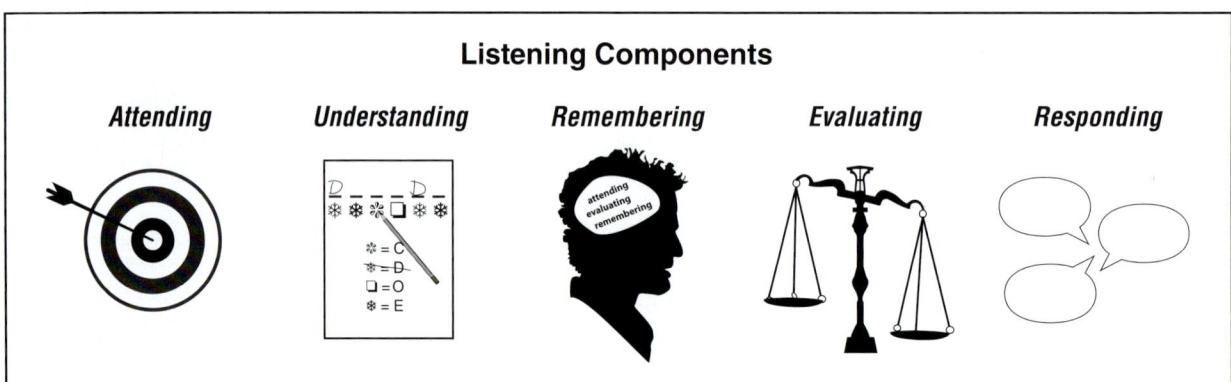

Figure 2.17 An instructor can serve as a role model by practicing the five components of active listening.

- **Attending** — Focus on the speaker while ignoring any other distractions. Instructors should ensure that the environment provides as few distractions as possible. Some suggestions for improving the attending step are as follows:
  — Look at the speaker when possible. Think about what is being said.
  — Visualize the situation or event that the speaker is talking about.
  — Wait until the speaker has finished delivering the message before responding.
  — Notice both the verbal and nonverbal messages.

- **Understanding** — Decode the message and assign meaning to it; this involves the following actions:
  — Organize the message into a logical pattern
  — Interpret nonverbal clues
  — Ask questions to clarify meaning
- **Remembering** — Retain information by taking notes, repeating lesson material back to an instructor, using **mnemonic devices**, and asking questions about unclear information.
- **Evaluating** — Critically analyze information to determine how accurate it is or to separate fact from opinion. To effectively evaluate a message, listeners must draw on their own personal experience, assess the credibility of the speaker, and interpret nonverbal cues.
- **Responding** — Indicate to the sender that the message or information has been understood or requires more explanation. Responses may be verbal, such as asking questions or requesting more information, or nonverbal, such as nodding to indicate that the information was received.

Practice is the best way to develop good listening skills. One way to practice active listening is to listen to recordings of speeches or stories, and repeat the main points. To improve listening and note-taking skills, an instructor should practice taking notes at meetings or in classes presented by other instructors. These exercises help overcome the barrier created by information overload and pinpoint the essential elements of the message. When instructors have the opportunity, they should attend speeches or presentations. While listening, instructors should assess the speaker's strengths and try to emulate those qualities. Instructors should also note any internal and external distractions that arise during the speech and try to minimize such distractions for their own students.

Factors That Affect Learning

Students who are struggling to learn new material may become frustrated, which further distracts them from the learning process. Instructors must realize that some students may experience additional stressors or situations that can undermine their success. Once instructors discover the underlying cause, they may be able to help students resolve some of these frustrations. Some students need assistance from the instructor to overcome learning obstacles or learning plateaus.

Learning Obstacles

Generally, obstacles to learning consist of external pressures and concerns that make the ability to focus on learning difficult and frustrating. Students may fear:

- Not knowing how to study appropriately
- Ridicule by the instructor or classmates
- Failure if they cannot perform as expected
- Leaving personal matters unresolved while spending long hours in a classroom setting

The learning environment can present obstacles for students as well. Examples of ways a learning environment can have a negative effect on learning include:

- Long stretches of time without a break to either sit or stand
- Poor lighting and ventilation
- Poorly organized training grounds or dangerous conditions

A student's personal feelings, such as anxiety or boredom, are obstacles to learning. Students will become anxious if they perceive that they are not prepared for the class or do not see the relevance to their jobs. Instructors should plan class intervals to allow opportunities to practice new skills, interact with training aids, and provide variety in teaching methods. **Table 2.2, p. 42,** lists some areas of student frustration.

Table 2.2 Areas of Student Frustration		
Fear or Worry	**Discomfort**	**Poor Instruction**
• Fitting in, acceptance	• Lack of personal strength and stamina	• Class too advanced
• The class situation	• Eyestrain	• Class too simple
• Failure	• Difficulty hearing	• Instructor unprepared
• Ridicule	• Classroom too hot or too cold	• No opportunity for participation
• Keeping up with requirements	• Uncomfortable seats or poor seating arrangements	• No variety in presentation
• Personal Problems		• Class too large
• Family	• Dangerous training conditions	• No direction
• Health		• Relevance not explained
• Money		

Learning Plateaus

A **learning plateau** can be compared to the landing on a flight of stairs — it is a break in upward progress (**Figure 2.18**). Some students stay there briefly, while others may have a difficult time moving past that point. Students sometimes create their own learning plateaus from emotional responses, such as boredom or fear of failure. These emotions may occupy their minds and interfere with their concentration and progress.

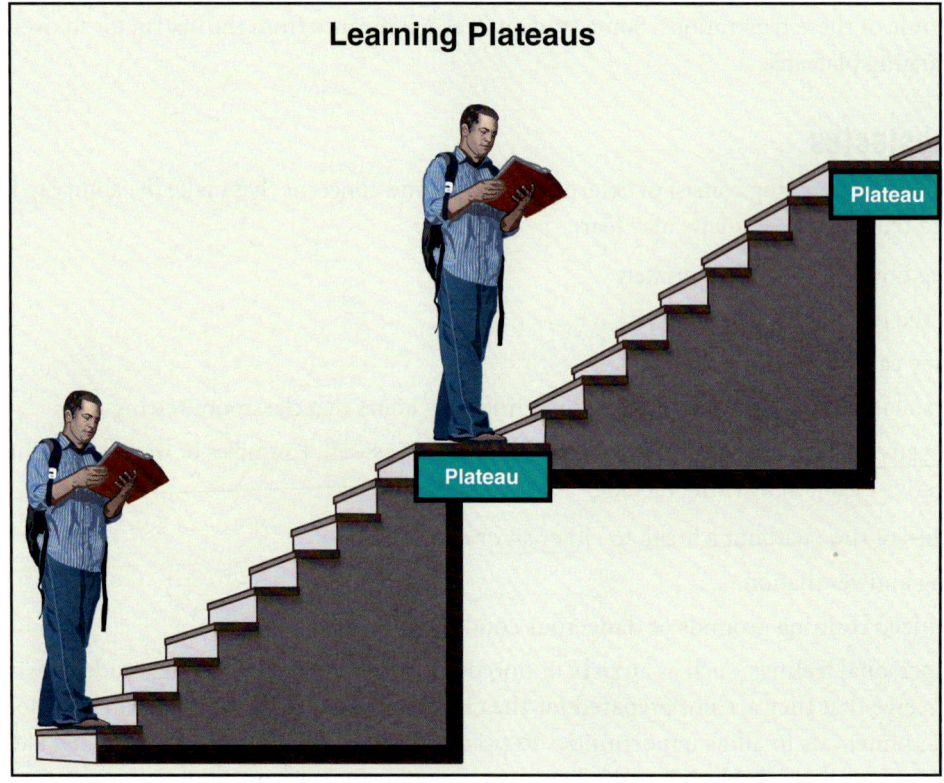

Figure 2.18 Learning plateaus can be frustrating to students but can also be overcome.

After students master the procedural steps of a skill, they need to practice until they meet a desired skill level. Once they reach this level, they will be exposed to more information and skills, and expected to progress to the next skill level.

Individuals may become discouraged if they have not been able to practice a task enough to feel proficient at a certain level, or they may find it more difficult to reach a particular skill level, especially if the rest of their classmates appear to have less difficulty with the task. At this point, further progress seems impossible, and an individual may feel like quitting. Athletes often experience plateaus in developing skills. Students, like athletes, must be coached and receive positive feedback as they practice.

Instructors should let students know that learning plateaus are normal, help them recognize signs of frustration, and work with them to overcome problems. One solution is for students to keep practicing until they thoroughly understand the skill and the procedures become automatic. Another solution is to take a break, direct students to review and think about the task for a while, then return to the task after some time away.

When students cannot get past plateaus, it may be that they have formed improper habits or tried to learn something beyond their abilities. But instructors should also consider that they may have failed to provide proper assistance. Instructors should review their methods of instruction to ensure that they are able to communicate and demonstrate effectively.

Motivation

Motivation is a key component to helping students to achieve their best work and meet learning objectives. Instructors can use the following techniques to help motivate students:

- **Provide relevance** — Tell students why the knowledge and skills they are learning are important and how they can be applied to real-life situations.
- **Set realistic and obtainable goals** — Identify criteria for successful completion of tasks, lessons, and courses.
- **Demonstrate enthusiasm** — Show as much interest in the course as is expected of students. Share ideas and receive positive comments or participate in reasonable debates.
- **Expect success and require outstanding performance** — Convince students that they are capable of mastering course goals. Encourage outstanding performance by guiding and coaching students to that level.
- **Incorporate motivators** — Make a conscious effort to determine student motivators. Provide external motivations such as rewards, recognition, and certificates that cause students to feel successful.
- **Generate interest and participation** — Use activities that include discussions to develop thinking skills and provide opportunities for participation in activities that hold attention and interest, stimulate thinking, develop thinking skills, and develop relationships with others (**Figure 2.19**).
- **Include instructional variety** — Use a variety of teaching strategies that match learning styles, learner characteristics, abilities, and needs. Use visual aids and demonstrations that relate to job requirements. Promote team work to share tasks and learn from one another.

Figure 2.19 Instructors can increase student interest by introducing activities in conjunction with classroom discussion.

Chapter Review

Answer the following questions to review the information provided in this chapter.

1. What is the sensory-stimulus theory?
2. What assumptions are made regarding adult learners for the theory of andragogy?
3. Describe Thorndike's Laws of Learning.

4. What is Maslow's Hierarchy of Needs?
5. What are the three domains of learning?
6. What are some examples of student diversity?
7. What are the differences between learning styles and learner characteristics?
8. What are some guidelines for engaging in verbal communication?
9. What are examples of nonverbal communication?
10. What are aspects of the listening process?
11. What are learning obstacles or plateaus?
12. What motivational techniques can Level I Instructors use to help students succeed?

Discussion Questions

The following questions are intended to generate discussion, expand your understanding of the chapter text, and allow you to think critically about what you have learned. Answers to these questions may vary.

1. What percentages do you personally believe you remember from the categories in the Cone of Learning?
2. Describe how at least one of Thorndike's Laws of Learning apply to you as an adult learner.
3. How can the domains of learning be used together to assist an instructor in presenting effective instruction?
4. How does self-confidence affect learning in the student's learning environment?
5. How does student diversity affect the learning environment?
6. How do you learn the information in a class if the instructor does not provide the content using your favored learning method?
7. How can verbal communications help an instructor be more effective in the learning environment?
8. How do you respond to nonverbal communication, both positively and negatively?
9. What are examples of ways you develop good listening skills?
10. How did you overcome the learning obstacles or plateaus you experienced in school, athletics, or other activities?
11. What motivational techniques do you use to be successful in your educational experiences?

Key Terms

Andragogy — Educational term referring to the study of adult education and its methods of teaching and learning.

Cognition — Concept that refers to all forms of knowing, including perceiving, imagining, reasoning, and judging.

Cone of Learning — Visual representation that depicts in what percentage of information human beings retain using their senses alone and in combination.

Learner Characteristics — The specific aspects that comprise an individual's ability to learn information; includes the individual's social/emotional, cognitive, and academic foundations.

Learning Domain — Distinct sphere or area of knowledge, such as the affective, cognitive, and psychomotor domains.

Learning Plateau — A break or leveling of a student's progress in a training course or class.

Learning Style — Learner's habitual manner of problem-solving, thinking, or learning, though the learner may not be conscious of his or her style and may adopt different styles for different learning tasks or circumstances.

Mnemonic Device — Any memory technique that is used to assist a learner in properly encoding and recalling vital information; is primarily used to associate seemingly dissimilar items or concepts.

Sensory Memory — Mental storage system for attention-getting sensory stimuli or input.

Instructional Planning

Chapter 3

SECTION A — INSTRUCTOR I

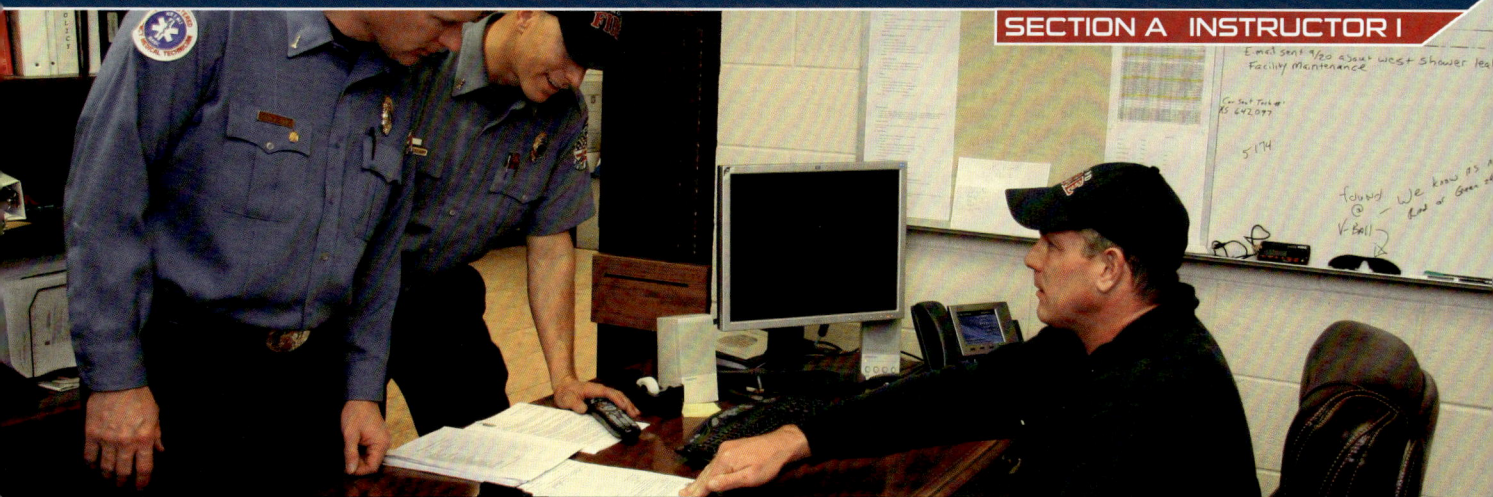

Chapter Contents

Planning to Teach **47**	Weather Variations 55
Organization 47	Instructional Resource Variations 56
Session Preparation 48	Differences in Learner Characteristics..... 56
Session Logistics 49	Differences in Knowledge Level 57
Training Aid Selection **51**	**Course Consistency** **57**
Learning Objectives and Lesson Content... 52	Safety Factors. 57
Class Size and Interaction 52	Training and Resource Materials 58
Learning Pace 53	**Chapter Review** **58**
Practice Factors 53	**Discussion Questions** **59**
Learning Environment Continuity **54**	**Key Terms** **59**
Instructor Changes 54	

JPRs addressed in this chapter

This chapter provides information that addresses the following job performance requirements of NFPA 1041, *Standard for Fire Service Instructor Professional Qualifications*, 2019 Edition.

4.2.1	4.3.1	4.4.2
4.2.2	4.3.2	4.4.3
4.2.3	4.3.3	4.4.4

Learning Objectives

1. Explain the importance of organization, session preparation, and session logistics when planning to teach. [4.2.1, 4.2.2. 4.3.2]

2. Describe factors to consider when selecting training aids for a lesson plan. [4.2.2, 4.3.1, 4.3.2, 4.3.3. 4.4.2]

3. Identify potential events than can impact learning environment continuity. [4.2.2, 4.3.2, 4.4.2, 4.4.3, 4.4.4]

4. Describe aspects of instruction that help achieve course consistency. [4.2.3, 4.3.2, 4.4.2, 4.4.3, 4.4.4]

Chapter 3
Instructional Planning

A good instructor does not simply appear in the learning environment and begin teaching, just as good students do not simply arrive for a class and begin learning. Both must prepare for the experience; both must have expectations and anticipations of what each wants to accomplish and how to accomplish it.

This chapter explains the following aspects of the instructional process:

- **Planning to teach** — Preparing to teach and preparing students to learn
- **Training aid selection** — Choosing appropriate training aids for the information to be taught
- **Learning environment continuity** — Tying each individual lesson plan together
- **Course consistency** — Presenting information that conforms to other information, standards, and accepted practices, especially within the training organization

Planning to Teach

Effective instructors take preparatory steps to ensure that the learning experience is worthwhile, relevant, and interesting. Taking time to properly prepare before the presentation will result in a positive learning experience that motivates students to think, question, and become involved in the learning process.

Planning time spent outside class is crucial to instructor accomplishments inside class. If instructors do not plan ahead, class time may not be used efficiently. While planning to teach, instructors must organize the learning materials, prepare for the class session, and ensure that all logistical needs are met.

Organization

New fire and emergency services instructors will benefit from proper organizational skills, which will lead to success and result in a variety of benefits such as:

- Increased credibility
- Improved efficiency of learning environment presentation
- Effective use of time, talent, and materials
- Reduced stress
- Meeting legal requirements for record-keeping (see Chapter 10, Records, Reports, and Scheduling)

Review procedures or guidelines to determine the appropriate methods for organizing classroom materials. When there are no procedures or guidelines available, you should identify and implement an organizational system that works best for you. You can find commonly used organizational systems in numerous books, magazines, and websites. Organize and review any materials needed for a particular session ahead of time and have easy access to them when they are needed during class.

Session Preparation

Instructors are often required to teach topics they have not taught before. Even a veteran instructor may be required to develop a presentation on an unfamiliar topic. To prepare to teach a new topic, instructors must first gain a thorough knowledge of the topic. This ensures that the instructor can perform the skills associated with the topic and answer any student questions. The instructor may already have this knowledge even if they have not taught the subject before.

NOTE: Even with adequate time to prepare, if the instructor is not competent with the topic, it would be prudent that another instructor be assigned well in advance to instruct the class. An unqualified instructor not only risks the safety of the students but also runs the risk of legal liability for providing incorrect or inaccurate information.

> **Instructor Credibility**
>
> An instructor should always understand the environment in which he or she will teach, both in terms of the resources available for the class, and in the students' perceptions of the instruction. For example, acknowledging that he or she is not prepared to teach a class risks students responding with lower confidence, perception of instructor credibility, and motivation. Good instructors will focus on the positives and overcome the limitations without calling attention to any of the shortcomings that they come across during the instructional process. Working this way will improve students' opinion of the instructor and the overall organization that offered the training.

An instructor should be careful to complete the following tasks ahead of every class session:

- **Read the lesson objectives** — Become familiar with what the objectives require students to know and perform.

- **Review the lesson plan** — Use an organization's standard lesson plan to determine what material must be addressed, what time frame is required, and what assistance, materials, and equipment are needed.

- **Check what equipment is needed** — Be familiar with training aids. Know how to operate and use various pieces of equipment for student activities **(Figure 3.1)**.

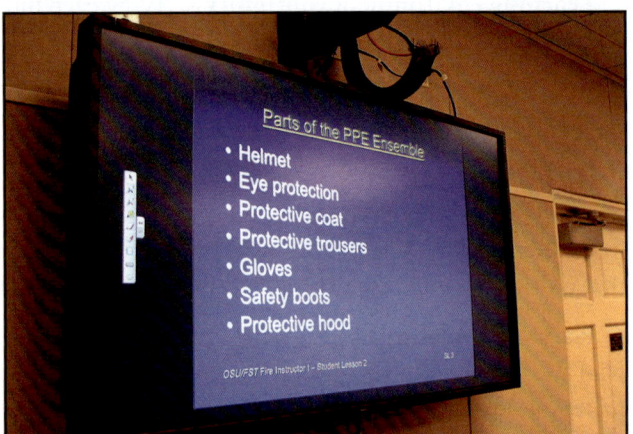

Figure 3.1 Instructors should familiarize themselves with the classroom's technology and equipment before student activities begin.

- **Locate required equipment** — Arrange for additional equipment that will be necessary but is not already in the classroom or training area; find avenues within the organization to acquire what is needed.

- **Determine what skills must be taught** — Practice the skill steps, or at least review the steps mentally by looking at pictures, equipment, or handouts. Any instructor who lacks recent experience or practice in a skill must take time to review it or determine a method of presenting it so that the skill is understandable to students **(Figure 3.2)**. Ideally, the instructor presents a skill at mastery level.

Figure 3.2 Review step-by-step skills before teaching them to students.

- **Review required lesson audiovisuals** — Preview audiovisuals to prepare for engaging students in related discussions and applications. This step means being familiar with the lesson plan and knowing when and how audiovisuals are used.
- **Check documentation requirements** — Check on, arrange for, and have available and ready all handouts, rosters, and other reports and records before class begins.
- **Arrive early** — Arrive early to make sure that the training area is set up appropriately for the lesson and that all needed elements are in place and functioning properly. This may require a significant amount of time if the instructor is not familiar with the facility.

The instructor must be familiar with the topic, familiar with the operation of all training aids, and be able to make logical, smooth transitions between sections of the material. For the new instructor, the best way to present an organized presentation is to practice the presentation. Practicing the presentation helps the instructor recognize any points where adjustments need to be made. It also helps to build personal confidence and create enthusiasm for the topic. Asking more experienced instructors how they approach teaching similar topics can be an excellent source of information for improvement.

Observe Other Instructors

Instructors should continuously and purposefully observe several other instructors' methods. Watch how others manage student groups, present information, and demonstrate skills. Look for advantages and disadvantages of their presentation methods.

Think about which methods accomplish learning objectives and which methods appear to create problems and cause disruptions in learning and lesson organization. Draw conclusions about what methods seem to work best, and adapt those as you develop your own teaching method and style.

Periodically, an instructor may need to teach a class on short notice. For an experienced and well-organized instructor, this is generally not a problem. A new instructor, however, may feel the stress of having to adapt to a situation rapidly. The incoming instructor must make every effort to quickly prepare and deliver a lesson that makes the time in class worthwhile. The incoming instructor should attempt to apply the preparation steps before beginning the class.

As instructors gain experience in a subject in a familiar course, the time it takes to prepare decreases. Inexperienced instructors should plan to spend one to three hours of preparation time for every hour of delivery time. For some courses, preparation time can extend to more hours or even days of scheduling, coordinating, and confirming that class and teaching needs are met.

What This Means to You

You may sometimes be put into the position of walking into the learning environment with short notice and little opportunity for preparation. One way to handle this situation is to make the class session a discovery zone where everyone learns something. For example, you can use each learning objective as an overview point and proceed from there. It may mean that the group learns or reviews the basic steps on whatever equipment or materials are available, such as getting sections of rope and practicing knots, or obtaining some manikins and practicing cardiopulmonary resuscitation (CPR).

Session Logistics

The term **logistics** as used in the fire and emergency services profession means the procurement, distribution, maintenance, and replacement of materials and personnel. Fire and emergency services organizations usually have a division or individual who manages logistics by providing the materials and equipment required to meet the mission of the organization and its various divisions. Within the training division, each instructor is usually responsible for acquiring the logistical support needed for any given lesson plan.

Logistics may also consume class time. The completion of some skills requires cleaning portions of the training area or restocking, refilling, or replacing certain items. The organization or the type of lesson plan may require that students perform some of the following duties:

- Cleaning a spill
- Returning the learning environment to its pre-training arrangement
- Parking vehicles in designated locations
- Cleaning manikins
- Refilling SCBA cylinders
- Reloading fire hose
- Restocking kits with supplies
- Recharging battery-operated equipment
- Cleaning fire hose **(Figure 3.3)**
- Inspecting, cleaning, and replacing tools

Instructors should calculate the time it takes to perform these duties. Students should perform some of these duties during class time, as part of the required skills training. Students should perform others after class time, so that the next class can begin promptly. Logistical needs may require several hours of preparation and restoration for classes with high equipment needs, such as live-fire training, EMT training, or driver training for groups of 25 to 30 students.

Figure 3.3 Logistics tasks may include cleaning fire hose before another set of students can practice the lesson.

The instructor is responsible for ensuring that all materials and equipment needed are determined and arranged for before the beginning of class **(Figure 3.4)**. Organized instructors will have a checklist of necessary materials and equipment to ensure that nothing is forgotten. This list can be retained as part of the lesson-plan documentation used for future class sessions.

For those instructors who work in a training facility, there may be staff members to assist with maintenance, inventory control, and scheduling of equipment. But the final responsibility always rests with the instructor to take the time to ask for assistance, to follow procedures when making requests or reservations, and to follow up on those requests.

Arriving early to get or assemble equipment and materials gives you the opportunity to perform the following steps:

Step 1: Check for missing items.

Step 2: Review operations.

Step 3: Arrange room layout.

Step 4: Find replacements or make repairs.

Step 5: Revert to a contingency plan (also known as Plan B). You should always have at least one contingency plan per lesson.

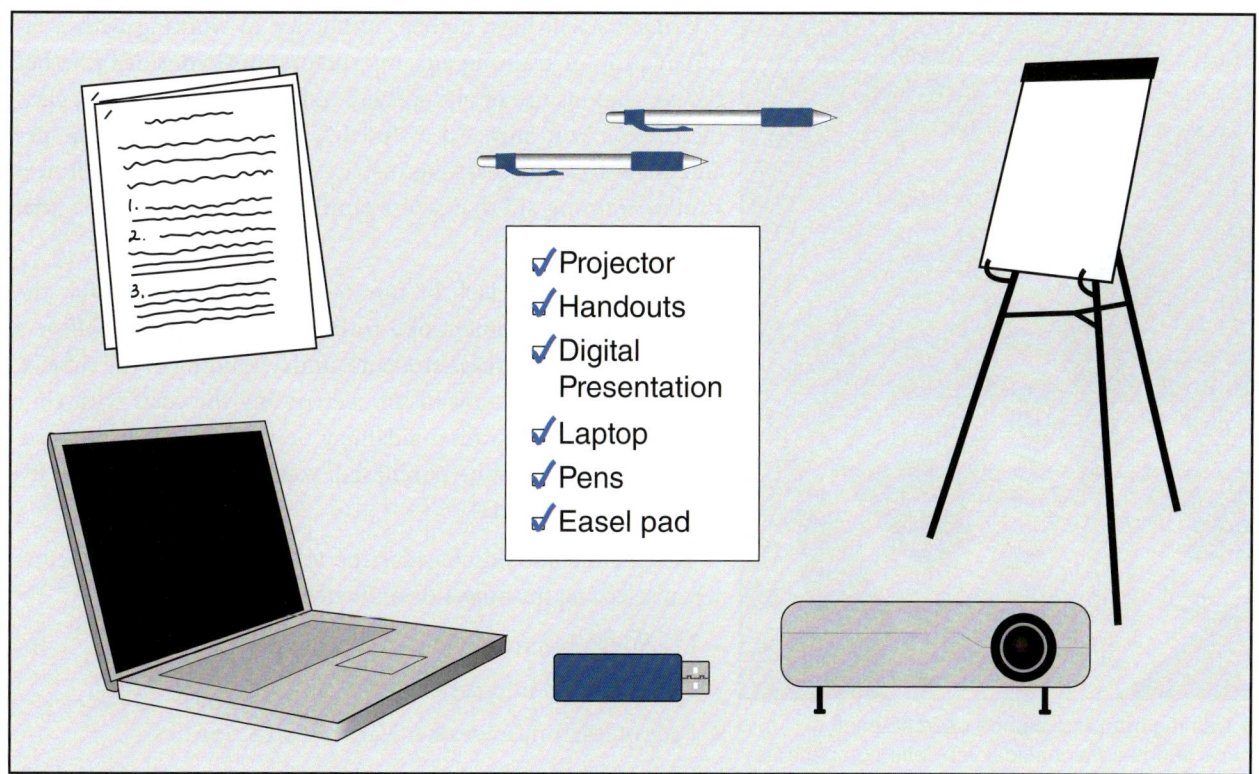

Figure 3.4 Instructors should confirm that they have all the materials and equipment they need before conducting a class session.

Training Aid Selection

Selecting the appropriate **training aids** to use with a lecture or demonstration requires planning. Procuring these aids is an important part of session logistics. However, the training aid recommended in a lesson plan may not be available. In these cases, instructors need to be able to adapt and select alternative training aids to meet the lesson plan's needs.

Based on the lesson plan, an instructor should select training aids that most appropriately illustrate or demonstrate the knowledge that the student must possess when the course is completed **(Figure 3.5)**. For instance, when a course is intended to help company officers become certified to command a multiple company operation, a computer-simulation program may be an effective training aid. Manikins for training and evaluation are appropriate when students are required to successfully perform CPR/AED or artificial respiration **(Figure 3.6, p. 52)**.

A further consideration when selecting the appropriate training aids is how a student's knowledge will be evaluated at the end of the course. In addition to other

Figure 3.5 When using training aids, ensure that they reinforce the lesson's learning objectives and content.

methods, the training aids themselves may be used as testing tools so students can demonstrate their proficiency. For example, the same CPR/AED manikin is usually used for the initial demonstration, student practice, and final evaluation. In a command course, technology-based training (TBT) can be used for both training and testing.

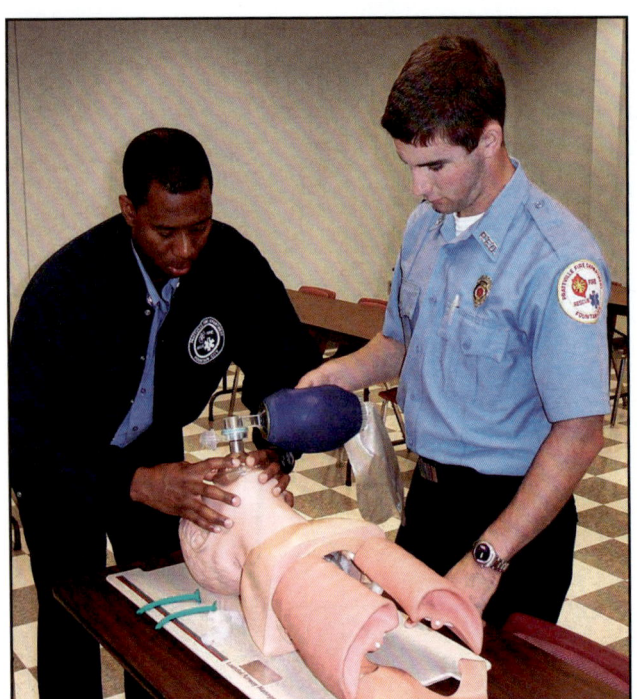

Figure 3.6 Training manikins can be used to teach a variety of skills such as mechanical ventilation on patients.

Students will have higher confidence in working with a familiar prop or training aid. Instructors must consider whether the continuous use of one specific training aid throughout class instruction will only test a student's familiarity with that specific item. In some cases, an instructor may want to substitute a similar training aid that is less familiar to students for the final evaluation.

Be aware that a lack of funding or time may prohibit the purchase or development of a training aid. But if you believe a training aid can increase students' understanding of the subject, you should discuss a means of overcoming the costs with your supervisor. Some training aids may be improvised or donated. If training aids must be purchased, you should follow the SOPs established by the AHJ.

Instructors should consider the following additional factors when selecting training aids or devices:

- Learning objectives and lesson content
- Class size and interaction
- Pace of learning and other learner characteristics
- Practice factors

Learning Objectives and Lesson Content

Any training aids that an instructor uses in a lecture or demonstration must reinforce the desired learning objectives and lesson content. Instructors should review the lesson plan, determine the content and objectives, and then select the training aid that will satisfy as many of the requirements as possible.

When the source for the training aid is a professionally produced curriculum or single item, compare it to the desired lesson plan outcome. You may need to alter the training aid or use only the relevant portion of it. When you do not have an existing training aid, it is then necessary to create one using the technology that is available.

For example, when the lesson content concerns fire behavior and the learning objective is to understand the fire tetrahedron, a simple digital image or a drawing illustrating this theory on a dry erase board would be appropriate. If the illustration is part of a longer video on fire development, the instructor can cue the video to the appropriate sequence before class in order to maximize class time. If no image exists in the purchased curriculum, the instructor may look for an image elsewhere or create one.

Class Size and Interaction

Teaching situations where instructors want to encourage student participation and interaction can influence training aid choice. With a large class, the class size and learning environment arrangement may limit interaction to a question-and-answer period at the presentation's end. Spontaneous diagrams on an easel pad or marker board can facilitate a highly participative coaching session in a small class.

Class size and seating arrangement have a direct relationship to the size of the image or training aid the class is expected to view. For example, using an easel pad in front of a 200-person audience seated in a large auditorium would not be effective. A large projection screen would be a better choice. On the other hand, an easel pad or marker board may be ideal when debriefing six trainees after a practical training evolution.

Learning Pace

The effectiveness of a training aid is affected by its safe application in an instructional environment. One factor in a training aid's suitability is the intended pace of learning. When the expected learning is fast paced, training aids can be integrated into the lesson plan to allow a class to move quickly through the information. When the lesson's content is primarily unfamiliar to students, training aids that allow an instructor to proceed at a slower, safer pace are more appropriate.

One strategy for teaching new material is to assign outside reading assignments from textbooks or self-study guides before the beginning of class. Guided discussions using marker boards, easel pads, or large illustrations used in class can reinforce new material. Students can then view a video for review and reinforcement.

When students represent a variety of knowledge levels or experiences, a student-centered strategy can include a mixture of the two scenarios. Those students who are not yet at the required learning level can have access to self-study materials before joining the more advanced students at the appropriate time. This could be monitored utilizing formative assessment tools, which will be expanded on in Chapter 9, Testing and Evaluation.

Practice Factors

To give students an opportunity to apply new knowledge and skills and to make the new application meaningful, include practice time and the training aids or props needed in the lesson plan (**Figure 3.7**). For example, you should provide a set of life safety rope lengths for a class on tying hitches used in technical rescue.

Figure 3.7 Time to practice using the correct equipment is key to any course.

The training aids required when applying concepts or skills such as incident management, accident investigation, personnel supervision, problem solving, or fire investigation may be more difficult to select. Training aids for prospective incident managers can include practice scenarios using the following resources:

- Paper, marker boards, or easel pads
- Tabletop models
- Technology-based training
- Interactive and virtual reality equipment

Similarly, investigation skills can be applied via written case studies, computer-based scenarios, structures (acquired or purpose-built) for practical evolutions, or staging an accident or incident scene for the student to physically investigate. Each of these applications requires that instructors carefully plan the use of training aids in the lesson in order to give students meaningful application of their new knowledge and skills.

Learning Environment Continuity

Continuity ensures that presented information flows in a logical and understandable stream. Many factors can affect **learning environment continuity**, but instructors can take steps to reduce their effects. Instructors should anticipate problems and prepare contingency plans for potential events such as:

- Instructor changes
- Weather variations
- Instructional resource variations
- Differences in learner characteristics
- Differences in knowledge levels

Instructor Changes

There are times when an instructor is not available to teach a scheduled class. Ideally, that instructor or the instructor's supervisor contacts someone familiar with and experienced in teaching the lesson. Instructors should carefully prepare for instances where other instructors may be involved in a class they are teaching. Trading control of instruction may affect the class continuity because all instructors have their own personal perceptions, views, beliefs, and methods of teaching. It is possible that their ideas and methods may be different from those of the lead course instructor. The lead course instructor can plan to maintain continuity by following these suggestions:

- **Know fellow instructors** — Instructors meet and exchange ideas through in-service training, meetings, conferences, and seminars. If an instructor must call on another to substitute or assist in teaching a class, it should be someone whom the instructor knows and trusts, who has similar ideas and methods, and who has similar or better experience.

- **Prepare the students** — Ideal situations do not always occur, and the desired instructor substitute or assistant may not be available. Students become dependent on and familiar with the methods, attitudes, and personalities of their instructors and may not relate as well to a different one. Instructors should tell students in advance to prepare for what may be a different teaching style.

- **Meet with substitute instructors or assistants to prepare for the class** — Every instructor involved in teaching a class must prepare for it. The lead course instructor who arranges for a substitute or for assistants should meet with the substitute or assistants and perform the following (**Figure 3.8**):

  — Outline the learning objectives and what must be accomplished in the lesson.

  — Assign specific duties or skills.

  — Show and orient each person to the teaching area or classroom.

  — Provide directions or assistance in locating and assembling equipment.

  — Coordinate rest break and cleanup times.

  — Perform any other duties needed to ensure a successful lesson.

Revisiting these considerations regularly also has the benefit of providing continuity in instructor preparation and delivery. Classes run more smoothly, instructors perform more effectively, and students have the continuity they need to participate successfully in the learning process and achieve the objectives or learning outcomes set for them.

Figure 3.8 The lead instructor should ensure that other instructors involved in a class are prepared for their tasks and introduce those instructors to the class.

> ### Preparing Students for a New Instructor
>
> Whenever possible, when another instructor will be conducting the lesson, the lead instructor should:
>
> - Give students some background on the additional or substitute instructor's experiences, knowledge, and teaching methods.
> - Introduce the additional or substitute instructor to the class when the person arrives or the day before he or she is going to teach a lesson.
> - Describe the lesson plan that a guest or substitute instructor will be teaching so that the students know what to expect.

Weather Variations

Weather changes can have an adverse effect on training activities and schedules. Instructors must build flexibility into courses that are scheduled during times of the year when inclement weather is possible. Extreme weather conditions may challenge a course's continuity as instructors attempt to teach the same skills during the summer heat and winter cold (**Figure 3.9**). You must be able to adapt the skills to the existing weather while still remaining faithful to the specific skill requirements.

When teaching psychomotor skills, the learning environment should not distract students from attending to the lesson's objectives. Even though students may later have to perform the same skills under severe conditions on the job, these conditions are not conducive to learning the skill.

Instructors should not expect their students to learn skills effectively in extreme weather conditions. On the other hand, instructors may decide to train or drill more experienced students in inclement weather conditions for the sole purpose of preparing students to face those conditions on the job. During inclement weather, instructors must keep student safety a high priority, ensuring safety measures account for adverse conditions. Students must be provided with the appropriate equipment, rehabilitation, and safety instruction that are applicable to the weather conditions.

NOTE: In some jurisdictions and situations, there might be specific policies or labor/management contract provisions regarding weather conditions and training. For instance, some labor/management contracts have heat-index provisions that limit training to days that have a heat index less than 100°F (38°C), or cold index provisions that relate to temperature and wind-chill to prevent injuries.

Figure 3.9 Instructors should closely monitor and inspect training facilities during inclement weather, as well as consult AHJ policies and procedures for training in extreme weather conditions.

> ### ⚠ Rescheduling Because of Unsafe Conditions
>
> Instructors should understand that there are conditions, such as ice or snow, where some evolutions cannot be safely performed. Even though the skills being taught in the evolution might have to be performed during the implementation of job duties under similarly adverse conditions, they do not have to be performed under those conditions while in training mode. If the instructor requires that the students perform such evolutions under unsafe conditions while in training mode, it could create liability issues for the instructor if a student gets injured. Sometimes it is best to postpone the class in the interest of safety.

Instructional Resource Variations

When scheduling equipment, use the same type of equipment in the learning sessions that is used in the testing session and on the job. Even though some brands may be less familiar to some students, using the same type of equipment is an important factor in maintaining the continuity of the course. When a group will be tested on tying knots using a certain type and size of rope, for example, students should also practice with that type and size of rope (**Figure 3.10**). When students use a certain type of SCBA on the job, train them on the same or generically similar equipment. This procedure is not only fair, but it also makes testing valid and reliable.

Instructional resources used in training include the information, product data, skill sheets, and references required to meet the lesson plan requirements. When this information is available to all course instructors, instructors can maintain continuity during the planning of their course presentations.

Course materials may also be substantially different from the latest training material being published in the fire and emergency services. The Level I Instructor should, whenever possible, verify that the information he or she is using for instruction is the most current information available. If the instructor discovers that updates to materials are necessary, he or she should consult with the individual responsible for purchasing and updating materials.

Figure 3.10 Standardizing equipment will help students learn to perform tasks using the same equipment they will use during an incident.

Differences in Learner Characteristics

Differences in learner characteristics lead to various learning styles and can create a challenge to instructors who are attempting to maintain class continuity. To maintain continuity, an instructor has to be flexible and adaptable within the structure of the course in order to appeal to the learning styles of all the students. Ideally, instructors adapt lessons that include a variety of teaching methodologies so that all types of students have the opportunity to learn through their preferred learning styles. This makes it more likely that students will gain the appropriate knowledge and skills and meet the lesson objectives.

Students may not be familiar with multiple learning styles or even their own learner characteristics, and the learning style they know may not be ideal for their success. By using a variety of teaching methods, instructors may introduce students to different and more successful ways of learning.

Historically, many training organizations have used traditional teaching styles that depended on a lecture-style format to provide cognitive information. Instructors generally asked questions or used short, written quizzes to check student understanding of the lectures or reading assignments. Students were expected to listen carefully to the instructor and take notes.

Over time, instructors have added visual aids to lectures to engage the sense of sight in the learning process. Some students are well served with the illustrated lecture format explained in Chapter 6, Classroom Instruction. Other students are better served with a demonstration or skill practice format as explained in Chapter 8, Skills-Based Training Beyond the Classroom.

There are also occasions when the abilities of students do not match the level to which the training material was written. Students may be either ahead of the material or not ready for it. Instructors should begin their course preparation by assessing their students' abilities and experiences and adjusting course materials accordingly.

Differences in Knowledge Level

Instructors may discover that their students are less knowledgeable (remedial) or more knowledgeable (advanced) about the topic for a daily lesson or for a course. You should be prepared to make adjustments in order to provide a successful, positive learning environment for both remedial and advanced students.

Remedial Knowledge Level

When the knowledge base of a student or a group of students is lower than the level of the course material, instructors have to adjust the teaching pace and expectation. You may need to budget more class time on review or practice. Students may need a review of basic skills or knowledge before they are ready to progress to the new material in the lesson. Reviewing previous material or skills may initially delay the lesson agenda or course time frame. However, it can allow students to feel more comfortable with their knowledge and abilities so that they feel ready to proceed. Assigning additional reading or study assignments for students to accomplish outside the class period may also help them advance in the class.

Advanced Knowledge Level

When the knowledge base of a student or a group of students is more advanced than the material, instructors should review the material with students, and then assign problems or exercises at the advanced level. When there is still time available, instructors have the following options:

- Preview the next lesson and involve students in discussions or exercises to determine their levels of readiness.
- Have the group create exercises or scenarios and plan for the equipment they need to perform the exercises in the next lesson. Then arrange to have that equipment available for the next lesson. This option can be used when it is not possible to move on to the next lesson because of equipment limitations.
- Have participants work together in groups to create test questions, complete with answer keys and text references. These tests can be used for study, debate, and discussion.

> **What This Means to You**
>
> Part of maintaining learning environment continuity is having contingency plans. You will, at some point, be faced with a situation where the equipment for a lesson will be suddenly unavailable or where a student's learner characteristics and learning style will be discovered part way through a course. Having a "Plan B" is important to address these situations and other unforeseen circumstances at the time of instruction. For example, the weather conditions could change on the day that a training evolution is planned. Students may have made special arrangements with their departments to be at the training, or they may have traveled some distance to be there. Having a contingency plan for such an occasion prevents the loss of a training day.

Course Consistency

Course consistency refers to the strategic use of accurate material that is developed at a definable learning level and shared between classes. Instructors and the training division lose credibility when the information has not been checked for accuracy and consistency. The following aspects of instruction should remain consistent throughout a course and between courses:

- Safety factors
- Training and resource materials

Safety Factors

Safety is the primary responsibility of instructors and students who are engaged in training exercises. Both instructors and students must consistently adhere to safety policies and procedures. You must be consistent in your use of safety equipment and procedures to ensure that the safety message is apparent to students.

Recognize and teach any safety issues that are appropriate to a lesson. When instructors or standards require that students wear PPE on the fireground, students need to wear it when they train. When the lesson requires that students wear a protective hood during live-fire training, all instructors involved in teaching live-fire training must also wear one.

You should plan skills and activities around the safest way to perform them. Instructors must inspect all learning environments and then eliminate or reduce potential hazards (**Figure 3.11**). In addition, encourage students to identify and report any safety concerns as soon as they are noticed. Teach students to perform psychomotor skills in a safe manner. Consistent, safe repetition of skills helps create a safety-conscious behavior pattern that students will continue to use after training is complete.

When a jurisdiction has strict procedures/guidelines for performing a particular skill, it is even more important to ensure that lessons consistently comply with those parameters. Although there may be multiple ways to perform a particular skill, such as raising a 24-foot (7.2 m) extension ladder for climbing, local SOPs or state/provincial performance objectives may be very specific in how the skill will be performed. In these situations, consistency is essential.

Training and Resource Materials

Consistency in the types of materials used in training is also necessary. Fire extinguishing agents, medical supplies, respiratory-protection equipment, and decontamination products should be identical to the types used in everyday and emergency activities. This consistency gives students the knowledge of how the product or material works and how it is packaged and looks.

Figure 3.11 Instructors should closely inspect training grounds for potential hazards.

Consistency also extends to the way that equipment and material is stored on the apparatus. Students perform more efficiently when they are familiar with a standard method of storing and accessing the equipment.

Instructors draw on a variety of resources, such as trade journals, the Internet, conferences, other training courses, and other instructors for their course materials. Instructors must ensure that all applicable sources can be cited to the appropriate credible resource or standard and are accepted by the sponsoring training organization. To ensure course consistency, the resources should support the course's learning objectives and should not contradict each other.

Chapter Review

Answer the following questions to review the information provided in this chapter.

1. Why is it important for instructors to use organization, session preparation, and session logistics when planning to teach?
2. What factors should instructors consider when selecting training aids for a lesson plan?
3. What potential events might impact learning environment continuity?
4. What aspects of instruction help achieve course consistency?

Discussion Questions

The following questions are intended to generate discussion, expand your understanding of the chapter text, and allow you to think critically about what you have learned. Answers to these questions may vary.

1. How much time do you think it takes for an instructor to successfully plan to teach a course?
2. How do training aids assist in learning new material?
3. How can instructors take steps to anticipate problems and prepare contingency plans for potential events that affect learning environment continuity?
4. Why is it important for instructors to keep aspects of instruction consistent throughout a course or between courses?

Key Terms

Course Consistency — Principle of instruction that states that information throughout a course should have the same level of accuracy, be presented with the same equipment and training aids, and maintain a similar level of learning.

Learning Environment Continuity — Principle of instruction that states that all information throughout a course should be in presented in a logical, understandable pattern; *also known as* Class Continuity.

Logistics — Process of managing the scheduling of limited materials and equipment to meet the multiple demands of training programs and instructors.

Training Aids — Broad term referring to any audiovisual aids, reprinted materials, training props, or equipment used to supplement instruction.

Chapter 4
Instructional Materials and Equipment

SECTION A INSTRUCTOR I

Chapter Contents

Purposes and Benefits of Lesson Plans ...63
- Lesson Plan Components............ 64
- Lesson Outlines and Learning Objectives 64

Copyright Laws and Permissions 69

Teaching Aids...................................71
- Nonprojected Teaching Aids........... 72
- Projected Teaching Aids.............. 75
- Simulators 79
- Training Props 80

Cleaning, Care, and Maintenance of Teaching Aids 81

Benefits of Teaching Aids and Props 82

Chapter Review 83

Discussion Questions 83

Key Terms 84

Skill Sheet 85

JPRs addressed in this chapter

This chapter provides information that addresses the following job performance requirements of NFPA 1041, *Standard for Fire Service Instructor Professional Qualifications*, 2019 Edition.

4.2.2   4.4.2
4.2.3   4.4.3
4.3.2   4.4.5
4.3.3

Learning Objectives

1. Describe the purposes and benefits of lesson plans. [4.2.2, 4.2.3, 4.3.3, 4.4.3]

2. Discuss copyright laws and permissions involved with instructional materials. [4.3.2, 4.3.3]

3. Describe the variety of teaching aids and instructional technology tools that an instructor may use in the learning environment. [4.4.2, 4.4.5]

4. Describe the cleaning, care, and maintenance of teaching aids. [4.4.5]

5. Summarize the benefits of using teaching aids and props in the learning environment. [4.4.2]

6. Skill Sheet 4-1: Adapt a prepared lesson plan provided by the authority having jurisdiction (AHJ). [4.2.3, 4.3.2, 4.3.3]

Chapter 4
Instructional Materials and Equipment

The Instructor I should teach from a prepared lesson plan that includes a list of any additional materials or equipment that may be needed to teach the lesson. Instructors need to understand the components of this prepared lesson plan and how to apply them to a training or classroom session. Also, instructors should be familiar with the equipment they may be required to use while teaching prepared lessons. This chapter discusses various materials that instructors will typically use in the classroom.

Purposes and Benefits of Lesson Plans

Instructors should never walk into a classroom and begin teaching without some plan as to what they will do, where they will go with the information, and how they will get there. Teaching without a plan gives no guarantee that course objectives will be met or that students will actually learn what is required by the course outcomes or NFPA **job performance requirements (JPRs)**.

An instructor should know how to build a lesson plan, and how to evaluate an existing lesson plan. A well-developed lesson plan will guide the instructor through the class session, and will indicate which instructional methods and **instructional materials** have been chosen to aid the process. When using a third-party prepared lesson plan, ensure that it will meet the needs and experience levels of the instructors and students.

Benefits that lesson plans provide to students, instructors, and administrators include:

- Standardizing the instruction. Enabling instructors to provide the same information in a similar format each time the lesson is taught makes teaching easier. For administrators, lesson uniformity also helps to ensure that a cadre of employees perform consistently and meet the requirements of the job.
- Giving a clear path for both instructors and students to follow. Sequential, orderly instruction makes learning more interesting and meaningful.
- Helping create consistency for administrators when more than one instructor must teach from the lesson plans. For instructors, lesson plans standardize the sequence in which the content should/will be taught in the following ways:
  — Introduce the material.
  — Guide students through practical training evolutions to meet performance objectives.
  — Guide students through a summary of the key learning objectives.
- Providing documentation for the training division and the organization's administration, which may include:
  — Teaching/learning information, methods and activities, and time frames for lessons.
  — The amount and type of materials, equipment, and other resources needed to teach the lesson, which in turn provides justification for budget requests involving material and equipment purchases.
  — Information for department or organizational accreditation by third-party agencies.

- Providing a document for developing test and evaluation requirements in order to:
  — Establish the testing criteria.
  — Show the material that was taught (based on the objectives).
  — Verify that the information presented is appropriate for testing.

Lesson Plan Components

An administrator may provide an instructor with a checklist of components for evaluating an existing lesson plan or for building a new lesson plan. In the most basic format, lesson plans consist of the following components:

- **Job or topic** — Short descriptive title of the information covered. The title briefly describes or indicates the lesson content. Topic titles usually come directly from the course outline.
- **Prerequisites** — List of information, skills, or previous requirements that students must have completed or mastered before entering this course or starting this lesson.
- **Time frame** — Estimated time it takes to teach the lesson. Lesson planners may set a time frame for each lesson objective so that the instructor has a better idea of how to pace the lesson.
- **References** — List of specific resources, textbooks, and other instructional materials on the lesson plan, along with page numbers for reference and review. References provide additional material to help instructors enhance their lessons.
- **Level of instruction** — Desired learning level that students will reach by the end of the lesson. It may be based on NFPA JPRs, or on an academically-established taxonomy.
- **Learning objectives** — Description of the minimally acceptable behaviors or skills that students must display or perform by the end of an instructional period.
- **Lesson outline** — Summary of the information to be taught (**Figure 4.1**). It may use the four-step instructional method, which will be discussed in greater detail in Chapter 6, Classroom Instruction.
- **Lesson summary** — Restatement or re-emphasis of the key points of the lesson in order to clarify variables, prevent misconceptions, reinforce learning, and improve retention.
- **Resources/materials needed** — List of all items needed to teach the number of students in the course. This section of the lesson plan includes other pre-lesson planning and preparation on the part of the instructor, such as:
  — Determining the appropriate training site and seating arrangement
  — Arranging for audiovisual equipment, teaching aids, props, and devices
  — Reproducing handouts
  — Acquiring tools, apparatus, and other equipment
  — Contacting guest speakers
  — Listing lesson resources; including information on instructor qualifications, textbooks and other instructional materials, and special equipment needs.
  — Determining specific instructional methods and learning activities required to meet objectives.
  — Selecting topic-specific training locations or changing training locations to meet the requirements of the topic being taught.
- **Assignments** — Readings, practice, research, or other outside-of-class requirements for students.
- **Evaluations** — Type of evaluation instrument the instructor will use to determine whether students have met lesson objectives.

Lesson Outlines and Learning Objectives

Level I instructors are normally provided a prepared lesson plan to guide instruction. However, they may encounter situations in which they have to create training exercises without the benefit of previously developed materials, or they may have to adapt existing materials to fit their needs. In these situations, instructors should realize that they do not need to produce formal lesson plans and curriculum like certified Level II and III Instructors. However, the ability to create a brief lesson outline with learning objectives, or to adapt previously developed content and components, are useful skills that meet the needs of the students and the objectives of the lesson plan. Whenever possible, the most experienced Instructor I available should be responsible for creating any needed lesson outlines. For additional information on the formal preparation of lesson plans, refer to Chapter 11, Lesson Plan Development. Steps for adapting a prepared lesson plan provided by the AHJ are presented in **Skill Sheet 4-1**.

Chapter 2
Cultural Change in the Fire Service

Lesson Goal

After completing this lesson, the student shall be able to understand the cultural heritage of fire fighting, the reason why it must shift to one that emphasizes safety for the benefit of all involved, and methods to implement this change.

Learning Objectives

Upon successful completion of this lesson, the student shall be able to:
1. Identify why cultural changes are needed in the fire service.
2. Describe the cultural evolution of the fire service.
3. Identify barriers to change.
4. Describe common change models.
5. Describe the benefits of cultural change.
6. Identify safety initiatives aimed at fire and emergency services.

Instructor Information

This is the lesson covering the cultural change towards safety in the fire service. This lesson describes the need, barriers to, and the benefits of change. The lesson also covers methods of implementing change and safety initiatives.

Important instructor information is provided in shaded boxes throughout the lesson plan. Carefully review the instructor information before presenting the lesson.

Methodology

This lesson uses lecture and discussion. The level of learning is comprehension.

Figure 4.1 The first page of a lesson outline can offer instructors a thorough summary of the content, including learning objectives, that will be covered during the lesson.

Basics of Learning Objectives and Lesson Outlines

When writing a basic lesson outline, the instructor should first consider the intended outcome of the lesson. This **learning outcome** determines both how to introduce the lesson to the students and the direction that the lesson will take. **Learning objectives** are the steps an instructor needs to follow in order to reach the learning outcome. Each objective should contain an action to indicate knowledge or skill, such as: "The student will identify a patient's pulse at three different locations on the patient." Some appropriate action verbs for use in learning objectives are shown in **Table 4.1**. For more information on learning objectives and lesson outlines, see Chapter 11, Lesson Plan Development.

**Table 4.1
Common Action Verbs for Learning Objectives**

Recognize	Identify	Design
State	Locate	Manage
Select	Explain	Estimate
Compare	Demonstrate	Evaluate
Determine	Analyze	Measure

A lesson outline is useful when training must continue but may not be offered as part of a larger course curriculum or at a formal training facility. When creating the outline, the instructor should begin with the learning objectives, then consider, in sequence, what points or steps should be followed to guide students toward meeting the objectives **(Figure 4.2)**.

One of the most widely-used learning outcome models, developed by Robert F. Mager in the early 1960s, continues to help define the role of learning objectives in training. According to Mager, learning objective statements should contain three components:

1. Performance (behavior) statement
2. Conditions description
3. Standards criteria

The *performance (behavior) statement* identifies what the student is expected to do. The behavior must be stated in observable terms and include a clear action verb such as *recall, identify, list, label, describe,* or *state*. Examples:

- Learning objective: "Describe the safety precautions used when ventilating a pitched roof with a power saw."
- Cognitive performance/behavior: "The student lists the safety precautions for ventilating a roof."

The *conditions description* describes the situation, tools, or materials required for a student to perform a specific action or behavior. Examples:

- Learning objective: "Given an adult cardiopulmonary resuscitation (CPR) training manikin, administer CPR."
- Condition: "The CPR training manikin must be accessible for the performance of administering CPR."
- Psychomotor performance/behavior: "The student will accurately perform CPR."

The *standards criteria* state the acceptable level of student performance. Standards provide measurable criteria for evaluating student performance and may include a statement about the degree of required accuracy or a time limit for completion. Examples:

- Learning objective: "Given photos of extrication tools, identify (label) 90 percent of the tools accurately."
- Criterion: "The standard requires 90 percent accuracy."

Brief learning objectives generally describe what performance is required of the students. If the instructor can pinpoint this piece of information, he or she can move on to outlining how to construct a basic lesson to guide students to the learning objectives.

Fire Instructor I   **Student Lesson 1**

3 min.
Lecture/
Discussion

     I.   **Preparation**

         A.   Administrative Details

             1.   Fire alarm.

             2.   Exits from the classroom.

             3.   Student roster.

         B.   Instructor Introduction

Easel Pad

> Print your name and contact information on an easel pad sheet. Reveal the sheet during your introduction so that students will have the correct spelling of your name and your contact information in case they need assistance following the lesson.

             1.   Name of instructor.

             2.   Experience in the fire service.

             3.   Experience with the use of fire hoses.

             4.   Training and certifications.

         C.   Motivation of Students

> It is critical that the students understand and appreciate the importance of being able to quickly deploy fire hoses, including coupling hoses. Review the following key points about the use of fire hoses. If possible, share a personal experience or an anecdote that illustrates the importance of being able to couple hoses without hesitation or coaching during an emergency operation.

Oklahoma State University Fire Service Training
2018

Figure 4.2 Instructors should follow a careful sequence of information while teaching from a lesson outline.

> **Comparing Job Performance Requirements (JPRs) to Learning Objectives**
>
> Fire and emergency services instructors who administer NFPA professional qualifications standards are familiar with job performance requirements (JPRs). They describe the skills required to perform a specific job and are grouped according to the duties of that job. The complete list of JPRs for each duty defines what an individual should be able to do in order to successfully perform that duty. Together, the duties and their JPRs define the job parameters.
>
> Annex C of NFPA 1041, *Standard for Fire Service Instructor Professional Qualifications*, explains the components of the JPRs and how to write them. The three critical components of JPRs, and their parallels to the Mager Model, are as follows:
>
> 1. Task to be performed — Performance (behavior)
> 2. Tools, equipment, or materials required to perform and complete the task — Conditions
> 3. Expected performance outcome — Criteria
>
> The following example, found in Annex C, shows how the three components of a JPR correspond to the Mager Model:
>
> *Ventilate a pitched roof (performance), given an axe, a pike pole, an extension ladder, and a roof ladder (conditions), so that a 4-foot × 4-foot (1.22 m by 1.22 m) hole is created, all ventilation barriers are removed, ladders are properly positioned for ventilation, and ventilation holes are correctly placed (criteria).*

NOTE: Because NFPA standards are written in the JPR format, this manual will frequently use the term "JPR" when discussing learning objectives. However, instructors and students should be aware that not all emergency services, such as emergency medical services (EMS) fields, utilize the JPR format for their objectives.

Lesson Outline Resources

Basic outlines may not provide enough information to adequately teach a topic or course. Well-prepared instructors often search for additional resources to reinforce information provided in lesson plans. While lesson outlines provide the theoretical concepts and key points of a topic, outside resource materials can provide practical real world examples. Some of the more common resources for fire and emergency instructors include:

- **National Fire Academy** — A division of the United States Fire Administration; provides training and training resources for emergency responders.
- **National Fallen Firefighters Foundation** — Created by the U.S. Congress to remember America's fallen firefighters and provide support and resources for their families; provides resources for firefighter safety education, including the Everyone Goes Home initiative and 16 Firefighter Life Safety Initiatives.
- **Firefighter Close Calls** — Compiles and publishes near-miss reports on firefighting incidents for training use.
- **National Institute of Standards and Technology (NIST)** — Produces research reports regarding emerging technologies in the fire service and new information about fire and explosion science.
- **National Institute for Occupational Safety and Health (NIOSH)** — Produces incident reports that describe fire and emergency service accidents and fatalities, as well as policy and procedure recommendations to help prevent future occurrences.
- **User-generated video websites** — Fire and emergency services organizations sometimes upload training videos to such sites. Eyewitnesses may also take video of accident and fire scenes and make this video public.
- **Supplementary texts** — Fire and emergency service periodicals and journals publish articles that can be used to reinforce daily lessons.
- **Fire and emergency services related websites** — There are many websites dedicated to the fire and emergency services community. Many feature community forums, blogs written by experts in the field, and articles that may or may not appear in print publications.

Copyright Laws and Permissions

When an instructor uses and distributes additional training materials, he or she must consider what is and is not permissible. Copyright laws provide legal guidelines for the use of published materials.

Copyright laws protect the work of artists, photographers, and authors, including but not limited to their literary, musical, graphic, audiovisual, and sound recording creations. They also give authors or developers the exclusive rights to publish their own works and/or to determine who else may publish or reproduce them. Since the *Copyright Act of 1976* was passed, the majority of U.S. copyright laws are federal statute and include the following provisions:

- All works published in the U.S. before 1923 are considered **public domain** (may be freely used by anyone).
- Works published between 1923 and 1977 are protected for 95 years from the date of publication.
- When the work was created (but not published) before 1978, the copyright lasts for the life of the artist, photographer, or author plus 70 years.
- For works published after 1977, the copyright lasts for the life of the artist, photographer, or author plus 70 years.

NOTE: A copyright holder can renew the copyright on material that he or she owns. Always check the status of the material in question's copyright, even if the item is old enough to be in the public domain.

Unauthorized use of copyrighted materials is considered infringement of the rights of artists, photographers, or authors. Infringement gives them a right to recover damages or gain profits from the use of their works.

Copyright laws may even apply to different parts of a single work. Instructors need to be aware that there may be copyrighted sections, which they do not have permission to reproduce, within public domain materials. For instance, the National Fire Academy (NFA) gains copyright permission to use certain materials in their courses. While the course materials are in the public domain, copyrighted portions of those course materials are not. Only the NFA has gained permission to use them. An instructor or organization not affiliated with the NFA would have to request (and receive) the same copyright permissions as the NFA in order to legally reproduce the material.

Instructors and students often copy materials from texts, journals, periodicals, and the Internet for use in class. Whether this replication counts as copyright infringement depends on how the material is used. Section 107 of the *Copyright Act* offers a statutory structure for **fair use** and lists the limitations on exclusive rights.

A few of the fair use guidelines are as follows (**Figure 4.3**):

- Instructors may make single copies of the following for scholarly research or when preparing to teach a class:
  — Chapter from a book
  — Article from a periodical, newspaper, or the Internet
  — Short story, essay, or poem

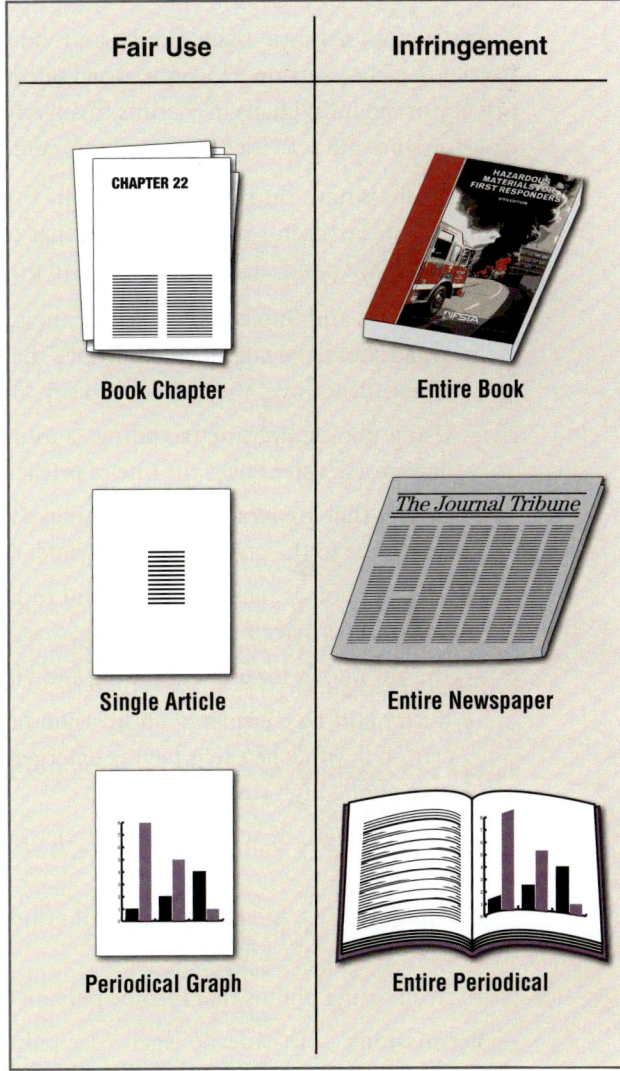

Figure 4.3 Instructors should ensure the materials they reproduce are fair use and take extra care to avoid infringement.

Chapter 4 • Instructional Materials and Equipment

— Chart, graph, diagram, drawing, cartoon, or picture from a book, periodical, or newspaper

— Video recordings of a television broadcast

- Teachers may make multiple copies of items for their students to use in class provided that the copied material is brief. Copying must be for the particular class being taught at the time and *not cumulative* or copied repeatedly for subsequent classes.
- Copying is not a substitute for the purchase of books, publisher's reprints, or periodicals. Students purchasing copies for class cannot be charged more than the actual cost of the photocopying.

Instructors must properly cite any sources they use and familiarize themselves with the applicable copyright laws regarding copying print, audio, video, or electronic materials they plan to use in class. Copyright laws that apply to the Internet or to digitally-reproduced material are constantly evolving, and instructors must monitor those changes.

Many instructors are tempted to search the Internet for supplementary material and reproduce that material for classroom use. They may assume that all materials on the Internet are public domain (free for anyone to use, regardless of copyright). However, materials on the Internet may be copyrighted even without notice. The rules of fair use apply to such materials, as do the copyright laws governing printed materials.

Instructors may use photographs and video recordings taken at emergency incidents to add relevance and impact to a class session. However, when selecting materials for this use, instructors must be careful to guard the privacy of the individuals or victims involved in the incident. **Invasion of privacy** is the wrongful intrusion into a person's private activities by the government or other individuals.

Individuals have the right to control the use of pictures of themselves and their property. If permission is *not* obtained, an instructor or organization that uses these pictures or films may be sued for invasion of privacy or libel. The legality and success of lawsuits to protect the right of privacy depend on several factors, such as:

- **Location(s) and Subject(s)** — Where the photograph or film/video was made and who was photographed or recorded should be documented. Events that are newsworthy and photographed or recorded for public interest take precedence over the right of privacy. Details:
  — Using photographs or recordings a month after an incident has lost its public relevance and is no longer newsworthy steps over the line of privacy.
  — Images that are retained as file photos and used in conjunction with a story of continuing interest may not be subject to the privacy requirements.
  — Taking photographs or video/audio recording is allowed when the setting is a public place rather than an individual's private residence.
  — The law allows for photographing and videotaping famous individuals who are seen in public.
  — It may still be considered an invasion of privacy to photograph or film any ordinary civilian, celebrity, or criminal, while he/she is having emergency procedures performed or being rescued from a serious accident on a public highway.
  — Using images of severely injured victims or fatalities may infringe on the rights of victims and/or their families.
  — Avoid making and using images of minors. Should this be unavoidable, confirm and document all permissions and applications.
  — Avoid using photos that include personally identifying information, such as license plates.
- **Permissions** — In order to legally use/publish photos or film from a public event, organizations must always obtain written permission (a **model release**) from any and all individuals who appear in those images. A model release for a minor must be signed by the model's parent or legal guardian.

- **Restrictions** — Organizations must explain how they plan to use a photo or video so those involved can make an informed decision to permit or restrict the use of their images. Educational and training organizations must also obtain model releases from any students they photo or videotape for:
  — Training sessions
  — Departmental/organizational web sites
  — Closed circuit or public broadcasts of organizational-sponsored television programs

  NOTE: Additional limitations are placed on the length of time that a copy of an on-air broadcast can be retained and used. The producer of the material grants permission to record a broadcast and determines how long it may be used.

- **Rights of individuals** — Individuals may ask to preview photos or recordings before they are used. They may also require anonymity while showing the photograph or recording, which the organization may accomplish by blocking faces with graphic overlays or shadows and altering voices. Instructors must honor these requests, if received.

Teaching Aids

Regardless of the type of instruction, teaching aids and instructional technology tools are integral to engaging student interest and encouraging active learning. The availability of teaching aids depends on an organization's needs, budget, and facilities. The following guidelines can help instructors use a variety of teaching aids more effectively:

- Illustrate a single concept or idea in each teaching aid, in order to avoid "overload" of too many concepts in a single aid.
- Introduce audiovisual teaching aids just before presenting them.
- Display individual steps in the sequence of an operation. Displaying them all at once can be confusing. After introducing all the steps individually, instructors can display them together so that students can see how they are related to each other.
- Avoid using multiple teaching aids simultaneously, except when their use has been carefully considered and strategically integrated into the lesson plan.
- Apply teaching aids in ways that emphasize the message, not the equipment.
- Maintain eye contact with students when using a visual aid.
- Ensure that all students can see the visual teaching aids.
- Ensure that all students in the presentation room can hear audio teaching aid devices clearly.
- Display projected or nonprojected visual teaching aids above the eye level of seated students.

Audiovisual teaching aids are also important during demonstrations. An instructor may use a model or simulation to show how a piece of equipment works. Students can view video recordings of step-by-step instructions in the training ground area while an instructor helps individual students. Props and equipment are also forms of audiovisual aids important to demonstrations.

Computer-based training (CBT), self-directed learning, and individualized instruction all benefit from audiovisual aids, as well. Instructors facilitating these instructional methods should become familiar with presenting audiovisual aids online. Aids may be located on a central website (streaming video), distributed from a database, or sent as attachments through email. Audio of in-class lectures may also be recorded and made available in digital formats. Handouts may be sent as files or downloaded. Also, instructors may share computer generated slides on web conferencing software to communicate with remote students.

The sections that follow categorize teaching aids and devices as nonprojected types, projected types, simulators, and training props. Each category includes descriptions of several types of devices or equipment. Also discussed are the cleaning, care, and maintenance of these teaching aids.

> ### Using Teaching Aids
> Instructors can create interesting and informative lectures by displaying illustrations or diagrams that they have purchased or prepared in advance. When displaying teaching aids, remember that they can also distract students, especially if the aids are displayed too early or too long. Avoid leaving illustrations and diagrams in view when they are not in use or relevant. While speaking, instructors should hold or position illustrations to one side, not directly in front of them. In addition, before using teaching aids, instructors should ensure that everyone in the class will be able to see them.

Nonprojected Teaching Aids

Popular and easy-to-use nonprojected teaching aids include the following:

- Marker boards and easel pads (**Figure 4.4**)
- Illustration or diagram displays
- Duplicated materials
- Models
- Audio recordings
- Casualty simulations

Figure 4.4 Paper from easel pads can be displayed on walls as part of learning activities or classroom discussions.

Nonprojected teaching aids offer several significant advantages over projected aids. For example, nonprojected teaching aids do not depend on high levels of technology or technical skill. As a result, they are easier to use and less likely to malfunction during presentations. Generally, nonprojection-type equipment is more economical, simple to use, and easy to maintain.

Marker Boards and Easel Pads

The easiest, most frequently used, and most versatile nonprojection-type equipment is a marker board. Dry-erase boards and easel pads can be fixed, mounted to a wall, or moved on wheels or stands. Some portable types can be folded to a compact size for travel. Instructors use marker boards for numerous instructional activities and rely on them for backup when more technical or complex classroom equipment fails.

While marker boards are considered low-tech, portable, versatile instructional materials and equipment, they possess the following limitations:

- Require consumable writing utensils
- Emit an odor that some may find offensive
- Present visual difficulty for some students
- Require continued cleaning

Instructors may also find that easel pads can be difficult to write on.

> ### Using Marker Boards
> When using marker boards, consider the following suggestions:
> - Lightly write any easily misspelled words or technical terms or lightly draw diagrams and other similar material on your easel-pad paper in advance with a pencil. The pencil marks are visible to the instructor standing at the front of the class but not to students. Quickly trace over the pencil marks with marker. This technique ensures accuracy and quick reproduction of material in class.
> - Draw complex or detailed diagrams in advance. If the classroom is shared with other instructors, indicate that the prepared materials are not to be disturbed. Keep diagrams covered until they are needed to avoid distracting the class. Use one of the following methods:

- Cover material with a sliding section of a marker board.
- Tape paper across material on boards.
- Cover material on an easel pad with blank sheets or title sheets.

- Avoid speaking when facing boards or easel pads. Instructors should pause speech when they turn away from the students to write.
- Write only what is necessary. It is rarely necessary to transcribe verbatim every spoken word of a lesson. Using key, concise points minimizes the writing time while keeping the board as uncluttered as possible.
- Choose marker pen colors that are easy to see. Black and other dark colors are recommended on light-colored boards or paper. Yellow or pastel-colored markers are difficult for students to see in a large classroom.
- Experiment with letter size in advance so that writing on visual aids is large enough to be read by students at the back of the room.
- Bring marker pens of several colors when scheduled to teach in an unfamiliar location and make sure that markers have not dried out prior to use.
- Use print, not cursive, when writing on marker boards.

Illustration or Diagram Displays

Illustrations and diagrams support visual learners and can help to clarify, organize, and emphasize key points of written or verbal instruction. Preparing these materials in advance saves class time. They can also be used to ensure consistency in the presentation of course material by different instructors. Some examples of illustrations and display boards are as follows:

- Technical diagrams, such as mechanical or electrical schematics, illustrate and help explain troubleshooting or repair procedures.
- Maps or plan diagrams illustrate routes, aid in preplanning, and/or assist in debriefing incidents.
- Anatomical charts assist in explaining human anatomy and physiology in emergency medical services (EMS) classes.
- Flowcharts help explain processes or procedures.
- Data charts, such as pie charts or bar graphs, illustrate common causes of injury, response types, or other statistical data.
- Photographs of incident scenes are invaluable in illustrating proper procedures.

Photos, illustrations, charts, and diagrams can be mounted or created on poster board. Maps may be mounted on rollers or hung flat from clips on marker boards. Easels may also be used to support mounted illustrations.

Duplicated Materials

Duplicated materials (handouts) include any printed matter that instructors distribute before, during, or at the end of a class. For students to gain the maximum benefit from handouts, instructors must strategically plan when and how to use them. Suggestions include:

- **Lecture review** — Distribute handouts of lecture material at the end of a presentation unless students need to refer to or take notes on them. If needed, distribute handouts just prior to the presentation so that students will remain focused.
- **Precourse material** — When handouts have the same content as the presentation, consider using them as precourse material. When students become familiar with the subject matter ahead of time, they can expand upon that learning/learn even more about the subject during class.
- **Self-study** — Give handouts that serve as self-study guides to help students work through textbooks and other learning materials at their own pace.

- **In-class assignments** — Hand out assignments to students for completion during the class session.
- **Take-home assignments** — Hand out outlines and requirements for take-home assignments, such as research projects, for practicing applications of theory learned in class, or reviews of material.
- **Note-taking guide** — Provide note guides with an outline of the lesson for students to fill with notes during class. These guides will encourage students to remain attentive to the lesson's major points.

CAUTION: Consult the copyright laws explained earlier in this chapter and apply them appropriately.

Using Class Handouts
When using class handouts, consider the following guidelines:
- Ensure that handouts are legible and complete for the intended purpose.
- Staple or clip together handouts with multiple pages. Students are less likely to lose or rearrange pages that have been stapled together in advance.
- Provide handouts with pages that have been three-hole punched in advance, unless they are bound otherwise.
- Include headers that contain the course name, instructor name, and date. When a handout has multiple pages, include page numbers.
- Provide sufficient copies for all participants. Always print more than are required, in case they are needed.
- Provide space on the handouts for taking notes.
- Be aware of and adhere to copyright laws.

Models

A model is an excellent tool for illustrating mechanical or spatial concepts. Students can clearly observe the types of relationships between parts of a model as they watch it function or manipulate it. Instructors can obtain or construct many types of models at low cost, but some types (depending on their size and complexity) require a large investment of time and money. Examples of models that are used as teaching aids include:

- **Tabletop miniatures** — Tabletop models showing a real or improvised city or building layout allow participants to enact a simulated incident in compressed time (5 minutes of simulation can represent 1 hour). Such models may allow for video monitoring from another location, simulated fires, and changing scenarios (**Figure 4.5**).
- **Cutaway models** — Models of equipment encountered during response are of great value when students are learning the inner workings of mechanical systems, such as valves or pumps. Instructors can often obtain cutaways at little or no cost by dismantling obsolete or surplus equipment. Some manufacturers may also have cutaway training models available (**Figure 4.6**).
- **Anatomical models** — Models of human physiological systems and structures are available in three dimensions, some of which have cutaway or take-apart features. These models benefit students who are developing EMS skills and

Figure 4.5 City miniatures (also known as tabletop models) offer flexibility during simulated training scenarios.

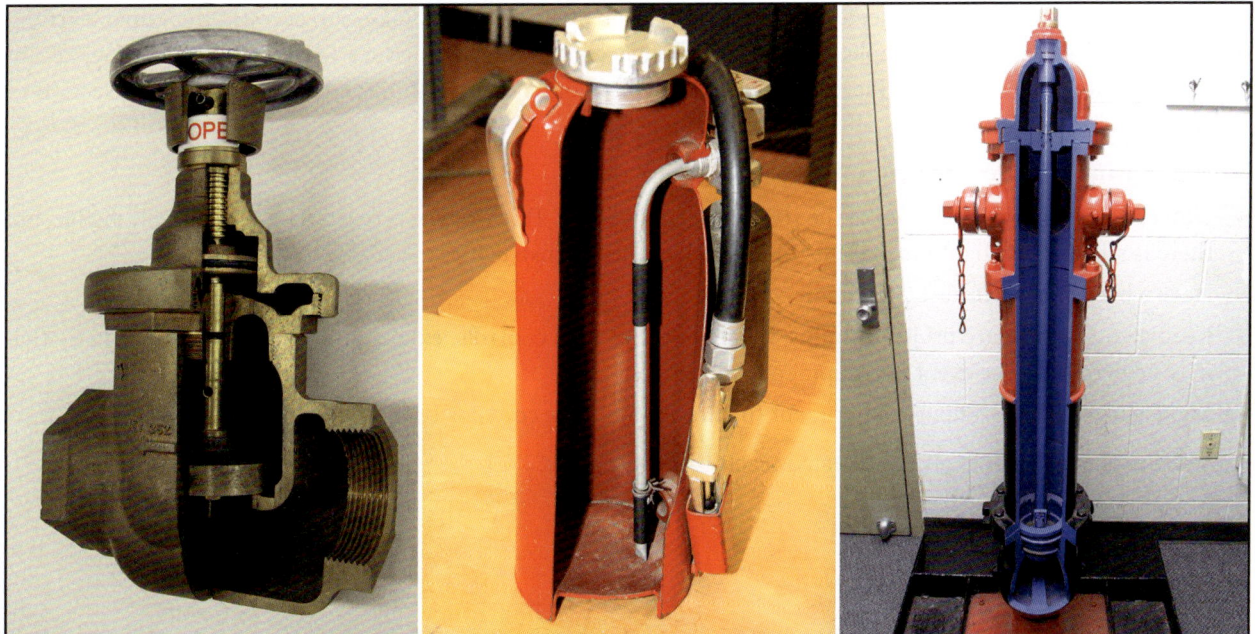

Figure 4.6 Cutaway models are useful for demonstrating how a piece of equipment functions.

knowledge, such areas as the mechanisms of internal injuries. A life-size or even small-scale skeleton is also a good anatomical model. Anatomical models are different from manikins in that they represent a specific system or function. Manikins are discussed later in this chapter.

- **Replicas or miniatures** — Replicas can demonstrate how actual equipment or devices are used. For example, replica cribbing and shoring assemblies may be based on actual designs, using stated equipment and resources. In contrast, manufacturers may offer smaller scale versions of equipment that, full-size, would occupy a significant amount of space. In both cases, students can see how the components are assembled and used before a practical training evolution.
- **Actual tools and equipment** — Actual equipment can also be used for demonstrations. For instance, an authentic self-contained breathing apparatus (SCBA) is a better teaching aid than an illustration of one.

Recordings

To add a level of authenticity and practicality to lessons, instructors may wish to include sounds that relate to the lesson. Examples of sounds that can be played for use in the classroom include:

- **Engine and pump sounds** — To familiarize students with normal operation and problem identification, such as pump cavitation
- **Dispatch radio traffic** — To highlight good practices in telecommunicator training or postincident critique
- **Heart, breathing, or blood-pressure sounds (for EMS training)** — To acquaint the student with physical conditions and body functions of victims in the field

Instructors may create their own digital-audio recordings of these sounds. Digital files are also easy to transfer to a student's personal electronic device or computer for listening outside of class, provided the use of the material does not infringe upon copyright.

Projected Teaching Aids

Using projected teaching aids, such as **computer-generated slide presentations**, is standard practice in today's fire and emergency services. Projected teaching aids are electronically displayed on a screen. They offer many advantages for classroom presentation, including vivid, multicolored images that are visible to a large audience. Most educational materials provided to Level I Instructors come with projected teaching aids, such as computer-generated slides, photographs, or training videos.

NOTE: When developing projected teaching aids, the instructor should take into account the visual learning issues of their students (for example, some students have contrast or color issues when looking at a PowerPoint slide or website). For more information, refer to the Web Content Accessibility Guidelines (WCAG) section of **Appendix B** in the manual.

When using projected aids, consider the following actions:

- Dim only the lights in the front of the room, if at all possible. This way the projected image is clearly visible, but there is still enough light for students to take notes and for instructors to see the students.
- Have a backup plan in case projection equipment fails.
- Keep a supply of spare projector bulbs, batteries for remote controls, extension cords, and anything else that may be needed in the event of equipment problems.
- Make sure that the computer equipment (including interface cords) is compatible with the projection device.

Some of the disadvantages of projected teaching aids include:

- Cost
- Upkeep
- Technological updates
- Immobility
- Necessity of a clean environment

> **Changing Technology in Presentation Aids**
>
> Instructors should always confirm the type and content of teaching aids, especially those they have not created, before using them in the classroom. The instructor should also ensure that the required equipment is available and in working condition.
>
> Some departments may still use video cassette recorders (VCRs) to play training materials recorded on VHS tapes. New materials are produced and made available predominately on DVD, Blu-Ray, or USB drive.
>
> Some departments may still have carousels of slide photographs and transparencies for use on overhead projectors, even though new materials are not being produced using this technology. Both slide machines and overhead projectors are relatively easy to use.
>
> Instructors who are required to use these resources and devices should read the instructions for their use and follow them accordingly. When possible, content should be updated to a more commonly accessible format to prevent its eventual loss if the equipment to play that content becomes unavailable.

Multimedia Projectors/Large-Screen Images

A projection system is a device for showing video, television, or computer images on a large screen. Projectors that are bright enough to display a quality video or computer-generated image in a classroom are affordable, portable, and versatile.

Projectors can be attached to a variety of media devices, such as DVD players, Blu-Ray players, laptops, desktop computers, and personal media devices (**Figure 4.7**). Projectors require a screen or wall to display an image.

NOTE: If no screen is available, a light-colored wall — with no decoration, minimal texture, and few architectural elements — will provide the best viewing surface possible.

Multimedia projectors are even more effective when used with computers. This combination allows instructors to display documents for editing or group discussion, play videos from the Internet or the computer's hard drive, or show simulation software to an entire class.

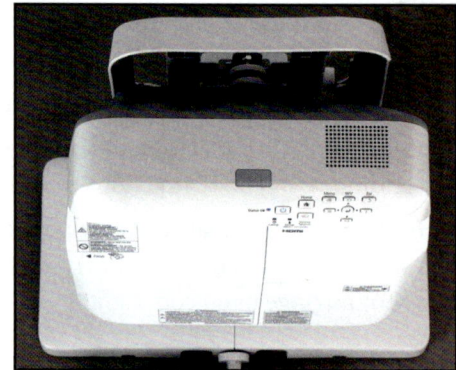

Figure 4.7 Modern multimedia projectors, such as this wall-mounted device, work well with various media devices and allow the instructor to display clear information on a large screen.

Keystoning

When a projector is not aligned perpendicular to the screen, the display will appear wider at the top than at the bottom **(Figure 4.8)**. This distortion is referred to as "keystoning" because of the shape's similarity to the top block in an arch. Keystoning is distracting because the top and bottom of the image is out of focus while the center is sharp.

Some portable projection screens have a keystone eliminator that allows the instructor to adjust the angle of the screen at the top. Most multimedia projectors include an electronic keystone eliminator, which allows the instructor to correct the display on either a portable or fixed projection screen.

Interactive Display Systems

An interactive display system, (for example, a Smart Board™) is a system that projects images from a computer and allows both students and instructors to interact with those images **(Figure 4.9)**.

The display system becomes a giant touch-screen that permits users to interact with the display, visit websites, and access the computer's applications and files. With interactive display system tools, the instructor can "draw" lines on the screen to emphasize portions of a photo or document **(Figure 4.10)**. An interactive display system can also be used to display computer-generated slides.

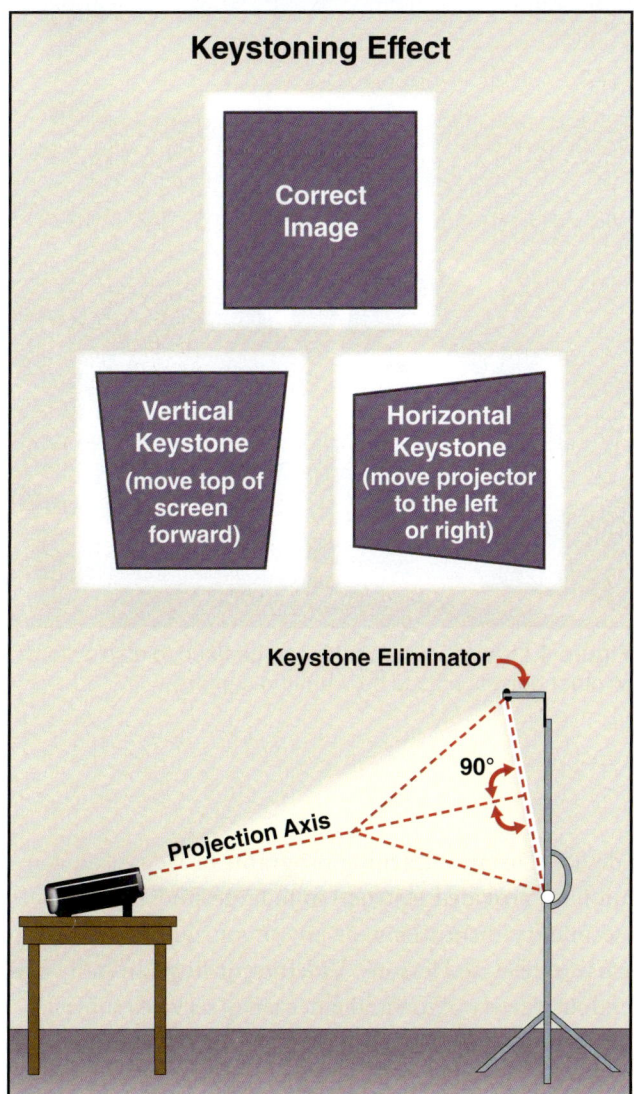

Figure 4.8 Instructors should adjust a projector's display to eliminate keystoning.

Figure 4.9 Instructors can use interactive display systems to create a giant touch-screen where information can be changed or manipulated.

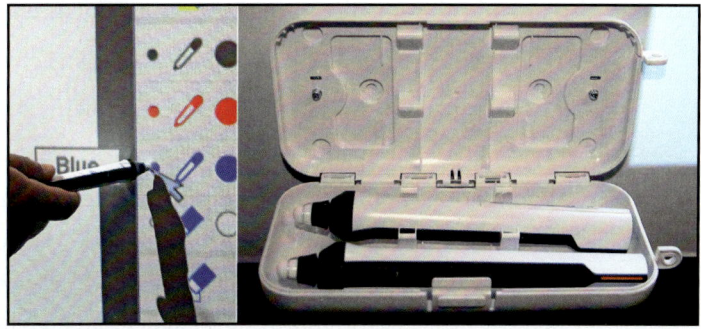

Figure 4.10 Multimedia projector tools, such as these interactive pens, can interface directly with a screen, allowing the instructor to "draw" lines and emphasize images, words, or photos.

Chapter 4 • Instructional Materials and Equipment

NOTE: There are nonprojected interactive board types that are similar to interactive display systems. Instructors are encouraged to consult their supervisors or the AHJ for the most suitable or available equipment for their course.

Visual Presenters

The visual presenter/document camera consists of a small video camera mounted vertically over a table-top platform **(Figure 4.11)**. Adjustable light units are mounted on either side of the platform. The platform also contains a light source to provide back lighting for transparencies. Some models may be portable.

The visual presenter displays images of objects, documents, or transparencies that have been placed in view of its camera. For large audiences, the image can be projected onto a television monitor, interactive display system, or screen so that all students can see it at the same time. Some models also allow for external audio that can be used as a public announcement (PA) system in a small classroom.

A visual presenter is helpful to instructors who want to display:

- Objects, such as tools or material samples
- Paper copies of documents
- Original photographs or illustrations
- Images or text in books or magazines
- Transparencies

Figure 4.11 A visual presenter can be used to display tools or other objects placed beneath the camera.

Video Presentations

Television has had tremendous impact as a teaching aid device. Distance learning programs broadcast on community-access channels via cable, satellite, or closed circuit have provided instruction to large student audiences over vast distances. State or provincial authorities, postsecondary institutions, or any organization wishing to reach a diverse group of students have also made effective use of televised lessons. Video recordings, initially created and marketed for televised learning, address a tremendous variety of fire and emergency services subjects.

Many older videos, as well as a wealth of new training material, are now available online. Many organizations also provide high-quality in-house training videos on their websites.

When using a video presentation in class, instructors should observe the following guidelines:

- Always preview a video privately before showing it publicly.
- Emphasize key learning points/objectives before showing a video.
- When showing only a portion of a recording, cue the video to the desired location in advance.
- Do not attempt to talk over the soundtrack. To discuss a particular point, pause the video. If the video must continue in order to illustrate the point, mute the sound.
- Remain in the classroom while students watch a video, as you may need to address technical problems, classroom disruptions, or other issues.

- Break up a long video presentation with instructor-class interaction or practical evolutions. A video presentation should not take more than half of the class time.
- Review key learning points/objectives after viewing a video.

> **Internet Access**
> Before showing/streaming online videos in class, instructors must confirm that their location has adequate Internet access. To avoid concerns about a stable connection, video content may also be downloaded before class starts.

Video Conferencing

Video conferencing allows two or more individuals or groups to communicate with each other at a distance through live, simultaneous video and audio connections. Video conferencing is similar to traditional instruction because instructors teach a group of students in a regular on-site classroom. However, the video conferencing classroom contains audio and video equipment that enables students from remote locations to interact with the group.

> **CAUTION:** Use of video conferencing must include careful class session preparation and attention to detail in interactions between students and instructors, or the experience may be negative for all parties.

Simulators

Simulators are teaching aids that represent systems, processes, or environments in which actual training would be unsafe, impractical, or prohibitively expensive. Simulators realistically imitate these environments for students without the safety hazards. The more realistic the simulation, the more effective the learning. The following types of simulators are available for fire and emergency services training:

- Computer simulations
- Virtual reality simulations
- Casualty simulations
- Anatomical/physiological manikins

Computer Simulations

Computerized simulations of burning buildings, command situations, casualty incidents, hostage scenarios, and other fire and emergency services applications are widely available. Computer simulations include three-dimensional (3D) rendered animations and interactive video.

Virtual Reality Simulations

Virtual reality simulations display fields of view as though students are part of the simulated environments. Environments can be manipulated in response to input from students. The simulations change in real time to react to a student's actions. Virtual reality simulations can be tremendously effective tools for reinforcing procedure-based or protocol-based skill sets.

Most of these simulations require an investment in equipment and/or software, which must be weighed against the instructional value of the product. Many simulations have built-in study materials and self-tests. Some types of simulators, and/or their additional programs, may be installed directly into an organization's computer network, enabling the instructor to keep centralized training records and statistics.

Disadvantages of virtual reality simulations include:

- Potential for light-induced seizures
- Possible motion sickness

- Extreme high costs
- Limited training scenarios due to pre-programming

Casualty Simulations

Simulated casualties provide tremendous benefits by increasing the realism of hands-on EMS training. Instructors can simulate injuries using commercially available **moulage kits** and prostheses, or applying Plasticine modeling paste, wax, and makeup (**Figure 4.12**).

Moulage kits typically contain plastic wounds that instructors can apply to a simulated casualty. Prostheses are also available, such as simulated amputated limbs, devices to simulate arterial bleeding, and other lifelike injury effects. Minimal training is required to use moulage kits. Plastic wounds are relatively quick to prepare and apply.

Instructors can also arrange to work with personnel who are qualified casualty simulators. Casualty simulators are typically EMS instructors and have been trained in the use of Plasticine, mortician's wax, makeup, prostheses, and simulated blood to produce realistic wounds and other special effects. The preparation process is more time-consuming than moulage, but the results are more realistic. For practical examinations or when filming simulations for later use, the time invested is worthwhile.

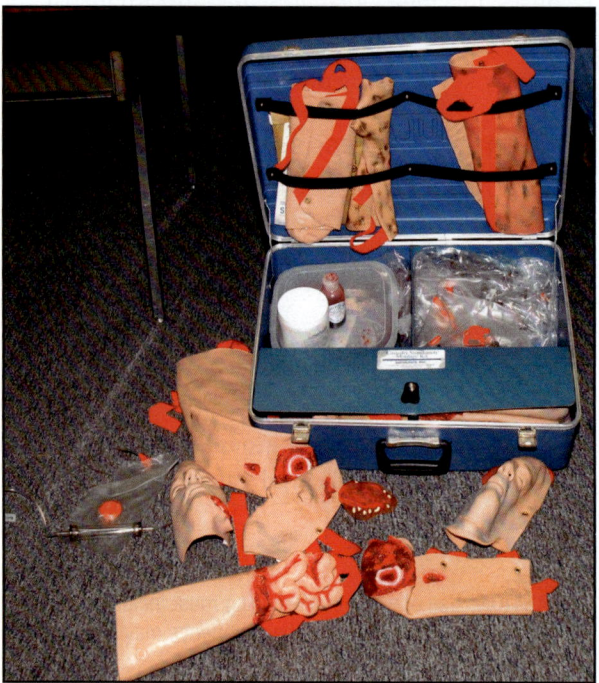

Figure 4.12 A moulage kit allows instructors to simulate an injury using realistic effects.

Anatomical/Physiological Manikins

Anatomical manikins are designed to offer types of simulation ranging from a simple representation of the human form to audible, visible, or palpable facsimiles of pulse, heart rhythm, reaction to defibrillation, or the ability to intubate or ventilate. Specific manikin models are designed to match any level of EMS training provided by an organization.

Training Props

Training props include permanent structures and portable devices. They simulate specific situations and assist in teaching subjects, such as:

- Technical rescue
- Vehicle extrication
- Spill and fire control of flammable/combustible liquids
- Transportation incident response and control

Designed for outdoor use, training props are often located at some distance from classroom facilities, which can present an obstacle for instruction. They are used for a variety of training scenarios, including rescue, property conservation, and fire suppression. Examples include:

- Trenches or collapsed structures (**Figure 4.13**)
- Motor vehicles (**Figure 4.14**)
- Railcars
- Ships
- Aircraft (**Figure 4.15**)
- Processing and storage facilities for flammable liquids

Figure 4.13 Search and rescue instructors can use prepared trenches to demonstrate environmental conditions and teach students proper shoring techniques.

Figure 4.14 Vehicle incident props allow firefighters to practice emergency responses to a vehicle fire or accident.

Figure 4.15 Simulated aircraft fires require extra containment measures and are designed for fire-suppression training.

What This Means to You

In order to use audiovisual aids successfully, instructors must take the time to learn about them. Find and fix any potential problems before instruction begins. Use the following guidelines to get accustomed to these teaching aids and to troubleshoot potential difficulties:

- Test the equipment before instruction begins.
- Use help menus and manuals to clarify issues.
- Ask questions of experts who have used the equipment, including the manufacturers.
- Explore any software that you will use as a facilitator in computer-based training (CBT) instruction.

Cleaning, Care, and Maintenance of Teaching Aids

Proper care and cleaning of teaching aids can prevent damage and extend their useful life. Instructors should regularly inspect all teaching aids before using them in class. Most manufacturers provide cleaning, care, and maintenance instructions when an item is delivered. File these instructions for reference and, if possible, attach a copy to the teaching aid or device.

Keeping teaching aids and instructional technology tools (devices, models, and equipment) clean is a fairly simple process. It is also an opportunity to periodically inspect the item and determine its maintenance needs. Cleaning suggestions include:

- Follow manufacturer-recommended procedures.
- Do not use abrasive cleaning agents on glass surfaces.
- Use a soft dust cloth to remove dust and fingerprints.
- Use a soft-bristled brush to dust hard-to-reach areas inside equipment.
- Do not use solvents.
- Clean dry erase boards completely after using them.
- Clean video and audio heads with an appropriate cleaning device at recommended intervals.

Suggested care guidelines for teaching aids and devices include:

- Follow manufacturer's recommendations for use and storage.
- Place dust covers over equipment when not in use.
- Store manikins properly in carrying cases or closed cabinets.
- Wrap power cords around carts, or remove and store them separately.
- Use lens caps on all optical lenses (cameras and projectors).
- Store class handouts in file folders.
- Allow projectors sufficient cool-down time after turning off the lamp. Many projectors have both soft and hard power buttons. Follow the manufacturer's guidelines carefully to prevent the cooling fan from disengaging prematurely and reducing the life span of the bulb.
- Do not leave electronic equipment in vehicles in direct sunlight, or when temperature extremes are expected.

Only qualified repair personnel should perform specialized maintenance of teaching aid devices. Instructors who are familiar with the equipment may perform routine maintenance, such as:

- **Inspecting teaching props** — Follow jurisdictional SOPs for inspecting props. Check for damage and deterioration before and after props are used in training evolutions.
- **Periodically cleaning air filters in multimedia projectors** — Clogged air filters cause cooling fans to work harder and decrease the life of the unit. Always clean the filter before replacing the projector bulb.
- **Replacing projector bulbs** — To prevent skin oil from contacting the bulb surface, handle the bulb by the porcelain base only (**Figure 4.16**).

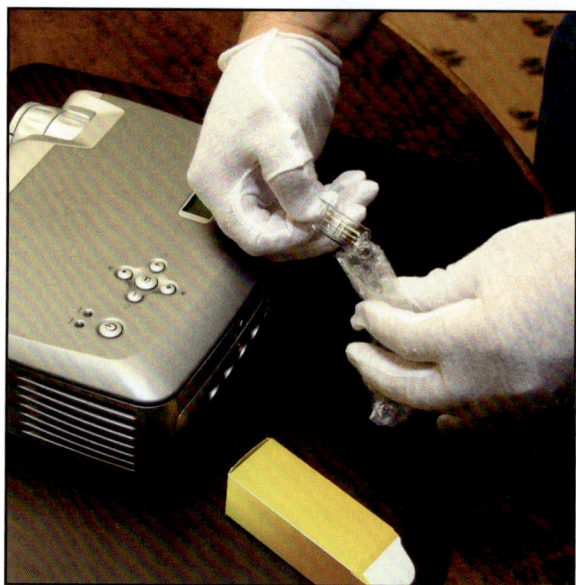

Figure 4.16 Clean, thin gloves can be worn to prevent skin oils from collecting on projector bulbs.

Benefits of Teaching Aids and Props

Teaching aids and props can improve the classroom experience for both students and instructors. Teaching aids help to:

- **Enhance student understanding** — Because a significant amount of what students learn is received visually, teaching aids increase the ability of students to understand the information in a lesson. Seeing an illustration or touching an item will help the item become more tangible to students, as opposed to simply hearing or reading a description of it.

- **Add interest to a lecture** — The additional visual or tactile stimulation provided by teaching aids adds variety to a lecture.
- **Clarify, prove, or emphasize a key point** — Seeing a key point projected for the class during an explanation or discussion can clarify and add emphasis to important information. Instructors can prove or validate a key point by using illustrations or examples of actual events.
- **Enhance memory** — Teaching aids visually reinforce the learning of information that is provided verbally, making it easier to remember and retain. Research indicates that people have better retention of information that they see and hear at the same time.
- **Help students organize ideas** — When dealing with complex ideas or processes, students benefit from seeing a list of components on a chart or board. Students follow a lesson better when they associate an illustration or outline with the words.
- **Gain and maintain students' attention** — Dramatic images or anecdotes help to capture attention and maintain student interest while the relevance or resolution of the topic is explained.
- **Illustrate a sequence of events or steps in a process** — Teaching aids help students to learn and remember steps in a sequence, stages in a process, or components of an assembly. They are especially helpful when a sequence cannot be easily explained, like the circulation of blood in the body.
- **Save lecture time** — Students can process more information when visual aids supplement a lecture. Less lecture time can make training more active and efficient.

Chapter Review

Answer the following questions to review the information provided in this chapter.

1. What are the purposes and benefits of lesson plans?
2. How are copyright laws and permissions involved when planning and/or using instructional materials?
3. What are some guidelines to help instructors use a variety of teaching aids and instructional technology tools more effectively in the learning environment?
4. What are some steps to clean, care for, and maintain teaching aids?
5. What are some benefits of using teaching aids and instructional technology tools in the learning environment?

Discussion Questions

The following questions are intended to generate discussion, expand your understanding of the chapter text, and allow you to think critically about what you have learned. Answers to these questions may vary.

1. Why should an instructor know how to construct a lesson plan?
2. Why is obeying copyright laws and permissions important in the learning environment?
3. What kinds of teaching aids or props have you found helpful as a student?
4. Why is it important for instructors to regularly inspect all teaching aids and instructional technology tools before using them in class?
5. How do teaching aids, instructional technology tools, and props improve the classroom experience for both students and instructors?

Key Terms

Computer-Generated Slide Presentations — Computer presentations that are sequenced and displayed like traditional slideshows that use a slide projector with a carousel. Popular software for creating and viewing these presentations include Microsoft PowerPoint and Apple Keynote.

Fair Use — Doctrine of the *Copyright Act* that grants the privilege of copying materials to persons other than the owner of the copyright, without consent, when the material is used in a reasonable manner.

Instructional Materials — Materials that an instructor may use to ensure and/or enhance a good learning experience for students; includes lesson plans, computer-generated slide presentations, lesson outlines, and student worksheets.

Invasion of Privacy — Wrongful intrusion into a person's private activities by the government or other individuals.

Job Performance Requirement (JPR) — Statement that describes a specific job task, lists the items necessary to complete the task, and defines measurable or observable outcomes and evaluation areas.

Learning Objective — Specific statement that describes the knowledge or skills that students should have acquired at the conclusion of a lesson.

Learning Outcome — Statement that summarizes what students will know or be able to do once learning is complete.

Model Release — Legal document that grants permission to use a person's image or voice in photos or recordings.

Moulage Kit — Makeup kit containing appliqué wounds and stage makeup; used during casualty simulations to imitate wounds on a manikin or actor.

Public Domain — Works of artists, photographers, and authors that were published before 1923 or are no longer covered by copyright ownership.

4-1
Adapt a prepared lesson plan provided by the authority having jurisdiction (AHJ)
[NFPA 1041 4.2.3, 4.3.2, 4.3.3]

Task Steps

Step 1: Review and analyze the prepared lesson plan provided by the AHJ.

NOTE: Analysis of the prepared lesson plan should include all available resources, facilities, and materials.

Step 2: Identify items in the lesson plan, learning environment, and/or resources that need to be adapted and/or adjusted for students.

Step 3: Adapt the lesson plan so that the needs of students and lesson objectives will be met.

Step 4: Confirm the adapted lesson plan is complete and well organized.

Step 5: Assemble and prepare materials for lesson delivery.

Chapter 5

Learning Environment

SECTION A INSTRUCTOR I

Chapter Contents

Classroom Environment 89
    Seating Arrangements 89
    Lighting 92
    Temperature and Ventilation 92
    Noise Level 93
    Audiovisual Equipment 93
    Other Learning Environment
        Considerations 94

Training Ground Environment 96
    Remote Sites 96
    Permanent Training Facilities 100
Chapter Review 101
Discussion Questions 101
Key Terms 102
Skill Sheet 103

JPRs addressed in this chapter

This chapter provides information that addresses the following job performance requirements of NFPA 1041, *Standard for Fire Service Instructor Professional Qualifications*, 2019 Edition.

4.4.2

4.4.5

Learning Objectives

1. Describe considerations for classroom learning environments. [4.4.2, 4.4.5]

2. Describe considerations for training ground learning environments. [4.4.2, 4.4.5]

3. Skill Sheet 5-1: Set up a learning environment (classroom, lab, or outdoor site). [4.4.2, 4.4.5]

Chapter 5
Learning Environment

Fire and emergency services training can occur in a variety of settings. Cognitive training can happen in a classroom setting or may be hosted online. Psychomotor training typically occurs in a training facility, such as a fire station, or at a remote site, such as an acquired structure **(Figure 5.1)**. Wherever training occurs, instructors must control the learning environment to ensure that students can concentrate on the lesson and remain safe. Before conducting training, instructors need to evaluate the environment to identify and remove potential distractions and hazards. Skills associated with setting up a learning environment are shown in **Skill Sheet 5-1**.

Classroom Environment

One advantage of working in a permanent classroom is that the environment rarely changes, making control of the room straightforward. Dedicated classrooms often have in-room controls for lighting and temperature. Rooms in fire stations or other non-educational facilities may not have similar controls. Instructors must be able to adapt to the location in order to create the best possible learning environment **(Figure 5.2)**. In any situation, the instructor should determine how much control he or she has over the following elements:

- Seating arrangements
- Lighting
- Temperature and ventilation
- Noise level
- Instructional technology tools (audiovisual equipment)
- Other classroom considerations such as power outlets, Internet access, and comfort facilities

Figure 5.1 Psychomotor training at a facility can utilize various structures such as this mobile live-fire training unit. *Courtesy of Chris Mickal/District Chief, New Orleans (LA) FD Photo Unit.*

Figure 5.2 Facilities such as this hangar can serve as the learning environment for both cognitive and psychomotor training sessions.

Seating Arrangements

Some seating arrangements are better suited to certain types of lessons. If chairs are not fixed to the floor, an instructor should know how to change the seating to make it most effective for the lesson. Instructors and students must respect the wishes and rules of the organization to return the room to its original arrangement when finished.

Common seating arrangement types include (**Figure 5.3**):

- **Fan or Chevron** — Permits students to easily see and hear an instructor and allows them work effectively in small groups.
- **Traditional** — Permits students to see, hear, and interact with an instructor. Student interaction can be limited and difficult, but the arrangement is applicable to any audience.
- **Auditorium or theater** — Fixed seating for a medium- to large-sized audience that faces a stage or lectern, allowing easy interaction between the instructor and students. It may require a sound system so the audience can fully hear an instructor and may lack a writing surface.

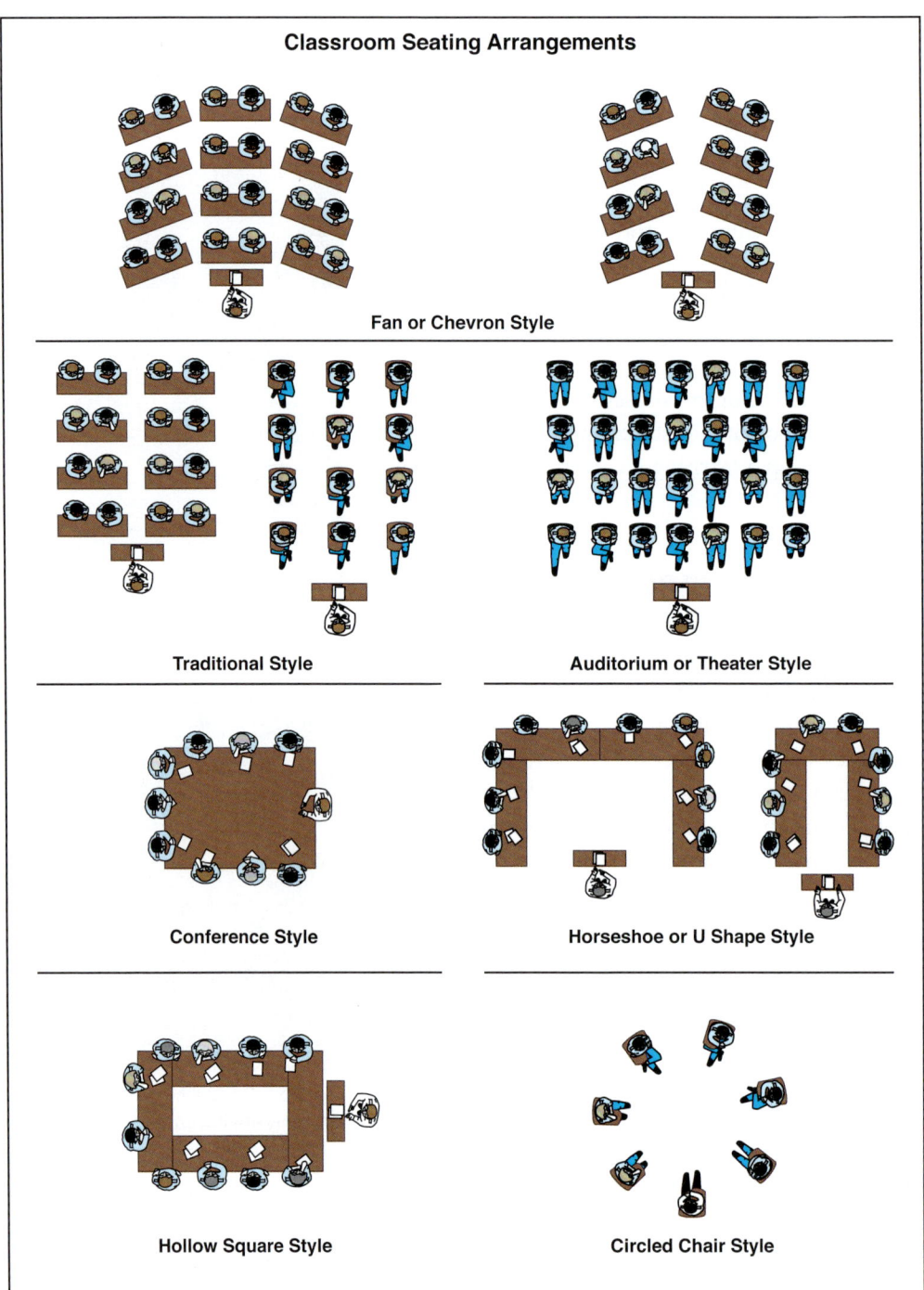

Figure 5.3 Each commonly used seating arrangement offers specific advantages.

- **Conference** — Allows for total group discussion where limited or no small-group activities are required. It is most effective for classes where students are seated around one table.
- **Horseshoe or U shape** — Provides a clear view of the instructor, who may be at either end of the U. Permits instructor presentation and total group discussion, but not small-group interaction. Used for small- to medium-sized audiences.
- **Hollow square** — Arranges tables into a square, with seating along the outside. It is effective for small- to medium-sized groups and permits instructor presentation and full discussion.
- **Circled chairs** — Arranges chairs facing the center, an open form designed to encourage group participation and discussion. The instructor (or facilitator) sits in the circle with the students. It is most effective for small to-medium-sized groups where discussion serves as the primary method of teaching. It is not useful when students are expected to take notes.

Instructors must determine how important it is for students to see and hear the instructor or audiovisual presentation, and how important it is that students can interact with each other. Based on these determinations, they should select the seating for the class. For example, if a lesson will rely heavily upon an audiovisual presentation, circled chairs is not the ideal seating arrangement, but other options may be suitable.

Another important consideration is the location of the table legs. Table legs can be both a physical obstruction and a learning distraction. When more seating is required, tables should be added. If the instructor adds more tables and seating, student safety should be a priority. Ensure that emergency exits and evacuation routes are not blocked.

Another issue related to types of seating is the provided work surface, desktop, or table. For example, a lesson that requires students to look at blueprints to complete preincident surveys will be difficult to complete if students sit at individual desks. Work surfaces vary and include the following (**Figure 5.4**):

- School desktops fixed to an individual chair
- Work surfaces that unfold across a student's lap from between auditorium-type seats
- Tables with varying depths that may accommodate students on one or both sides or are arranged in a U or square shape
- Long conference tables
- Round tables

Figure 5.4 Some learning environments will offer a variety of work surfaces for students, such as moveable desktops, fixed desktops, or round tables.

Instructors may not have the opportunity to choose the type of table or desktop surface for the classroom, and some training organizations may provide only desks or tables with small writing surfaces that do not accommodate an open notebook. Instructors should inspect the physical setting before beginning the class and perhaps arrange to have students place unneeded items in a storage area to prevent cluttering the work surfaces (**Figure 5.5**).

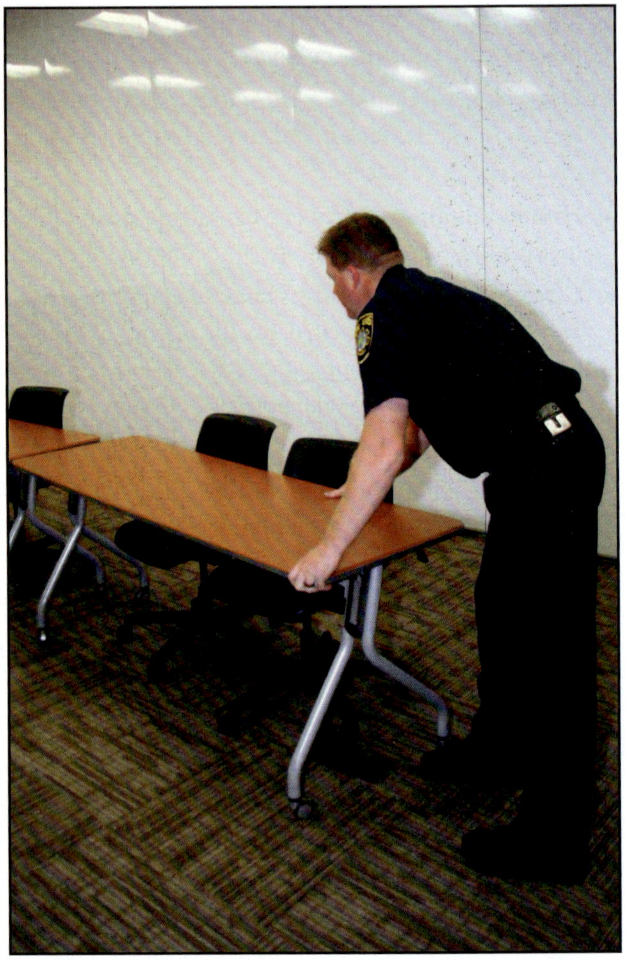

> **Seating Comfort**
>
> Comfort is important to learning. Students should be able to see the instructor comfortably from their seats as well as interact with each other when appropriate. Frequent rest breaks, usually after every 45 to 50 minutes of instruction, allow students to stand, move around, stretch, and attend to other comfort needs. Delaying a break to finish a segment of instruction may be counterproductive, because students lose interest in learning when they are distracted by comfort needs.

Lighting

Permanent classrooms have lighting designed to enhance the learning experience. Typically, this is a mix of incandescent, fluorescent light, or LED (light emitting diode) lighting. The primary light source is usually fluorescent, because it causes less eyestrain, does not glare on reflective surfaces, and is more energy efficient. The trend is moving toward LED lighting due to the increased energy efficiency, long component life, and the ease of dimming, compared with fluorescent lighting.

Figure 5.5 Inspect the classroom setting before students arrive to make any necessary adjustments and decisions about work spaces.

Incandescent lighting is often controlled with a dimmer switch. The lower lighting level allows students to see images on projector screens or computer monitors but have enough light to take notes or read handouts. This effect may be duplicated in a room with fluorescent lighting if separate switches allow the instructor to turn off the front lights and leave on the back lights.

When training must be conducted in a non-classroom setting, instructors should inspect the room or area before class. If the learning environment does not permit the use of audiovisuals, instructors may have to adapt their instruction.

Whatever lighting is used, the instructor must always consider student safety as a priority. A proper balance of lighting will allow students to clearly see audiovisual components while giving them safe movement within the training facility.

Temperature and Ventilation

The temperature of the learning environment can distract students and instructors. When temperature becomes a distraction, instructors may need to provide more frequent rest breaks for students. Learning environments that are too hot or too cold tend to preoccupy students as they try to become comfortable. Instructors should familiarize themselves with the HVAC systems at classroom facilities as follows:

- Determine the location and setting of climate controls before teaching in the facility.
- Determine whether controls can be adjusted and how to adjust them.

- Contact facility personnel if the instructor does not have access to the building's environmental controls.
- Make any climate adjustments within adequate time for the temperature to change prior to instruction.
- Advise students on how to dress for comfort when the systems cannot be adjusted.

Buildings or areas that do not have HVAC systems may be equipped with openable windows. In warm weather, consider whether the fans available will be sufficient to create a cooling air flow in the room. In cold weather, air flow should be minimized. Apparatus bays may need ventilation to disperse exhaust before instruction. When conducting training, monitor air quality within the learning environment for the safety of the students, yourself, and other instructors.

NOTE: In colder climates, training organizations may have indoor facilities large enough to host apparatus training. These facilities must have adequate ventilation systems to ensure that apparatus exhaust does not present a hazard (**Figure 5.6**).

Figure 5.6 Some AHJs have indoor facilities large enough to conduct apparatus operations. Such facilities should be well ventilated and have adequate lighting. *Courtesy of Spokane Valley Fire Department.*

Noise Level

When inspecting a learning environment, instructors should attempt to eliminate potential sources of noise. Depending on the type of facility, they may be able to turn off or lower the volume of loudspeakers, radios, and pagers. They might also close classroom doors or windows to reduce outside noise.

Students who are in-service or on call may have to respond to radios or cell phones during class and may inadvertently cause noise interruptions during a training session. Instructors should consult with these students before sessions to determine how they will be contacted if called to duty. Instructors can then prepare the rest of the class for this potential interruption, so that in-service students can leave with as little distraction to others as possible.

If the noise within the training environment exceeds maximum noise exposure levels (90 decibels in the U.S., 85 decibels in Canada), the instructor should provide safe and approved hearing protection.

Audiovisual Equipment

Instructors can also take steps to eliminate distractions before using instructional technology tools. Take time to ensure that the equipment is arranged properly to allow all students to view presentations. Also, take the following steps to avoid distractions when using audiovisual equipment:

- Do not stand between the audience and the projected image.
- Locate the projector so that it does not obstruct students' views.
- Ensure that the projected image fills the screen area without extending over the edges or keystoning.

- Ensure that the projected image is not distorted.
- Locate the projector to minimize motor noise.
- Prior to the day of instruction, preview any audiovisual aids in the classroom from the point of view of the students so that you can understand transitions between all media types and ways to present the lesson as smoothly as possible.

When using a computer and projector system, instructors should test the presentation before class. The initial image may be used to welcome students, or the system may be turned off until it is needed.

When using unfamiliar technology tools, the instructor should ask the facility staff how to use the equipment. Instructors should become familiar with the operation of the unit being used.

Interaction and cooperation with support staff is essential when providing a course on closed-circuit television, interactive television, or computers (computer-based training). A test of all the equipment and remote receiving sites must be performed in advance in order to meet all training objectives. Broadcast airtime is valuable, and instructors should not waste it trying to resolve system problems.

Other Learning Environment Considerations

In addition to the learning environment considerations already discussed, instructors should consider other features, such as:

- Power outlet access
- Internet, phone, and cable television access
- Visual distractions
- Comfort facilities
- Safety hazards and emergency exits

Power Outlet Access

Before beginning a presentation in a remote location, instructors should locate necessary electrical outlets, determine the need for extension power cords or power strips and know where power cords are located. Use extension power cords that meet the local fire code, usually 6 feet (approximately 2 m) and adequate size for the electrical load being attached. Never plug too many pieces of equipment into a single outlet or power strip, as this can overload the circuit. Be sure to use appropriate electrical outlets, extension power cords, and adapters. It is a good habit for instructors to unplug all electrical equipment from outlets at the end of the class each day.

Safe use of electrical equipment can prevent accidents or injuries. This includes protecting extension power cords and eliminating tripping hazards. For example, extension power cords should be taped to the floor or encased in a cover strip to prevent tripping. Instructors should also evaluate the following:

- Location of adapters for grounded plugs or power strips
- Condition of extension power cords (whether cords are frayed or worn or plugs are damaged)
- Location of the main breaker panel and identification of the circuit breaker that controls classroom receptacles
- Locations of timer-controlled automatic power switches that may turn off lights in the classroom during a presentation

> **Student Power Access**
>
> As society becomes more technologically savvy and more classes encourage students to bring their electronic devices for use in the classroom, the instructor needs to be cognizant of the need for students to charge or provide power for these devices. The importance of providing power for student devices increases as the instructor plans on incorporating these devices into the learning environment. For example, if an instructor is certain that all students in a class will have access to a Web browser, the class textbook may be assigned in eBook format.

In more modern classrooms, access to a power source may be incorporated in the desk or table **(Figure 5.7)**. However, this is not common and most times the instructor will need to plan to provide extension cords and outlet devices, such as power strips. This can pose significant safety/trip hazards in the room, and provision needs to be made to safely secure these cords. An instructor needs to be mindful of what is being plugged in and the power it will draw. Be careful not to overload the electrical equipment being used. Use multiple outlets when possible. Work with the facility representative to address these issues.

Figure 5.7 Power sources may be incorporated into the surfaces or frames of some desks or tables.

Internet, Phone, and Cable Television Access

Internet access is becoming more widely available in classroom facilities, either via wireless networks (Wi-Fi) or wired Ethernet ports. If wired Ethernet ports (wired, broadband Internet connections) are available, the instructor should locate these ports and arrange the classroom accordingly. If Wi-Fi is available, the instructor should ensure that all equipment can connect to the network before class begins and sufficient bandwidth is available for the instructor and students. It may be necessary to contact an information technology representative at the facility before instruction to obtain security passwords for logging into the Wi-Fi network or for assistance with connection issues. When using distance technology, the instructor must ensure that all participants have the proper equipment, login information, and appropriate access to the technology.

If needed, the instructor should also locate telephone and cable television outlets in the presentation room. When they are not conveniently located, it may be necessary to rearrange the seating, move the television or lectern, or arrange for an extension cable.

Visual Distractions

Students can be distracted by posters, photographs, maps, and other wall decorations, as well as whatever may be visible outside the classroom window. Before class begins, an instructor should remove those decorations and close the window blinds. If training is conducted in work areas, seating should face away from the apparatus, other equipment, or working personnel. The only thing that should be visible behind the instructor is a blank wall or projection screen.

Comfort Facilities

During the introduction to a session, instructors should inform students of the location of restrooms, water fountains, and applicable tobacco/drug use policies. When refreshments are provided, they should be located away from the students to reduce the temptation for students to move around the room **(Figure 5.8)**. Instructors must adhere to site rules for bringing refreshments into the training facility.

Safety Hazards and Emergency Exits

During the introduction to a session, instructors should inform students of the location of storm shelters and exits, including emergency exits. Information about potential disaster emergencies should be provided when necessary.

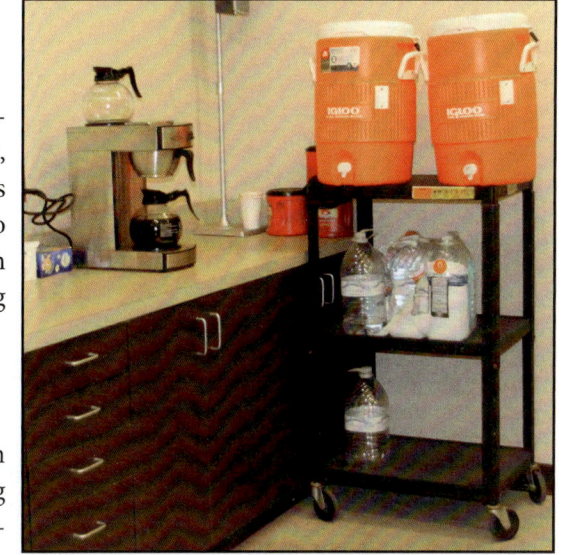

Figure 5.8 Locate refreshments away from students to inhibit distractions during lectures.

> **Online Environments**
>
> In the traditional face-to-face classroom environment, the instructor controls most of the physical environment that students experience. In the online environment, the ability to control that setting is eliminated. The instructor needs to verify at the outset of the course that a student understands all expectations. Allowing learners access from almost anywhere is one of the great advantages of the online learning environment; however, it is also one of its greatest challenges due to the lack of instructional control and the potential for distraction.
>
> Consider the following three distinct environments:
>
> - The instructor's environment — Take care to minimize noise and distraction that may affect you or distract a student.
> - The online environment — This environment is typically a commercially produced package that may allow for significant interaction between a student and the instructor. Do not let one student monopolize the session and distract the other students.
> - The learner's physical environment — This environment can be highly varied and very distracting. Take care to ask questions to see if students need more help.

Training Ground Environment

The training ground (sometimes referred to as the *outdoor learning environment*) can be a challenging place to teach. The locations used for psychomotor training evolutions can be as varied as the types of evolutions. The training ground can include any of the following facilities:

- Props
- Permanent facilities
- Mobile facilities
- Remote sites
- Acquired structures

Regardless whether training occurs in a classroom or at a training ground, the lead instructor is responsible for providing a safe training environment for all instructors and students. The requirements for providing this level of safety are found in national laws, state/provincial laws, local ordinances, government rules and regulations, international consensus standards, organizational policies and procedures, and even case law. The instructor must ensure safety requirements are met before, during, and after training **(Figure 5.9)**. Remote sites and permanent training facilities are detailed in the following sections.

> **Training Ground Safety Regulations and Standards**
>
> Instructors must adhere to the following requirements when planning evolutions or selecting training locations:
>
> - NFPA 1403, *Standard on Live Fire Training Evolutions*
> - Occupational Safety and Health Administration (OSHA), federal, state/provincial, and local laws and ordinances that pertain to environmental protection
> - All jurisdictional training policies and procedures
> - Any facility or site-specific guidelines, including SOPs on travel to and from the site

Remote Sites

When developing a list of possible remote training sites, instructors should collect data that includes location, name of owner/representative, availability (access and time), water supply source, and types of training evolutions that the site could support.

INCIDENT ACTION PLAN SAFETY ANALYSIS			
1. Incident Number: 0036	**2. Location:** Worlds of Fun	**3. Date:** September 4th – 6th	**4. Time:** 0900 – 1530

Incident Area	Hazards/Risks							Mitigations (e.g. PPE, buddy system, escape routes)
	Type of Hazard: Rope entanglement	Type of Hazard: Missed eddy recovery	Type of Hazard: Trip and fall	Type of Hazard: Impact forces	Type of Hazard: Slip and fall wet surfaces	Type of Hazard: Submersion	Type of Hazard: Impact	Type of Hazard: Pulled in water	
Throw bag deployment	x								Tech **shall not** secure rope to person
Throw bag deployment						x		x	Tech will use self rescue techniques
Throw bag deployment	x		x	x	x			x	Keep work area clear of trip hazards and caution at edge work
Throw bag deployment		x					x		Down stream safety with throw bag deployment capabilities
Throw bag deployment						x			Tech will wear swift water PPE
Throw bag deployment							x		Tech will direct projectile up stream of rescuer

Prepared by (Name and Position)
Brad Clausing - Instructor
Clint McLelland - Instructor

Figure 5.9 An incident action plan safety analysis is important for ensuring safety when training at a remote site. *Courtesy of Overland Park Fire Department.*

⚠ Safety at Remote Sites

For any type of remote site, the instructor must verify that an incident action plan and site safety plan in accordance with agency policies has been developed and make sure that the site is suitable for any training conducted there. The instructor must also verify that permission to use the site is in place with proper insurance/liability coverage prior to the activity.

Examples of remote sites and their potential training uses include:

- **Parking lots** — Used for driver/operator training, supply and attack hose deployment, vehicle extrication, and EMT training **(Figure 5.10, p. 98)**.
- **Subdivisions under construction** — Used for driver/operator training and building construction training.

- **Acquired structures** — Used for live-fire evolutions, ventilation training, and forcible-entry training. Buildings that have been in a fire may be used for fire pattern analysis and origin and cause determination training. Buildings under demolition may be used for collapse and confined-space rescue training. Structures intended for use in live-fire training must be in compliance with NFPA 1403, *Standard on Live Fire Training Evolutions*; see Chapter 8, Skills-Based Training Beyond the Classroom, for more information. Structures intended for use in training other than live-fire evolutions must include a complete safety assessment per the AHJ before training can begin.

Figure 5.10 Parking lots can serve as remote sites for driver/operator training scenarios, vehicle extrication, EMT training, and other operations.

- **Military or government-owned reservations** — Used for wildland fire suppression general skills training, off-road driver/operator training, and joint military fire department training.

- **Airports** — Used for aircraft crash/fire/rescue training, driver/operator training, and foam fire-suppression training. When training at an airport, the instructor should contact the tower and airport operations to obtain clearance **(Figure 5.11)**.

- **Grain elevators/silos** — Used for technical and rope rescue training.

- **Industrial sites** — Used for technical and rope rescue training, hazardous materials spill control, fire-suppression training, and joint-training evolutions with local industrial fire brigades.

Figure 5.11 Some airports offer training areas where students can practice aircraft crash/fire/rescue training techniques or foam fire-suppression evolutions.

- **Open wildlands** — Used for wildland fire-suppression and off-road driver/operator training. This may involve joint department training with controlled burns.

- **Vehicle salvage yards** — Used for vehicle extrication training.

- **Parking garages** — Used for standpipe operations and high-angle rescue training.

- **Warehouses and aircraft hangars** — Used for large area search training, rapid intervention team/crew training, hoseline deployment training, and tactical simulations.

> ### ⚠️ Live-Fire Training
>
> NFPA 1041 requires that a Level II Instructor be present to supervise any high-hazard training, such as **live-fire exercises**, **controlled burning**, and any fires set for training at acquired structures. Whenever possible, live-fire training should be conducted with props and burn buildings at permanent training facilities. Typically, this ensures that a Level II Instructor from the facility will be available to oversee the evolution.
>
> But in practice, it is often impossible for a Level II Instructor to be present, especially if training is conducted at a remote site. A Level I Instructor will sometimes have to oversee a live-fire evolution at these sites. Before doing so, the instructor should consult with local, county, or state/provincial environmental officials, and must review the requirements in NFPA 1500™, *Standard on Fire Department Occupational Safety and Health Program*, and NFPA 1403, *Standard on Live Fire Training Evolutions*.

> This is the most effective way to ensure the instructor is fully aware of all the issues related to this high hazard training and is able to take necessary safety measures. Failing to take these steps may put students and the instructor in danger.
>
> For more information on planning training evolutions at the Instructor I level, refer to Chapter 8, Skills-Based Training Beyond the Classroom. In addition, more information on planning, leading, and monitoring training evolutions can be found in the Instructor II section of this book, Chapter 12, Training Evolution Supervision.

Some remote sites can be used repeatedly, but others may only be available for a single training course or session. Organizations must get permission from the property owners or their agents before training at a remote site. Again, instructors should also thoroughly inspect potential sites to ensure that it is appropriate for the desired training. It may be helpful at that stage to take photographs, record video, and/or make a site map. Potential considerations for the inspection and planning processes at remote sites are as follows:

Weather Conditions

Instructors should try to lessen weather's effects on students. Examples:

- Provide rehabilitation for students exposed to temperature extremes and high humidity.
- Provide shielded observation and waiting areas to protect against strong winds and rain.
- Provide cleats to add to the soles of shoes or boots on icy ground. Sand, salt, and/or ice melt can be spread to help melt the ice or provide foot traction.
- Prohibit some operations such as the use of aerial devices or ground ladders during high wind or thunderstorms. Lightning storms and thunderstorms can develop quickly and are unsafe to train in even if miles away.
- Follow regulations and local policies that dictate under what conditions (temperature, humidity, wind, etc.) students may train. One resource for instructors is the Humiture Chart and wind chill charts.
- Training should be cancelled, rescheduled, or changed to meet conditions if student safety cannot be maintained with the original plan.

Site Environment Considerations

Considerations involving the site environment include:

- **Terrain** — The initial inspection should determine how the terrain will affect ladder and apparatus use, site access, and water runoff. If the terrain creates safety hazards, instructors should mark these locations on the site map and inform students of areas that are either off-limits or potentially hazardous.
- **Site space** — Make sure the site is large enough for the planned training. If it is not, training will be less effective and less safe.
- **Exposures** — Before live-fire training begins, identify exposed structures and personnel. Consider the terrain, water runoff, wind direction, and wind speed.
- **Environmental laws and codes** — Training must comply with national, state/provincial, and local environmental laws, in addition to building and fire codes and zoning ordinances.
- **Water supply** — Ensure that a site's water supply is adequate for the training.

Vehicle Traffic

Vehicle traffic may affect training operations conducted on public streets or in parking lots. Take the following actions when training near vehicle traffic:

- Limit or prohibit public access to the training area whenever possible **(Figure 5.12, p. 100)**.
- Enforce the use of safety cones and vests during training so students are familiar with their proper use at emergency incidents.

- Involve either the local department of public works, the department of transportation, or law enforcement personnel in the training scenario to provide scene support such as closing roadways or controlling traffic. This can serve as a good opportunity for these agencies to become familiar with interagency cooperation at emergency scene operations.

Training Ground Noise

Fire and emergency services personnel often work in environments with high noise levels from vehicle engines, pumps, sirens, radio transmissions, and shouting. Incidents simulated for a training incident may be as loud as unscripted incidents. Take the following precautions to protect students from training ground noise:

- Follow appropriate guidelines for hearing protection. Use earmuffs or earplugs in environments with noise levels exceeding 90 decibels in the U.S., 85 decibels in Canada **(Figure 5.13)**.
- Turn off vehicle and machinery motors when instructors are giving instructions or explaining procedures.
- Bring extension cords to place noisy generators and compressors out of the immediate training area.
- If noise cannot be controlled, use a microphone and speaker system so students can properly hear the instructor.

Figure 5.12 Whenever possible, training areas should limit or prohibit public access and vehicle traffic to ensure maximum safety and reduce distractions.

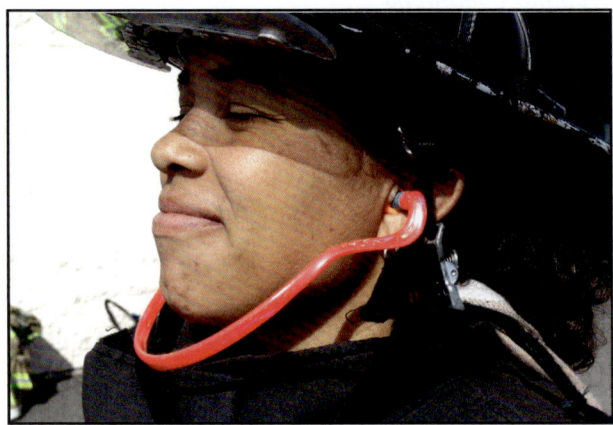
Figure 5.13 Instructors should ensure that all students follow appropriate guidelines for hearing protection on the training ground.

Lighting

Instructors on the training ground should provide initial instruction in a well-lit and comfortable area. Light the immediate training area and staging areas where tools and other items are kept during training. Make sure the terrain is well lit to help prevent tripping and falling. When applicable, use fluorescent markings to highlight potential hazards.

Access/Egress

The site map should indicate access and egress points, traffic flow, and the relationship between staging and incident areas. This is especially important if students are in-service during training. Other considerations:

- Additional personnel may be required to control access to the site.
- Whenever possible, provide at least two means of egress.

Permanent Training Facilities

Many fire and emergency services organizations have access to permanent facilities within their state/province. Permanent training facilities usually contain permanent and portable props required for a variety of training, such as **(Figure 5.14)**:

- Drill towers
- **Burn buildings** and **smokehouses**
- **Flammable/combustible liquid pits**
- **Vehicle driving courses**

Figure 5.14 Permanent training facilities can consist of drill towers, flammable/combustible liquid pits, and burn buildings, as well as other structures and props.

An instructor must be familiar with the training area to ensure training is conducted safely. Before teaching at a training facility, instructors should inspect the area to determine the condition of the facility and training props; identify and mitigate any safety concerns; and locate simulated incidents, student parking lots, apparatus staging areas, and observation seating (usually fixed position bleachers).

At the beginning of any outdoor training session, instructors should give students an overview of the training scenario which includes safety issues, expected outcomes, and assignments, whether at the unit, company, or individual level. Any skills that the instructor must demonstrate should be presented near the beginning of the session. Instructors should then conduct a walk-through of the area or structure that emphasizes exit routes, control zones, rehab facilities, and accountability practices.

Instructors should develop a list of potential facilities, including those owned by the jurisdiction, local colleges and vocational/technical schools, and regional and state/provincial training agencies. After compiling a list of facilities, the types of training props available, and the names of facility representatives, instructors can plan the types of evolutions that can be performed at each location. Next, instructors can arrange with individual representatives to schedule and coordinate the use of each facility.

Chapter Review

Answer the following questions to review the information provided in this chapter.

1. What must an Instructor I consider when organizing a classroom learning environment for effective learning?
2. What are some challenges an Instructor I encounters when using training ground environments as a place to teach?

Discussion Questions

The following questions are intended to generate discussion, expand your understanding of the chapter text, and allow you to think critically about what you have learned. Answers to these questions may vary.

1. What classroom learning environments have you had that facilitated learning? Why?
2. What classroom learning environments have you had that impeded learning? Why?
3. You are teaching an indoor class on community risk reduction that has twenty students. What would you do to help set up an effective learning environment?

4. What remote sites in your jurisdiction might be potential classroom training sites?
5. What outdoor learning environments have you had that facilitated learning? Why?
6. What outdoor learning environments have you had that impeded learning? Why?

Key Terms

Burn Building — Structure designed to contain live fires for the purpose of fire suppression training.

Controlled Burning — Any burn that is safely set and controlled for the purposes of fire and emergency services training.

Flammable/Combustible Liquid Pit — Training prop designed to provide controlled burns of flammable or combustible liquids; used in training for the extinguishment of flammable/combustible liquid fires.

Live-Fire Exercises — Training exercises that involve the use of an unconfined open flame or fire in a structure or other combustibles to provide a controlled burning environment. *Also known as* Live-Burn Exercises.

Smokehouse — Specially designed fire training building that is filled with smoke to simulate working under live fire conditions; used for SCBA and search and rescue training. *Also known as* Smoke Building.

Vehicle Driving Course — Permanent or temporary training course used for training in apparatus and vehicle driving and operation.

5-1
Set up a learning environment (classroom, lab, or outdoor site).
[NFPA 1041, 4.4.2, 4.4.5]

Task Steps

Step 1: Obtain and organize course materials, instructional media, resources, and equipment.

Step 2: Adapt materials and resources, if necessary, to the learning environment.

Step 3: Organize the learning environment so all considerations are addressed and needs are met:

    a. Lighting **(Figure 5.15)**

    b. Distractions

    c. Climate/weather (temperature)

    d. Noise

    e. Seating **(Figure 5.16)**

    f. Audiovisual

    g. Teaching aids

    h. Safety **(Figure 5.17)**

Step 4: Confirm audiovisual equipment is arranged to allow unobstructed viewing by all students.

NOTE: Prior to beginning the lesson, make sure all audiovisual equipment is in working order. When inspecting a projector, ensure the lens is clean, check the air filter, and replace the bulb if necessary.

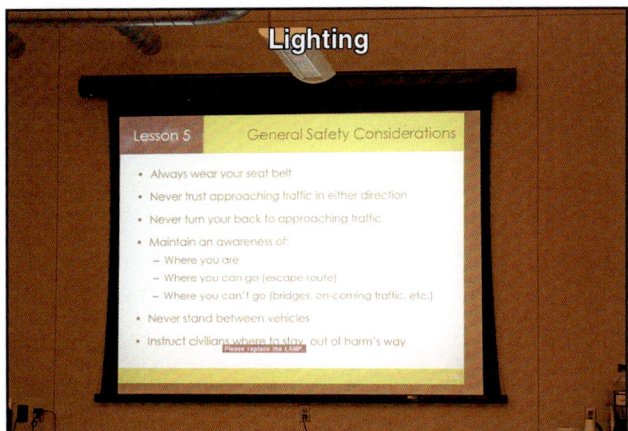

Figure 5.15

Figure 5.16

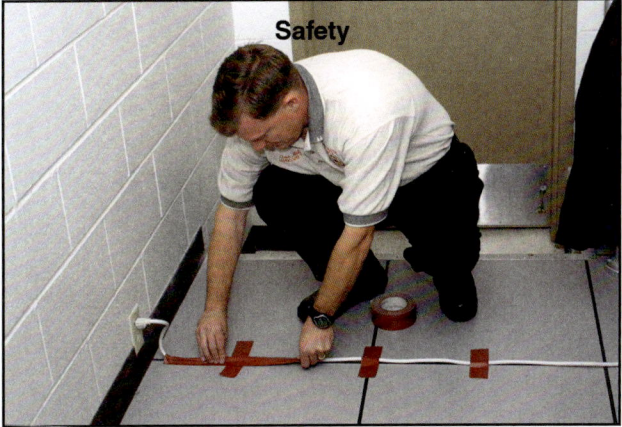

Figure 5.17

Classroom Instruction

Chapter 6

SECTION A INSTRUCTOR I

Chapter Contents

Presentation Techniques **107**
    Characteristics of Effective Speakers... 107
    Presentation Planning 108
    Organizing the Presentation.......... 109
    Methods of Sequencing 110
    Transitions 110
Four-Step Method of Instruction **112**
    Preparation........................112
    Presentation......................114
    Application 114
    Evaluation........................114
Instructional Methods...................... **115**
    Giving an Illustrated Lecture 115
    Providing Demonstration.............116
    Leading Class Discussions............118
    Asking Effective Questions.......... 121
    Psychomotor Skill Instruction 124
Structured Exercises **125**

Case Studies..................... 125
Role Playing 126
Simulations...................... 126
Field and Laboratory Experiences 127
Competency-Based Learning in the
    Fire and Emergency Services **127**
Teaching Strategies **128**
    Traditional Instructor-Led Training (ILT) 128
    Blended or Hybrid Learning 131
    Student-Centered Learning 131
    Flipping the Classroom.............. 131
    Computer Simulation 131
    Self-Directed Learning 133
    Individualized Instruction............ 134
Chapter Review **135**
Discussion Questions **135**
Key Terms................................... **135**
Skill Sheets **137**

JPRs addressed in this chapter

This chapter provides information that addresses the following job performance requirements of NFPA 1041, *Standard for Fire Service Instructor Professional Qualifications*, 2019 Edition.

4.3.2   4.4.4
4.4.2   4.4.5
4.4.3

Learning Objectives

1. Describe presentation techniques that most effectively communicate information to students. [4.3.2, 4.4.3]

2. Describe the four-step method of instruction. [4.4.3]

3. Discuss instructional delivery methods. [4.4.3, 4.4.5]

4. Describe the use of structured exercises. [4.4.3, 4.4.5]

5. Describe competency-based learning in the fire and emergency services. [4.4.3, 4.4.4]

6. Identify teaching strategies that encourage active learning. [4.4.2, 4.4.3, 4.4.5]

7. Skill Sheet 6-1. Give a prepared classroom [cognitive] lesson. [4.4.3, 4.4.4, 4.4.5]

8. Skill Sheet 6-2. Give a prepared practical [psychomotor] lesson. [4.4.3, 4.4.4, 4.4.5]

Chapter 6
Classroom Instruction

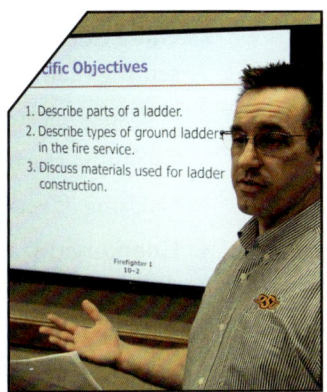

Classroom instruction is a complicated and multifaceted skill. This chapter outlines significant resources and skills required in classroom teaching. For example, classroom instruction requires that an instructor practice public speaking and teaching to become confident in the skill. After building this confidence, instructors can focus on turning public speaking into effective lecturing and engaging students in active learning. The purpose of this chapter is to introduce various aspects of classroom instruction:

- Presentation techniques
- Four-step method of instruction
- Instructional methods
- Structured exercises
- **Competency-based learning** in the fire and emergency services
- Teaching strategies

Presentation Techniques

Presentation is the art of clearly and concisely explaining information in ways that an anticipated audience can understand. It requires both an understanding of interpersonal communication and the ability to apply that understanding to public speaking. For instructors, public speaking usually occurs in a classroom setting with an audience of students **(Figure 6.1)**. Presentation techniques can also apply to presentations given to superiors, administrative bodies, collaborative agencies, or the general public.

The sections that follow will describe the characteristics of effective speakers; explain how to plan, organize, and sequence a presentation; and describe how to effectively use transitions. The instructor needs to expand on this knowledge and practice each of these methods in order to become proficient in them.

Figure 6.1 Instructors must grow accustomed to public speaking before an audience of students.

Characteristics of Effective Speakers

The first step toward becoming an effective public speaker is to identify the characteristics displayed by effective speakers. Characteristics that apply to the instructor in the classroom include:

- **Audience-centered** — The speaker knows the audience and adapts his or her topic, speech organization, presentation style, and personal appearance to this audience. In the classroom, this involves matching the instructor's presentation style to the students' learning styles.

- **Good development of ideas** — Effective speakers create interesting, appealing, and memorable ways of presenting their information. In a classroom setting, this may include:
  — Using relevant examples
  — Telling stories to which the audience can relate
  — Using effective metaphors
- **Good organization of ideas** — Effective speakers organize their material so that their audience is never lost during the presentation. Persuasive speeches include an attention grabber, necessary background information, an illustration of the problem or situation and, finally, solutions for the situation. An informative speech may be organized either topically or by level of complexity.
- **Best choice of words** — Tailor a presentation's wording to its intended audience. Explain any industry jargon to an introductory level audience. A more experienced audience will appreciate the correct usage of technical terms.
- **Good delivery skills** — Effective speakers use the following communication techniques to enhance the words that they have chosen:
  — Keep appropriate eye contact with the audience members.
  — Speak to the entire audience, not just one section or one side of the room.
  — Use appropriate gestures to illustrate mental pictures or emphasize key points.
  — Pause periodically so students can think about what they have heard and ask questions. Do not cut a pause short just because no one is responding.
  — Refrain from adding too many personal anecdotes to presentations. A few anecdotes add interest and relevance to a class, but too many can be a distraction.
- **Good vocal characteristics** — Major elements are:
  — *Pronunciation.* Pronouncing each word correctly, stressing the right words or syllables, and pausing where appropriate.
  — *Good grammar.* Using correct tense, possession, and pronoun agreement.
  — *Vocal variety,* also known as *inflection.* Refers to changes in loudness, pitch, and rate of speech. Speakers use these changes to emphasize important points and to keep the audience's attention.
  — *Enunciation.* Clearly emphasizing each syllable, accent, and pause. Avoid slurring or mumbling when speaking in front of a group.
  — *Projection.* Speaking loudly and clearly enough to be heard in the back of the room or auditorium.
  — *Rate of speech.* Speed at which words are spoken. Effective instructors will speak more slowly when presenting new information or emphasizing important points, or when students need to take notes. As students become more familiar with the material, instructors can speak more quickly.
- **Conversational tone** — A relaxed tone makes listeners feel at ease and ready to receive information.
- **Positive attitude** — Effective speakers display a positive attitude about the subject matter they are presenting.
- **Appropriate use of humor** — Appropriate humor can create a relaxed atmosphere and get the attention of the audience. Avoid inappropriate humor that may offend members of the audience.
- **Personal style** — Effective speakers use a personal style, capitalizing on their own unique experiences and abilities.

Presentation Planning

Instructors can plan for presentations so that they are prepared and at their most effective. The following suggestions should improve any presentation:

- Practice the delivery of a presentation. Attempt to put into practice the characteristics of effective public speakers.
- Make a video recording of the presentation and review it for distracting actions and speech patterns (**Figure 6.2**). Making a recording also enables instructors to experiment with different ways to present materials, which can increase instructional effectiveness.
- Check the presentation materials to ensure that they are complete, in order, and correct for the topic.
- Analyze the presentation to ensure that it is logical in its sequence.
- Get plenty of rest the night before a presentation.
- Relax before a training session.
- Select comfortable clothing and always dress appropriately.
- Anticipate potential problems and prepare to resolve them should they occur. Have a backup plan for unanticipated technology failures, including hard copies of the presentation.

Figure 6.2 A video recording of a classroom presentation can reveal the instructor's distracting actions and increase overall instructional effectiveness.

Distracting Behaviors to Avoid

- Needless or excessive pacing around the floor
- Playing with or tapping pens, pencils, and other items
- Jingling keys or change
- Chewing gum, fingernails, matchsticks, or toothpicks
- Using electronic devices like cell phones
- Excessively getting off topic
- Focusing on a single student
- Overusing slang and verbal pauses like "um" or "okay"

Organizing the Presentation

Oral presentations generally consist of three parts: opening/introduction, body, and summary/conclusion. This format tells the listener or student the topic of the presentation, gives details about the topic, and then restates the main points.

All presentations should follow this general format. Descriptions of the three parts are:

- **Opening or introduction** — Use the opening of the presentation to get the attention of students. Introduce students to the topic and purpose of the presentation, and tell them how it relates to them or their jobs. Also present a summary or outline of the main points to help students remain focused.
- **Body** — Present the information in a logical sequence, along with supporting facts and information. Separate the body of a long presentation into smaller, more easily understood segments. Ensure that each segment conveys a single point or idea and has its own opening, body, and summary. Use transition phrases to link the segments.
- **Summary or conclusion** — Review the objective of the presentation and how it is relevant to the overall goal of the course. Emphasize the main points and introduce the next lesson or the demonstration that is associated with the presentation.

Methods of Sequencing

Experts in the fields of teaching methodology and speech communication have established ways of effectively sequencing information in a presentation. The sequence depends on the topic and the organization of the lesson plan. Generally accepted sequences for instructional delivery include:

- **Known-to-unknown** — Begin with information that students are familiar with or already know before introducing unfamiliar or unknown material. This method is effective because students base their learning experience on something they already recognize.
- **Simple-to-complex** — Begin teaching the basic knowledge or skill, then introduce more difficult or complex knowledge as the lesson progresses. Basic knowledge and skills are necessary foundations for mastering more complex knowledge and skills. For example, the instructor would teach basic rope skills before teaching rigging and hauling techniques.
- **Whole-part-whole** — Begin this sequence with an overview of the entire topic or a demonstration of the complete skill in real time. Next, divide the topic or skill into subsections or steps and describe or demonstrate each of them. Close with a summary of the entire topic or a demonstration of the complete skill.
- **Step-by-step** — Teach each individual step in the correct order and then have students practice the steps in the same order. A variation on this sequence is called progressive-part, in which steps 1 and 2 are learned before progressing to step 3. After mastering 1, 2, and 3, the student learns step 4, and so on. Finally, the student must perform all steps sequentially in a single skill.

Instructors commonly use all of these sequences to present new material because they provide a solid foundation for learning. These sequences also help instructors outline the points that are essential to understanding the topic and mastering a skill.

Educational research recommends that instructors evaluate students' understanding during the lesson. A short list of evaluation techniques that can be integrated into a lesson plan for this purpose include:

- Gauge student insight by showing physical examples of unfamiliar objects or demonstrate unfamiliar processes.
- Measure comprehension by diagramming a complex, structured set of ideas on a chalkboard or handout, or through other visual aids **(Figure 6.3)**.
- Use demonstrations and modeling where possible, particularly cognitive modeling in which the instructor recites steps or processes out loud while students demonstrate a skill.
- Reinforce aspects that interest students and increase student motivation by initiating discussion on the interest value or application possibilities of the new material.

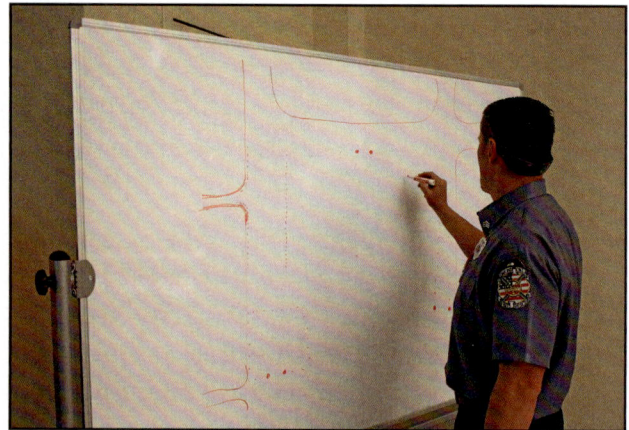

Figure 6.3 Diagramming ideas should help students understand concepts.

In all lessons, instructors must ensure that they introduce students to the key points; explain why the information is important; emphasize these points in the related parts of the lesson; and review and summarize them at the end of the lesson. Even if instructors present only a lesson segment, they should still ensure that the segment flows logically using one of the sequential methods.

Transitions

For continuity and consistency, instructors use transitions to keep students' attention between portions of the lesson. Transitions preview what will happen next, or relate an upcoming concept or skill to a previous one. Effective transitions can create interest among students, keep their attention, and help them to make logical connections between portions of the lesson. The lesson plan should indicate necessary transitions for the benefit of both experienced and inexperienced instructors. Effective transitions help instructors to:

- **Maintain interest** — Keep the audience interested in the overall class topic.
- **Pace the lesson** — Keep the information flowing steadily, without interruption.
- **Maintain consistency** — Help ensure that topics and lessons throughout a course are taught in a similar manner with recognizable transitions.
- **Establish relationships** — Show how parts of the topic are related to each other.
- **Provide previews** — Give the audience an idea of what to expect in the next portion of the material.
- **Provide summaries** — Conclude the previous idea or topic.

Knowing when to use transitions is a question of timing. Transitions can be used effectively when:

- Ending one topic and beginning another
- Ending a complete lesson within a series or course
- Starting a new lesson within a series or course
- Moving from one teaching method to another
- Providing rest breaks for students and instructors

The length of time a transition requires will vary according to its use. Announcing a rest break for the class takes only a few seconds, while summarizing a complex topic may take several minutes. Including transitions in a lesson plan helps establish the time required and prevents the lesson from continuing too long.

Speech communication professionals teach two types of transitions: verbal and nonverbal. They may be used separately or together. With practice, the instructor can learn how to use them effectively and with variety.

Verbal Transitions

Verbal transitions provide a summary and/or preview within a single sentence or two. Types/examples of verbal transitions include:

- **Summary statement and preview** — Example: *Now that you understand how to operate the components of the SCBA, our next step is to learn how to assemble them into a working system.*
- **Review of the lesson or course agenda** — Example: *Today we saw a demonstration and received some practical training over how to use Class A foam to extinguish a flammable liquid fire.*
- **Change of media** — Example: *In order to illustrate what we have been discussing in the slide presentation, we will now view a video clip that shows how rapidly a fire can develop in a controlled environment.*

Words or phrases that may be useful as transitions include "in addition to," "in other words," "as well," "therefore," "in summary," and "not only" **(Figure 6.4)**. Use of the words

Figure 6.4 Using transitional statements during a lecture can help the instructor summarize information, review the lesson, or cue changes in media.

"finally" and "in conclusion" should be avoided in oral presentations, as they give students permission to stop listening. Other ways of using specific words or phrases to make verbal transitions include:

- **Repeating key words or their synonyms** — Repetition emphasizes the importance of the word or phrase.
- **Sequencing the parts to ensure continuity** — Use ordinal numbers, such as first, second, or third, and other sequencing words like "next" to establish the relationship between parts of an idea or process.
- **Using rhetorical questions** — Rhetorical questions (ones that do not require an answer from the audience) help to establish a relationship between the information that has been provided and the information that follows.

Nonverbal Transitions

Nonverbal transitions also help an instructor to emphasize a point within a topic. These transitions may consist of a change of facial expression, a pause, a gesture, or physically moving from one place or position to another.

Nonverbal transitions may also be used to move from one teaching method to another. This kind of transition may disturb student concentration because it involves an obvious change. Altering the light level, turning on audiovisual equipment, or assembling a model takes time and cannot be accomplished effectively while the instructor is lecturing. To make these transitions less noticeable, the instructor may have an aide assemble equipment while the instructor lectures, or call for a break when new equipment needs to be set up for the next portion of the lesson.

> **Media Transitions and Animations**
>
> Computer-generated presentations offer a variety of options for audio and video transitions between slides. While an Instructor I rarely has to create a large number of slides for classroom use, he or she may want to tailor a few slides for a particular lesson. When adapting slides, instructors are encouraged to use the same type of transition style used in the prepared slides. This minimizes the potential for distraction. Playing with different animation styles during a slideshow may be visually appealing, but it can keep students from focusing on important content.

Four-Step Method of Instruction

The **four-step method of instruction**, taken as a whole, is a widely accepted structure for teaching a lesson. As the name implies, it consists of four steps: preparation, presentation, application, and evaluation **(Figure 6.5)**. For instructors to be successful in completing the stated learning objectives, they must have knowledge of lesson plan terminology and definitions and be able to apply the four-step method of instruction. The sections that follow describe each step in greater detail. Skills associated with giving a prepared classroom (cognitive) lesson are shown in **Skill Sheet 6-1**. Skills associated with giving a prepared practical (psychomotor) lesson are shown in **Skill Sheet 6-2**.

Preparation

In the preparation step, instructors introduce the lesson and show how the material is relevant to students' needs. An instructor can accomplish this by:

- Introducing the topic.
- Gaining the students' attention.
- Stating the learning objectives.
- Explaining how the information or skill in the lesson is directly relevant to the students' jobs.
- Motivating the student to learn the information.
- Stating the lesson's key points so that students are prepared to listen for them.

Four-Step Method of Instruction

STEP 1 PREPARATION

Purpose	How to Accomplish
To prepare the students to learn	Tested methods for preparing students to learn
1. Prepare the mind of students by creating: • Attention • Curiosity • Interest • Desire 2. Create a foundation for learning: Begin associating students' experiences with the lesson's contents.	1. Generate curiosity by asking rhetorical questions or questions that cause students to relate personal experiences to the topic. 2. Create attention by including a personal experience, analogy, or topic-related story. 3. Generate desire by citing the personal benefits associated with mastering the knowledge and skills. 4. Create interest by presenting new concepts, procedures, or equipment. 5. Create continuity by reviewing previous lessons. 6. Determine student knowledge by conducting diagnostic quizzes or pretests.

STEP 2 PRESENTATION

Purpose	How to Accomplish
To communicate content developed to change the behavior of students	Tested methods of presenting knowledge and skills
1. Present knowledge, new skills, concepts, or procedures to students. 2. Instruct, motivate, and educate students.	1. Select the appropriate presentation style for the audience, subject, and desired outcome. 2. Present lectures, demonstrations, and activities. 3. Use appropriate visual aids and props. 4. Explain procedures. 5. Emphasize key points. 6. Explain concepts, philosophies, principles, and implications. 7. Proceed from known to unknown and simple to complex. 8. Use textbooks and other reference materials. 9. Apply active learning principles. 10. Summarize key points and concepts at the end of the presentation. 11. Require students to take notes.

STEP 3 APPLICATION

Purpose	How to Accomplish
To provide the opportunity for students to apply theory, critical thinking, critical decision-making, or psychomotor skills to practical situations	Creative, organized, and tested methods for presenting and practicing practical skills
1. Demonstrate skills-based knowledge through appropriate means. 2. Provide students with the opportunity to perform under supervision. 3. Involve students actively in the learning process. 4. Provide the opportunity to practice and master critical skills in a nonemergency learning environment.	1. Have students perform the task or activity under supervision. 2. Observe performances closely. 3. Check and correct errors. 4. Instill correct habits in students. 5. Check key points and safety points. 6. Develop discussions based on theory, decision-making, or skills application. 7. Conduct periodic skills tests. 8. Assign projects and activities. 9. Assign problems for students to resolve.

STEP 4 EVALUATION

Purpose	How to Accomplish
To evaluate the learning process	Tested methods for evaluating the learning process
1. Evaluate student understanding. 2. Evaluate teaching effectiveness.	1. Have students perform tasks unassisted. 2. Conduct performance tests. 3. Ask prepared questions. 4. Have students demonstrate and explain tasks. 5. Have students observe and critique other student performances. 6. Conduct final examinations. 7. Evaluate notebooks, projects, assignments, and activities. 8. Have students complete course and instructor evaluation forms. 9. Have instructors complete course evaluation forms.

Figure 6.5 The four-step method of instruction covers the important components of instructional delivery.

Each of these actions will help to create a foundation from which the instructor can make the presentation. By relating the topic to previously learned information or past student experiences, instructors can show students why the topic is important and how it will benefit them.

Figure 6.6 In the presentation step, the instructor explains the objectives and structure of the lesson.

Figure 6.7 During the application step, students can practice skills.

Figure 6.8 Practical tests allow instructors to evaluate students' skills.

Presentation

In the presentation step, the instructor follows an orderly, sequential outline to present the lesson content. (**Figure 6.6**). Key points on this outline include:

- Teaching methods
- Learning activities
- Demonstrations and practices
- A list of instructional support materials needed for the lesson, such as audiovisuals, worksheets, and handouts
- Summary sections given at logical stopping points and at the end of the outline. The summary gives the instructor an opportunity to emphasize critical information throughout the lesson and provide closure at the end.

By following the presentation outline, the instructor enhances the students' ability to achieve the desired learning objectives and outcomes.

Application

In the application step, the instructor provides opportunities for students to learn through a variety of activities, including exercises, discussions, work groups, skill practices, and practical training evolutions (**Figure 6.7**). Most learning takes place during this critically important step.

Instructors can combine the application and presentation steps so that, as students learn the lesson content, they can participate in activities that require them to think, correctly manipulate tools, or safely demonstrate skills. During an exercise, the application step is often associated with performing the operations or steps of a specific task. However, students may also be asked to:

- Give a presentation
- Lead a group discussion or brainstorming session
- Apply research methods
- Demonstrate outlining and writing techniques

Evaluation

In this fourth and final step, students demonstrate how much they have learned through a written, oral, or practical examination. Written and oral tests are typically used to evaluate cognitive learning, while practical tests are used to evaluate a student's ability to perform a specific skill within safety parameters (**Figure 6.8**). The purpose of the evaluation step is to determine whether students have achieved the lesson objectives.

Instructional Methods

An instructor is encouraged to become comfortable with a number of different methods to give them the most flexibility to present and/or adjust the material in the most effective manner. The following sections describe various instructional methods, their benefits, and recommendations for accomplishing them.

> **Instructional Method Terminology**
>
> In academic environments, "pedagogy" refers to the profession and practice of teaching, while "andragogy" "refers to the teaching and training of adult learners. For the purposes of this chapter, the phrase "instructional methods" refers to strategies an instructor can use to share information based on learning objectives, and to evaluate student feedback and progress during instruction.

Giving an Illustrated Lecture

In the illustrated lecture format, the instructor explains a topic with the help of audiovisual aids, such as:

- Computer-generated slide presentations, such as PowerPoint or Keynote
- Illustrations on dry-erase boards or chalkboards
- Drawings and photographs
- Recorded video

The illustrated lecture format is an effective method for providing facts, rules and regulations, clarifications, examples, and definitions. It allows one speaker to reach an audience of any size, from a single student to a full auditorium. Many students can be taught at the same time while the instructor only prepares one presentation. Another advantage is that students are familiar with this format, so they are aware of what to expect and what is expected of them.

Computer-generated slide presentations are the most widely accepted visual aid that instructors use to accompany their lectures. Although these presentations can be a valuable asset, an instructor must remember that they are merely tools to help illustrate key points or generate discussion. PowerPoints can overwhelm and distract if they last too long, contain too many words, or contain too many special effects. When giving an illustrated lecture, consider the following recommendations:

- Incorporate time for questions into your lesson plans. Pose questions to students throughout the lecture, and allow them to ask questions either during the lecture or at the end of the session.
- Be prepared to ask questions extemporaneously when it becomes clear that the students are losing interest. Direct questions to the students who are paying the least attention.
- Use effective listening skills to pay attention to student feedback.
- Avoid presenting too much information at once. Students need time to process new material, especially if they are also taking notes.
- Provide supplemental information using handouts and reference lists.
- Break lectures into smaller segments of about 12-18 minutes. Intersperse these lecture segments with discussion groups or skill practice time.
- Provide a note-taking guide so students can take notes on the verbal portion of the lecture without having to also write down information from the slides.
- At the end of each segment, have students transition into pairs or small groups to compare notes, ask each other questions, and discuss the lecture material.
- Give students 3 minutes at the end of the class to write down everything they remember from the lesson.
- Provide a clear preview of the information that will be contained in the lecture.

- Include only essential and relevant information in the lecture.
- Review frequently, after each lecture segment and at the end of the lesson, to ensure that the stated objectives or learning outcomes are achieved.

> **Student Engagement in the Learning Process**
>
> Students are increasingly distracted with evolving technology. In addition, more students are entering the fire service without applicable skills, such as how to use hand or power tools. In order to keep students interested and participating actively in the course, instructors must understand these challenges and find ways to accommodate them.

Providing Demonstration

Demonstrations are an effective way to teach manipulative skills, physical principles, safe techniques, and mechanical functions. In the cognitive domain, demonstrations are used to illustrate theoretical or scientific concepts that students are not expected to perform. In the psychomotor domain, they are used to model a task or skill that students must learn to perform. This is the most common use of demonstrations for training in the fire and emergency services. **Table 6.1** provides some general guidelines for demonstrating skills.

> **Emphasize Safety**
>
> Because of the hazardous nature of work in the fire and emergency services, instructors must emphasize the importance of safety while demonstrating every step of a skill or task. Many students want to be able to perform a skill quickly when they first learn it, but skill and speed come only with practice. Trying to perform a skill without having carefully learned the steps or developed coordination can be a safety hazard. Instructors should always stress the importance of safety when demonstrating a procedure, during practice time, and in final student evaluations. Students can be frustrated by the disparity between watching an instructor perform a skill with apparently effortless fluency and learning to perform it themselves at a much slower pace. Consider teaching them the phrase: "Slow is smooth, smooth is fast."

The instructor safely demonstrates a task while explaining how and why it is performed **(Figure 6.9)**. Students absorb this information through sight and hearing. Whole-part-whole is a helpful sequence for demonstrating psychomotor skills. It consists of three steps:

Step 1: Perform the skill at normal speed so students can see an overview of the skill.

Step 2: Perform the skill at a slower speed, emphasizing each part individually, so that students can see the details of the skill.

Step 3: Perform the skill a third time, at normal speed, with explanation during performance.

When students practice the skill, they use psychomotor skills and add the sense of touch to their learning experiences. The following advantages easily outweigh any disadvantages to using demonstrations in both the cognitive and psychomotor domains:

- Students can receive immediate feedback.
- Instructors can readily observe behavioral changes.
- Students have a high level of interest when participating.
- Instructors can easily determine whether students have achieved the learning objectives.

Figure 6.9 Demonstrating a task is an effective training method.

Table 6.1
Skills Demonstration

Preparing for a Demonstration

Know clearly what is to be demonstrated and its learning objective.

Be proficient in every step of the demonstration by practicing in advance with all instructors who will be involved.

Acquire all equipment and accessories, ensure that they work, and arrange them for use.

Arrange the room or demonstration area so that all students can see and hear the demonstration.

Demonstrating the Skill

Begin the demonstration by linking new information with the students' current knowledge.

Explain what the demonstration will show the group how to do.

Explain why the skill is important.

Demonstrate the skill once at normal speed.

Repeat the demonstration step by step while explaining each step slowly.

Repeat the demonstration again while a class member or the group explains each step.

Consider using a video camera and large-screen monitor when the group is large in order to allow students to see the process up close or observe small details.

Allow students the opportunity to ask questions and clarify any misunderstandings.

Ask for a student volunteer to demonstrate the skill while explaining the steps. Give reassurance by coaching and guiding the student through the process. Offer suggestions or corrections during the demonstration.

Provide the opportunity for students to practice, and allow them to supervise and correct each other as they become skilled. Again, closely monitor student activities when students practice potentially dangerous skills for the first time.

Reassemble the group and demonstrate the skill one more time at normal speed and/or one more time slowly as the group explains the steps as a summary. Relate the skill to the learning objective and performance on the job.

- Learning skills correctly reduces risk when conducted in a safe environment and under careful supervision, thus giving students the confidence to perform the same skills on the job.

Potential drawbacks of the demonstration method include the following:

- Instructors must plan for extensive preparation and cleanup times, especially when using such items as power tools, hose, breathing apparatus, and cardiopulmonary resuscitation (CPR) manikins.
- Careful lesson planning is important because assembly and practice can use much of the class time.
- Large groups of students require extra equipment for practice, as well as additional instructors for supervising, coaching, and enforcing safety regulations. Instructors must closely monitor students who are practicing potentially dangerous skills for the first time.
- Skills that must be performed or practiced outside depend on the weather. Instructors must have a contingency plan available in the event of inclement weather conditions.

Leading Class Discussions

The discussion method allows for interaction between instructors and students. The instructor talks to the overall group, and invites the group to reply. Group members talk to the instructor and to each other, either in small groups or as one large group **(Figure 6.10)**.

Lesson plans may include directions for structured discussion sessions in either large or small groups. These directions usually provide discussion topics and can also specify the most optimal arrangements for the groups. As instructors become more experienced, discussions may result spontaneously as a response to student questions. Instructors should remember that discussions are less predictable than lectures in terms of the amount of class time they require.

During a discussion, instructors and students can:

- Exchange views and ideas
- Ask questions and receive answers
- Provide examples based on experiences
- Arrive at conclusions
- Form a consensus

For this method to be effective, students must have a basic knowledge of the subject before the discussion begins. The discussion method is not a good format for introducing new material to inexperienced students.

Figure 6.10 Group discussions allow instructors and students to interact in ways that lectures do not.

Discussion as Active Learning

Group discussions are an example of active learning, a form of instruction in which students participate in classroom activities and are forced to think about what they are doing. Research has demonstrated the benefits of active learning at every level of education. As a form of active learning, classroom discussion:

- **Fosters student understanding** — Group discussions give students an opportunity to reflect on the lecture material.
- **Improves student communication skills** — Small group interaction allows students to learn to listen effectively, develop their positions on topics, and logically discuss information.
- **Improves cooperation within a group** — Group discussions help to create a sense of teamwork and cooperation between students.
- **Places the responsibility for learning in the hands of the student** — Discussions increase the student's sense of ownership of the learning process.

Whole Group Discussions

In whole group discussion, the lesson plan contains the basic information for the discussion. The instructor should determine whether or not a specific topic can generate enough interest for a whole group discussion. A whole group discussion can help students to accomplish the following learning objectives:

- Share information and knowledge.
- Apply theories and critical thinking skills.
- Express personal views and ideas.
- Collaborate and work as a team.
- Clarify attitudes, values, and beliefs.

An instructor should establish the time required for the discussion and ensure that it is available in the class period. The instructor should also develop an opener for the discussion, which may consist of a short narrative, **case study**, hypothetical situation, or problem. These openers and topics may be provided in prepared or adjusted lesson plans. Because discussions can be time-consuming, the instructor must ensure that the time is used efficiently and the topic is specific enough to help students stay focused.

When planning for whole group discussions, instructors should select the type best suited for the topic, time frame, and lesson objectives. The two most common categories of the whole group discussion format are as follows:

1. **Guided** — The instructor presents a topic to a group, and the members of the group discuss ideas in an orderly exchange controlled or guided by the instructor. The intent of this type of discussion is for students to gain knowledge from other group members, modify their own ideas, or develop new ones. As facilitator, the instructor's role is to guide the discussion and meet the lesson objectives in the following ways:

   — Keep the discussion on the topic and on schedule.

   — Add pertinent details.

   — Ask thought-provoking questions.

2. **Conference** — A conference discussion is less controlled than a guided discussion. In this method, instructors are facilitators not teachers. They do not tell the group how or what to think. The intent is for the students to understand how they view a topic rather than being influenced by their instructor. The instructor's responsibilities in this format include:

   — Providing background information on the topic.

   — Stating or restating problems, asking questions, or clarifying students' comments.

     Other than this, the instructor should allow the students to control the discussion and should not actively participate.

   — Controlling or eliminating bickering and irrelevant discussion, reconciling differences of opinion, and uniting students.

NOTE: The term *conference* is used for both a discussion format and a type of meeting that has the same purpose, only on a larger scale.

Ahead of the class session, the instructor should develop a means of closing the discussion. The closer may include a summary of the problem and a list of the solutions developed. Students may also be asked to write their own summaries — an exercise that requires them to rephrase and restate the major points generated in the discussion.

Small Group Discussions

Instructors do not actively participate in small group discussions. Instead, they select a student to facilitate or lead the discussion in each group. The main advantage of this format is that students express their ideas and opinions more openly with their peers than they do when an instructor is present. Small group discussions are most effective when:

- The task is structured.
- Students are experienced in working with others.
- The learning outcome is clearly defined.
- Students have time to prepare for the discussion.

The instructor's role in this format is to define group goals, establish a time frame for discussion, and monitor the groups to make sure students stay on task. When the discussion is finished, instructors may want to have students write summaries of their discussions or present their findings to the rest of the class. Prepared lesson plans will generally indicate when small group discussions are desirable in the lesson plan, how much time should be provided for the discussions, and what prompts or topics should be provided to the small groups.

Leading Discussions

Both large and small group discussions require that instructors demonstrate leadership to ensure that students achieve course goals. Instructors should monitor how involved students are in the discussion, who is and is not participating, and whether the discussion stays on topic. The instructor should assume, or delegate, the following roles:

- **Director** — Keep the discussion moving forward. Make sure students stay focused on the topic.
- **Gatekeeper** — Ensure that all students have an opportunity to speak, that no one dominates the discussion, and that any cultural differences are considered.
- **Timekeeper** — Remind students of the time remaining for discussion or summary.

To ensure that the students understand the intention of the discussion, the instructor should perform the following tasks:

- Open the discussion by stating the topic or problem to be solved.
- Paraphrase students' contributions to ensure that they understand the material.
- Ask questions to make sure that students understand their own positions, as well as those of their classmates.
- Act as a resource for additional information and statistics.
- Summarize the results of the discussion.

NOTE: Instructors who regularly plan small group discussions should coach students in the leadership skills they will need to serve as group facilitators.

Discussion Techniques

A variety of techniques may be used to direct the outcome of a discussion. These techniques apply to both large and small group discussions and include:

- **Brainstorming** — Students try to generate as many ideas as they possibly can, operating under the principle that there are no bad ideas. The group then evaluates the ideas and decides which ones have the most merit. Brainstorming requires students to use creative thinking to propose a solution to a problem based on their knowledge and experience.
- **Nominal group process** — In this format, the discussion closely imitates an organizational decision-making process that students will encounter in their jobs. This technique is more structured than brainstorming and requires that ideas be more realistic (**Figure 6.11**). Steps:

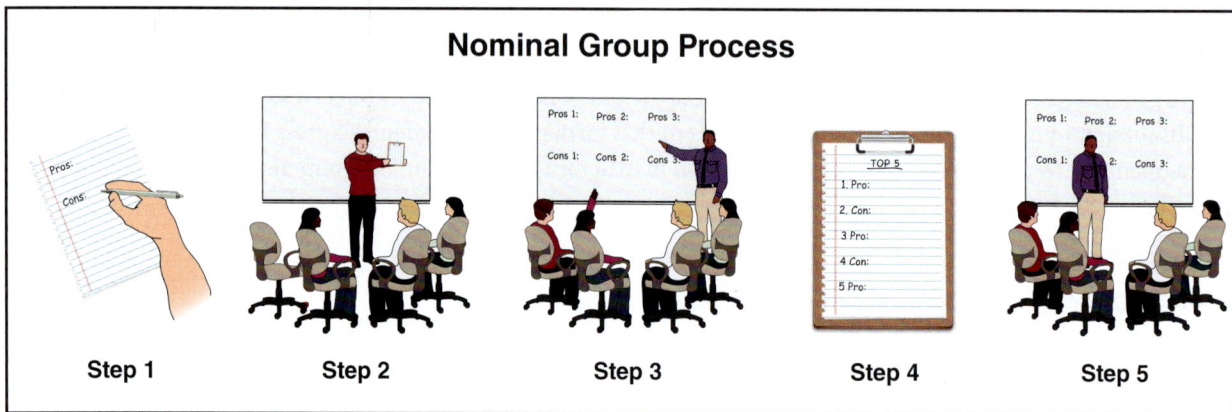

Figure 6.11 Instructors can use the nominal group process to imitate real organizational decision-making methods.

1. Begin the session by having students write a list of the pros and cons of the topic.
2. Have students present their lists to the group, each speaking in order until all have commented.
3. Correlate, examine, discuss, and rewrite comments presented.

4. Have the group select the top five considerations.

5. The instructor summarizes the findings.

- **Agenda-based process** — The instructor provides the students with an agenda of topics or key points. Students then do research and prepare reports to give to the group. In the discussion, students may ask questions or express opinions on the reports.

Asking Effective Questions

Instructors use questions for a variety of reasons, including to receive feedback on how instruction is progressing and to promote critical thinking. Questions may also be used to achieve the following objectives:

- Promote discussion.
- Encourage interest and curiosity.
- Motivate students to acquire knowledge on their own.
- Assess students' level of understanding.
- Control the behavior of disruptive or nonparticipating students.
- Provide an opportunity for students to openly express their ideas and opinions.
- Stimulate interest that generates related questions.
- Review and summarize information.
- Assess whether students have achieved the lesson's learning objectives.

Students' responses to questions can also help instructors recognize the need to alter a lesson plan or adjust their teaching style. For example, instructors may realize that they need to incorporate techniques more appropriate to their students' learning styles.

Some curriculum developers include prepared questions in their lesson plans, but instructors should know how to create effective questions of their own. This skill also enables the instructor to evaluate prepared questions so that the questions can be adjusted or improved, if needed. Follow these guidelines for developing and asking effective questions:

- **Plan and/or review questions in advance** — Questions should help students achieve desired course outcomes or learning objectives, and be appropriate for their location within the lesson plan.
- **Write and arrange questions in a logical order** — Start with questions that request basic information before moving on to questions that require critical thinking.
- **Phrase questions clearly** — Whenever possible, use clear, simple language so that students focus their attention on the answer, not the question. When a question is complex enough to require explanation, develop the explanation in advance.
- **Ask only one question at a time** — Avoid asking questions in succession without waiting for a response to each one.
- **Be sure that the wording of the question doesn't make the answer obvious** — Students will recognize the intended answer and think that the question is, at best, hypothetical, and, at worst, a waste of time.
- **Allow a wait time** — When questions are directed to the entire class, wait until there is a response. It may take time before a student raises his or her hand to answer. An appropriate wait time would be 3-5 seconds.
- **After waiting, call on a student directly** — Following the wait time, address a student directly by name. Give the student time to respond and do not hurry, especially when the student is shy.
- **Never use questions to intimidate, embarrass, or humiliate students** — Intentionally intimidating, embarrassing, or humiliating a student is inappropriate and unprofessional.
- **Distribute questions evenly** — Make sure to call on as many students as possible.
- **Ask questions at a variety of levels and of a variety of types** — Instructors should use all types of questions that are appropriate to the topic and the students' learning styles.

- **Adapt questions to students' ability level** — Matching the questions to individual student abilities ensures that most students will be able to answer at least some of the questions.
- **Ask appropriately challenging questions** — Questions should never be so easy that no thought is required to answer them.
- **Avoid asking questions too soon** — For questions to be effective, students must have the knowledge to answer them. Questions that are intended to determine a student's current level of knowledge or opinion can be asked early in the session.
- **Follow up on student answers** — Instructors can elicit further response by using techniques such as inviting elaboration, encouraging other class members to respond, or remaining silent. Examples:

— Once a student has answered the question, the instructor may ask, "Could you expand on your answer?" as a way of encouraging the student to go into greater detail.

— The instructor may also want to involve other students by asking them to add an idea, fact, or experience to the first answer.

— While instructors may be uncomfortable with silence during a class, being deliberately silent can be an effective tool when asking questions. The instructor's silence can motivate a student to elaborate on the initial answer.

New instructors should practice developing questions and including them in their lesson plans or outlines. When there is extra time in a session, the questions can be used to help students focus on the presentation's key points. When time allows, questions can also be used to start group discussions.

> **Waiting After Asking a Question**
>
> Inexperienced instructors are often uncomfortable with the silence that follows their questions, as students try to formulate their answers. To avoid this silence, instructors may quickly answer their own questions, which defeats the purpose of including questions for students in the lesson plan. Instructors should wait for students to respond, even if the wait time seems lengthy. Another tactic is to encourage students to answer by giving them a hint about the intended answer, or offering a motivating remark, such as "I know you're all listening," "Somebody knows," or "Give it a shot." Instructors can also politely redirect the question to another student if the first one takes too long to respond or gives an incomplete answer.

Question Types

Different types of questions produce different kinds of answers. When considering which kind of question to ask, instructors should consider what they want to accomplish at that particular point in the lesson. Instructors can use the following types of questions to start discussions, stimulate thinking, provide feedback on how training is being received, and enable students to both assess their own learning and manage any gaps (**Figure 6.12**):

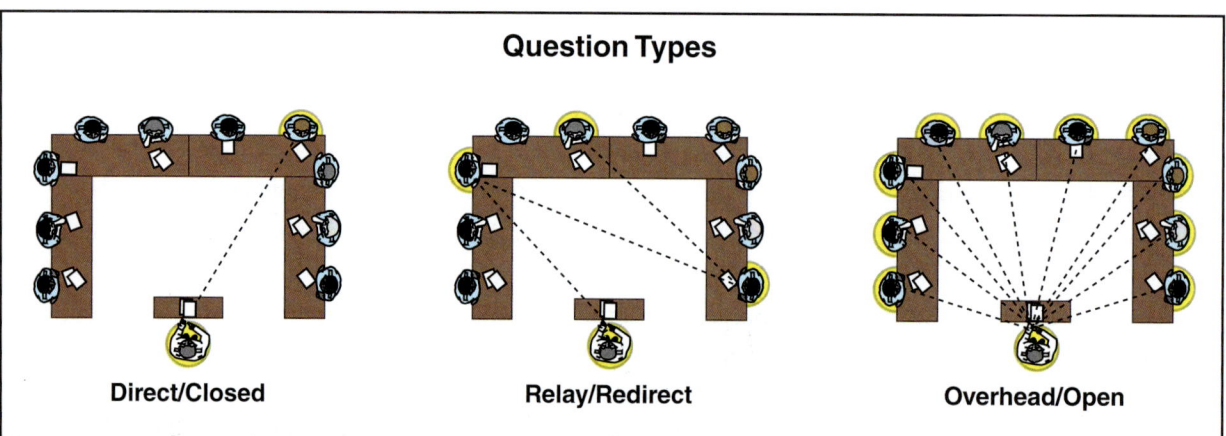

Figure 6.12 An instructor should tailor questions based on what he or she wants to accomplish during the lesson.

- **Rhetorical** — Rhetorical questions are used to stimulate thinking. They do not necessarily have one correct answer, and often do not call for a spoken response. For example, an instructor might open a safety lesson by saying: *What are the most important pieces of equipment you will use on an emergency scene to protect yourself from injury or exposure? By the end of this lesson, you will know how to answer this question.*

- **Closed** — This type of question has a limited number of possible answers. The instructor is able to anticipate and judge the accuracy of student responses. Example: *What is the definition of flashover?*

- **Open** — This type of question has many acceptable answers. The instructor has general criteria for judging the accuracy of an answer, although students' answers may be unexpected. Example: *What information can you gain during size-up?*

- **Direct** — The instructor directs a question to a student, who must then respond. This type of question is not frequently used with adult learners because it can make students uncomfortable. However, asking direct questions can be an effective way to encourage bored or disengaged students to participate.

- **Overhead** — The instructor asks a question of the entire class, not just one student. Any member of the class is free to respond, either by calling out the answer or by raising a hand and waiting to be recognized. This technique is helpful in starting discussions or offering ideas or opinions. If no one answers, the instructor can then direct the question to an individual student. Instructors may also allow students to consult with each other in groups to produce the answer.

- **Relay** — Instead of answering a student's question, the instructor asks the rest of the class to answer. Relay questions are a good way to open a discussion or stimulate interest, but instructors should avoid this technique if they do not know the answer. In these circumstances, it is better to say, "I don't know."

- **Redirected** — This type of question is useful when a student asks for an answer that the instructor believes he or she should already know. The instructor can ask the student to provide small amounts of information that, taken together, will answer the original question.

Responding to Students' Answers

When instructors ask questions, they should prepare for a variety of student responses **(Figure 6.13)**. Instructors should learn appropriate ways to respond when students provide an answer or pose a new question as their reply. Some basic guidelines for instructors are as follows:

- Use positive reinforcement. If a student answers a question correctly, the instructor might say, "That is absolutely right" or "You seem to have a good understanding of this topic."

- When a student answer is only partially correct, positively reinforce the correct portion, then redirect the question back to the student or ask another student to complete the answer.

- Always provide correct answers if the students do not. Providing correct answers benefits the entire class. Wrong answers are to be expected as part of the learning process. An appropriate response could be "That is close. Maybe I didn't ask the question clearly. What I am asking is . . ."

Instructors are sometimes reluctant to discourage responses, so they accept all answers while tactfully trying to direct the group toward the right conclusion. But taking a wrong answer and asking either why it is wrong or where it may fit appropriately in the lesson is a technique that provides students with opportunities to:

- Think and analyze problems.

Figure 6.13 While posing questions to students, instructors should be prepared to respond appropriately to answers or offer questions for follow-up.

- Compare facts and ideas, and apply them to different situations.
- Critique their responses and find correct solutions.
- Explore and discover new methods of application.

Answering Students' Questions

Answering students' questions is one of the most difficult things for instructors to do. Some students may pose questions that appear to be logical, but are actually complex, illogical, or off topic. An instructor can respond to these questions in the following ways:

- Be aware that some questions can be controversial or distract the class. Instructors should only provide answers to questions when they have the knowledge, experience, or resources to do so. If this is not the case, they should defer the question in order to research the answer or to consult a more knowledgeable source, such as a senior instructor or administrator.
- Redirect the question to another student who is likely to respond correctly. This approach can also be used to generate group discussion.
- Defer questions that are beyond the scope of the lesson, or tell students that they will learn the answer later in the course.
- Always answer truthfully. Never bluff students by providing false or misleading information. Doing so can destroy an instructor's credibility.

Psychomotor Skill Instruction

Much of what is done in the emergency services directly relates to performing an action. An action often involves complex thought and deliberate response to observations made in the field. It may also need to be altered or adjusted to fit the exact circumstance.

The development of psychomotor skills typically occurs in three phases. In the **cognitive phase**, the student is a beginner and is developing the basic procedure of the skill through verbal and visual stimulus. Students use trial and error to strengthen their understanding of the skill, a process that requires a great deal of effort. In this phase, the instructor has a great deal of influence and must demonstrate as much patience as possible to encourage the student and stay positive. A step-by-step skill sheet is helpful and will keep the student on track to fully meet the lesson objectives.

In the **associative phase**, students grow more comfortable performing the skill. They associate cognitive knowledge with the muscular movement needed to master the skill. In this phase, students realize and overcome initial errors, and they develop and strengthen connections between the steps of the skill. Fluidity increases, but the student still needs to think about the steps to be completed (**Figure 6.14**). In this phase, the instructor is more of an evaluator and a coach than a teacher.

In the **autonomous phase**, the student's actions become smoother, and the performance of the actions becomes quicker. Students have moved toward the point where they no longer need to think about the action. The instructor has now become the coach and is there to help students refine

Figure 6.14 In the associative phase of psychomotor skills development, students acquire fluidity while strengthening connections between steps.

their efforts. This phase is associated with an effortless approach to the task and the ability to adapt the task to varying environments.

Throughout the three phases of psychomotor skill instruction, the instructor should always consider the students' emotional, mental, and physical safety. Teaching and practicing safe procedures engrain best practices that will carry on throughout the students' training and service careers.

An instructor must use a systematic and structured approach to psychomotor skill instruction. A sequential approach includes:

- **Preparation** — The instructor is ready to teach the skill and is familiar with the skill level of the student.
- **Conceptualization** — The student is given the cognitive elements of the skill. This is where the student learns what to do in a larger context that includes possible problems, common errors, and potential risks.
- **Visualization** — The instructor demonstrates the skill in its entirety, providing a model for student performance.
- **Verbalization** — The instructor verbalizes the skill to the student, breaking down the task into subtasks in the correct order. The student should be able to verbalize the steps and assist in placing the steps in the correct order.
- **Practice** — Deliberate and conscious effort to refine the skill performance and accomplish the skill safely is critical for ultimate success and eventual mastery.
- **Feedback** — Instructors use feedback to convey whether students are meeting the instructor's expectations, and to correct errors or observed problems. Students use feedback to indicate whether they feel successful in achieving an objective.
- **Mastery** — Both instructors and students will strive to demonstrate their ability to perform the skill routinely without error. Instructors may indicate to students which skills cannot be mastered over a single class session but should be strived for over the course of a career.
- **Autonomy** — Ultimately, the instructor wants students to be able to adapt the skill in practical situations without error.

> **Teaching Psychomotor Skills**
>
> Psychomotor skill instruction follows the sequence of the adage: "I do, We do, You do." First, instructors demonstrate the skill to their students, modeling the way they want to see it performed. Next, students perform the skill, step-by-step, with their instructor. Finally, the instructor evaluates each student as he or she performs the skill independently.

Structured Exercises

Structured exercises include a variety of instructional methods that actively involve students in the learning process. Prepared or adjusted lesson plans may include structured exercises for instructors to lead. The following sections provide brief descriptions of structured exercises that instructors may encounter:

- Case studies
- Role playing
- Simulations
- Field and laboratory experiences

Case Studies

A case study, sometimes referred to as a scenario, is a description of a real or hypothetical problem that an organization or an individual has dealt with and may face in the future. Typically, a case study reviews and discusses detailed accounts of past events. Students then analyze the situation and synthesize possible answers to the problem. The purpose of studying past incidents is to be prepared for similar circumstances in the future.

Case studies provide students with the opportunity to discuss ideas and solve problems. These discussions develop their ability to examine facts and analyze situations in order to reach a conclusion or determine a course of action. The instructor provides students with time to review, research, and discuss the situation. Face-to-face or electronic communication between students must also be established to encourage student interaction. Students must be willing and able to communicate and defend their suggestions to other members of the group.

Role Playing

In role playing, students act out the role of a character in a scenario to prepare for situations they may encounter while fulfilling their duties. Role playing can be used when training personnel for a variety of tasks or situations, such as public information officers who interact with members of the community or public safety telecommunication personnel who must receive and dispatch calls during emotionally intense incidents.

At the end of the role play the instructor debriefs participants and observers about the objective of the activity. This debriefing also gives students an opportunity to explain their feelings and actions during the role playing activity. Some role playing scenarios may be emotionally charged, and the instructor should allow time in the debriefing for students to successfully separate themselves from the role they have played. The instructor should also summarize the scenario and reinforce the importance of any positive behavior that was exhibited.

A structured activity with many advantages, role playing:

- Encourages application of knowledge and safe practice of skills.
- Permits students to practice under conditions that simulate an incident without the danger of fatal consequences.
- Improves understanding of critical features of interpersonal relations.
- Allows students to identify multiple approaches to a problem.
- Increases the development of empathy as students discuss their own perspectives and listen to their classmates' opinions.
- Helps develop critical awareness.
- Prepares students for emotionally challenging events, such as dealing with trauma patients.

Instructors should also be aware that role playing has the following disadvantages:

- Preparation can be complicated and time-consuming. Consider using case studies as the basis for role playing scenarios, which reduces the preparation time for both.
- Role playing can take time to perform and may result in students digressing from the topic when they are bored, uninterested, or see no value in the activity. Instructors can combat this attitude by explaining the importance and relevance of all activities and how they relate to the students' work-related duties.
- Role playing requires students to identify, at least superficially, with the characters and scenario. Some students will be more comfortable with that process than others. Remind students that role playing is the closest simulation possible, in a controlled setting, to some types of stressful encounters; and that they should not take anything personally when they are in character for the activity.

NOTE: When debriefing a role playing scenario, instructors should remind students that actions taken during the scenario should not be considered part of a fellow student's personal character, but rather part of a larger training situation designed to ingrain situational behavior and skills.

Simulations

Training simulations also allow students to participate in scenarios that represent situations they are likely to encounter on the job. These simulations may take many forms, from practical training evolutions using demonstration devices to technology-centered activities such as **computer-based training (CBT)**. Simulations permit students to experience a situation, make decisions, and see the results of their decisions without the negative consequences that can occur at an actual emergency. The key to all simulations is to ensure that they effectively reflect the equipment, procedures, protocols, and situations that students will encounter on duty.

Simulations may be tailored to the type and scope of the training. For example, a tabletop emergency management drill is an economical and effective simulation **(Figure 6.15)**. Confined-space rescue training, that permits actual operations in a simulated hazardous environment, is an example of a practical training evolution. CBT permits individual students to operate apparatus pump panels, simulate command of structure fires, and even attack a computer-created structure fire. Most simulations include role playing elements, such as students taking on assigned duties and interacting with one another.

Field and Laboratory Experiences

Field and laboratory experiences provide students with demonstrations and simulations. These experiences allow students to inspect, use, test, and evaluate equipment or processes, either in actual installations or in laboratory settings. Prior to learning in a laboratory or field environment, students should learn safe and applicable procedures to ensure a safe learning environment.

Figure 6.15 Tabletop simulation models are a cost-effective way to conduct certain types of training.

In the field, students are typically given a tour of an installation. They may be permitted to observe a fire detection and suppression systems test or see the steps required to replace a component. Field experiences tend to be less controlled than laboratory settings, so students may not be able to handle components. On the other hand, they will be able to see systems functioning on larger scale than is often possible in a laboratory.

In a laboratory, students can see models of equipment, such as cutaways of apparatus engines, pumps, or sprinkler control valves **(Figure 6.16)**. They may perform chemistry experiments to simulate fire behavior or fire spread in an enclosed space. In the controlled environment of a laboratory, the instructor explains the equipment or process, demonstrates the steps required, and observes students as they repeat the skills. Students may work independently or in groups. Instructors may also choose to give students a challenge, such as providing them with a defective SCBA regulator and asking them to repair it.

Figure 6.16 Models allow students to study the inner workings of equipment, fixtures, or machinery.

Competency-Based Learning in the Fire and Emergency Services

In the fire and emergency services, students should be competent with one set of requisite information and skills before they progress to a new one. Competency-based learning (CBL), sometimes referred to as **mastery learning**, requires that each student successfully master the learning objectives or outcomes of the lesson or course. In this approach, instructors are provided with the specific criteria (standards) that students must meet during testing for competency. This information is stated in the learning objectives.

Competency-based learning:

- **Focuses on competency** — Primary focus is on the successful and accurate completion of skills; also known as *performance-based*.
- **Meets individual needs** — Training is individualized or adjusted to meet the learner characteristics of the student.
- **Provides immediate, specific feedback** — Instructors provide feedback to the individual student when the student performs the skill.

- **Employs modules and multimedia** — Courses and lessons are divided into similar blocks that are supported by a variety of audiovisual training aids.
- **Depends on instructors** — Instructors must help students learn the skill and become proficient at it.
- **Meets specific objectives** — The ultimate goal of every course is for students to master a set of skills that meet specific learning objectives.
- **Uses criterion-referenced testing** — Success is based solely on a predesignated level of competency, usually 70 to 100 percent.

The competency approach uses criterion-referenced teaching, learning, and assessments; and focuses attention on learning objectives. Students who have problems meeting the desired criteria on their initial efforts get additional instruction, time, and opportunities to perform to the acceptable level. The learning objectives are written to establish the criteria for competency as follows:

- Identify and clearly describe the learning outcome (behavior). Example: *The student will don an SCBA.*
- Define the important conditions under which the students will perform (conditions). Example: *The student will don an SCBA **while wearing full personal protective equipment**.*
- Define the criterion of acceptable performance (degree). Example: *The student will don an SCBA while wearing full personal protective equipment **within 45 seconds**.*

When the material is difficult or complex, instructors may need to schedule additional instruction time so that students can assimilate complex concepts. No student should proceed to new material until he or she has become competent in the basic requisite material.

The competency approach has advantages and disadvantages, although with proper planning, instructors can overcome the disadvantages. Some advantages of the competency approach to teaching include:

- Students are prepared to advance to more complex knowledge or skills.
- Prior knowledge is used as a building block for new skills, which can make gaining competency easier for the student.
- Students are made aware of the learning objectives from the beginning so that criteria for passing the course are never in question.
- Time is given to tailor learning to the student's individual learner characteristics and learning style, which assists the student in gaining competency.

Disadvantages include:

- Instructors must plan for and provide extra time to ensure that all students become competent on the subject. This may interfere with lesson planning or make it difficult for an instructor to stick to a schedule.
- More effort is required on the instructor's part to determine students' learning pace and to match it.
- Faster students may feel that slower classmates are holding them back.
- A wide variety of training materials must be available to meet the learning needs of all students.

Teaching Strategies

As instructors learn more about educating people, they often come to the realization that people learn differently depending on the topic. With these variations in mind, instructors can use different teaching strategies to adapt to changing learning environment and challenges.

Traditional Instructor-Led Training (ILT)

Instructor-led training (ILT) is the most prevalent approach to teaching in the fire and emergency services. ILT has the advantage of being flexible, economical, and familiar to both students and instructors. Three features of instructor-led training are detailed in the following sections:

- Multiple instructors
- Generating and maintaining student interest
- Reinforcing learning

Multiple Instructors

Instructor-led learning environments traditionally have only one instructor serving in multiple roles. A variation on that training style is to split some of the responsibilities among multiple instructors. In addition to sharing the workload, multiple instructors combine their knowledge and experience so that students are better able to meet the course learning objectives or outcomes. This strategy can be an effective educational technique and a unifying force between organizations. In one variation, known as *team teaching*, multiple instructors teach the same topics at the same time with different groups of students. Team teaching provides a more effective use of class time and has smaller student-to-instructor ratios **(Figure 6.17)**.

The instructor with the most knowledge or experience in a particular topic usually teaches the cognitive and demonstration portions of the lesson to all classes. That instructor then joins with other instructors in supervising the practical training evolution. Multiple-instructor or team-teaching learning environments:

- Provide an effective use of resources from multiple areas and specialties.
- Allow an instructional delivery method that works well when the topic is broad.
- Provide students with an exposure to a wide variety of teaching methods and skills training.
- Keep the attention of the group by utilizing each instructor's different voice, pace, and personality.

To prepare for multiple-instructor presentations, consider the following suggestions:

- Make an extensive course plan that outlines each instructor's role and time commitment.
- Maintain communication between all instructors to ensure course continuity and consistency.
- Choose instructors whose teaching styles contrast, yet balance one another.
- Decide in advance who will teach which topic.
- Review all lesson outlines together so that each instructor knows what the others are doing.
- Agree on and use the same format for all instructional technology tools.
- Designate one lead instructor per topic.
- Determine time commitments and require each instructor to adhere to them.
- Agree that all instructors will be present for the entire course, not just when it is their turn to teach.
- Meet with other instructors to review session results.

Figure 6.17 In team teaching, several instructors can work with groups of students simultaneously. Doing so helps to maintain a low student-to-instructor ratio.

Generating and Maintaining Student Interest

Instructor-led training relies on students' interest for the overall class session to be successful. Interested students are open and responsive and want to concentrate on what they are learning. They are also more willing to actively participate in the learning process.

Generating interest is only the first step. To help maintain interest, instructors need to show students how they have a personal connection with the lesson. The following are strategies to help maintain student interest:

- **Relate learning to student interests** — Material relevant to student experiences illustrates the overall usefulness of the information presented.

- **Offer material that pertains to students' professional goals, duties, and tasks** — Students are more likely to be curious about information that has particular importance to their professional futures.
- **Use humor appropriately** — Use humor spontaneously and courteously.
- **Stimulate emotions** — The experiences that students bring to training sessions evoke a range of emotions. Instructors can call on these experiences to promote student interest.
- **Explain and illustrate with examples, stories, analogies, and metaphors** — Providing examples can make abstract concepts more relatable to students.
- **Use questions to stimulate interest** — With practice and experience, instructors can learn to ask appropriate, well-timed, thought-provoking questions that encourage participation and promote understanding. The instructor should be careful not to force a specific viewpoint on the class.
- **Use unpredictability and uncertainty** — Anticipating the unexpected is exciting, and students who do not know what will happen next are more likely to pay close attention. As long as students feel safe and know that no one will be hurt, this technique can be an effective way to keep students interested.

> **Words to Teach By**
>
> Your students have entrusted you with their most valuable possession – time. Don't you dare give them anything less than your best.
> - Be Prepared. Know your material, prepare your classroom, master your audiovisuals (including the machines), and have a backup plan.
> - Preview your video presentations. Use only those portions that are important to your lesson.
> - Be excited about what you teach. Excitement is contagious!
>
> *Courtesy of Rod Smith, Assistant Chief, Lane County Fire District No. 1, Veneta, Oregon.*

Reinforcing Learning

Instructors reinforce learning through repetition and behavioral reinforcement. Repetition involves repeating, re-emphasizing, and reviewing key points with the students. Behavioral reinforcement encourages students' attempts to learn by acknowledging their successes and correcting their mistakes.

Intentional repetition is an important part of organizing a presentation. For example, in a lecture, elements are introduced, explained, and then summarized. In a skills demonstration, the entire skill is demonstrated, and then repeated in smaller segments.

Through emphasis and repetition, instructors help students to recognize the importance of key points. Instructors should provide opportunities for students to apply knowledge and practice important skills in a safe learning environment.

Behavioral reinforcement is based on a psychological theory that connects a change in behavior with the consequences of that change. Behavioral reinforcement establishes rewards (positive reinforcement) for students' successes or good behavior and negative consequences for their failures or bad behavior. In the fire and emergency services, rewards for success include promotion or commendations. In the learning environment, positive reinforcement may be as simple as an instructor's compliment or encouragement after a student properly performs a skill or masters a theoretical concept. Educational research has demonstrated that students learn faster and make permanent behavioral changes when their successes are met with meaningful positive reinforcement.

> **Negative Reinforcement**
>
> While positive reinforcement praises good performance, negative reinforcement punishes poor performance. Negative reinforcement is intended to make students so afraid of failure that they stop performing poorly. In the classroom, negative consequences include failing to pass an exam or skills test. On a training ground or at a scene, unsafe behavior may result in the negative consequence of a reprimand from an instructor or officer.
>
> Educational research suggests that negative reinforcement, including natural consequences of poor performance, is not as effective as positive reinforcement. Students may be better motivated to learn when they recognize that their success in a class or on a scene is directly tied to their own motivation and ability.

Blended or Hybrid Learning

In contrast to instructor-led training, blended or hybrid learning combines traditional (face-to-face) and online teaching methods. Ideally, the instructor or course designer has critically evaluated which parts of the course would be better suited to face-to-face interaction and which would be enhanced by independent, online learning.

Instructors who use the hybrid learning environment need to be very organized and disciplined in order to maintain a high level of student satisfaction and performance. With the reduction of a formal structure, it is easy for both the instructor and the student to become frustrated and/or confused about what is happening and what needs to happen in the course.

Student-Centered Learning

In the traditional academic-type learning environment, students spend most of their time watching and listening to the instructor. In the student-centered learning environment, the instructor shifts the focus to the students. The instructor provides students with active learning activities, such as answering questions, problem solving, and formulating questions of their own. In student-centered learning, the instructor becomes more of a facilitator than a traditional teacher. This method of instruction, more than the traditional method, leads to an increased depth of knowledge and understanding of the material.

Flipping the Classroom

In the traditional classroom model, the instructor lectures over new material in class, and the student is assigned exercises and activities to complete at home. If students have questions or difficulties with the homework, they have to wait until the next class period to ask the instructor for help. The flipped classroom reverses that traditional teaching model. Instead of students coming to class for a lecture, they read and study the new material at home. When they come to class, students have more time to ask questions, discuss the material, and engage in activities that reinforce their learning. By bringing active learning activities into the classroom, the instructor can reduce student frustration and provide a more positive learning environment.

Computer Simulation

As mentioned above, many training programs use instructional technology tools such as CBT modules and programs that allow students to work at their own pace (**Figure 6.18**). CBT may also be used in a more structured format to provide **distance learning**. CBT programs minimize the interaction between a student and an instructor. In some cases, the instructor may present an illustrated lecture remotely over the Internet or

Figure 6.18 Computer-based training permits more opportunities for independent study and remote instruction.

closed-circuit television. In other cases, the instructor may only be involved in answering questions that arise as students read information on their own and complete assignments.

A comprehensive list of CBT presentation methods is beyond the scope of this manual. The AHJ should inform the instructor as to the products and instructional technology tools he or she will be using to deliver instruction. Instructors should use whatever resources are necessary to familiarize themselves with the instructional system. Psychomotor skills demonstrations and practical training evolutions will still take place in person on the training ground, with the exception of any parts of the skill that can be taught online.

> **What This Means to You**
>
> Computer-based training is a broad term that incorporates many specific types of virtual learning environments. The following is a list of common software packages, known as learning management systems (LMS), designed to assist in the virtual learning environment:
>
> - Blackboard™ and Web CT™
> - Moodle™
> - D2L (Desire 2 Learn)™
>
> **NOTE:** This list will change as the marketplace matures.

CBT platforms require software to function effectively, and each may have unique requirements. Instructors must make sure that students' computers have necessary Internet access, sufficient speed, and enough internal memory, as well as:

- Internet browsers
- Computer operating systems (for example, Windows™ or Macintosh™)
- Word processing programs
- Document readers, such as Adobe Acrobat Reader®
- Unique software needed for the course

When facilitating CBT, instructors sometimes take on the role of computer support specialist. Instructors may be able to answer some questions. However, if they cannot promptly resolve a student's technical problems, instructors should contact someone else who can do so. No student should be penalized because a piece of equipment, software application, or computer does not work properly.

> **Security in Distance Learning**
>
> One significant concern with all types of CBT is security. Security begins when students enroll or are assigned to a course or program. Each student is assigned a unique login that provides the necessary level of access to the course website or database. This access should not include the test bank of questions and answers, other students' grades, or archival material such as assignments or tests from previous classes. Instructors should follow the security protocols established by the AHJ when delivering CBT.
>
> The instructor must take precautions when corresponding with students through e-mail. First, instructors should follow all e-mail policies and procedures in their organization. In the absence of these policies, instructors should use good judgment in e-mail correspondence. An instructor should always be careful when writing an e-mail message because of the possibility that the message may reach the wrong party or be shared with others without the instructor's knowledge. Instructor and student e-mail lists must be strictly controlled to ensure that they are not distributed to unauthorized persons or groups. E-mail communications may be misdirected, forwarded, altered, or distributed to unauthorized persons or sites. Confidential information such as student test results must never be transmitted via e-mail.

> **CAUTION:** After an email or electronic file has been deleted, traces still remain on the user's computer system.

Instructors should familiarize themselves with all the software used in the CBT system. Many of these tools are useful for any distance learning scenario, and instructors may wish to augment a distance learning program by using these tools. Software used in CBT programs could include:

- **Wikis**, **blogs**, and **file sharing** services
- **Social networking** systems
- Live streaming audio and video
- Chat sessions
- **Teleconferencing** and **web conferencing**
- Course management system (CMS)
- **Learning management system (LMS)**

Instructors using CBT should be aware of the ways that this format may interfere with their ability to effectively deliver a message. Feedback from students is minimal, and the technology presents an additional barrier between the instructor and the students. To overcome this interference, instructors may need to make the following adjustments to their instructional delivery:

- Be sure to take a thorough roll call prior to the session. This not only ensures that all registered participants are accounted for, but also helps to confirm that students are not experiencing technical difficulties that would hinder their ability to hear or see the session.
- Wait longer for students to answer questions. Whether the CBT uses a text/chat-based platform or streaming audio/video, there is a time lag when asking questions over the Internet.
- Limit movement. Live streaming video will be broadcast more clearly if there is not a great deal of motion being captured. Stay within the filming area of the camera. Arrange components to easily reach any computer controls without disrupting the presentation.
- Connect locations via interactive television (ITV). Direct questions to individual locations participating in the lesson rather than asking questions to the entire group. This ensures that each location is present and participating.
- Give clear instructions for media transitions. Students may have to switch cameras for views, visit specific websites, or access particular files during a presentation. They must be told to do so at the proper time.
- Allow time to deal with technical issues that may occur during the lesson. If a computer support specialist is not readily available, have a contact number you can call for support, if needed. This contact can be a member of your department or organization, or a representative of the software publisher or hardware manufacturer.

> ### Interactive Television (ITV)
> Interactive television (ITV) is used to link multiple classroom sites and permits one instructor to reach more students **(Figure 6.19, p. 134)**. It is a popular approach to distance education. Each site is able to see, hear, and talk to the other sites.

Self-Directed Learning

In **self-directed learning** or independent learning, individual students work at their own pace to accomplish course objectives, including the completion of tasks and skills in a predetermined format and the completion of work within a defined timeframe. Students are solely responsible for achieving these objectives, which may be

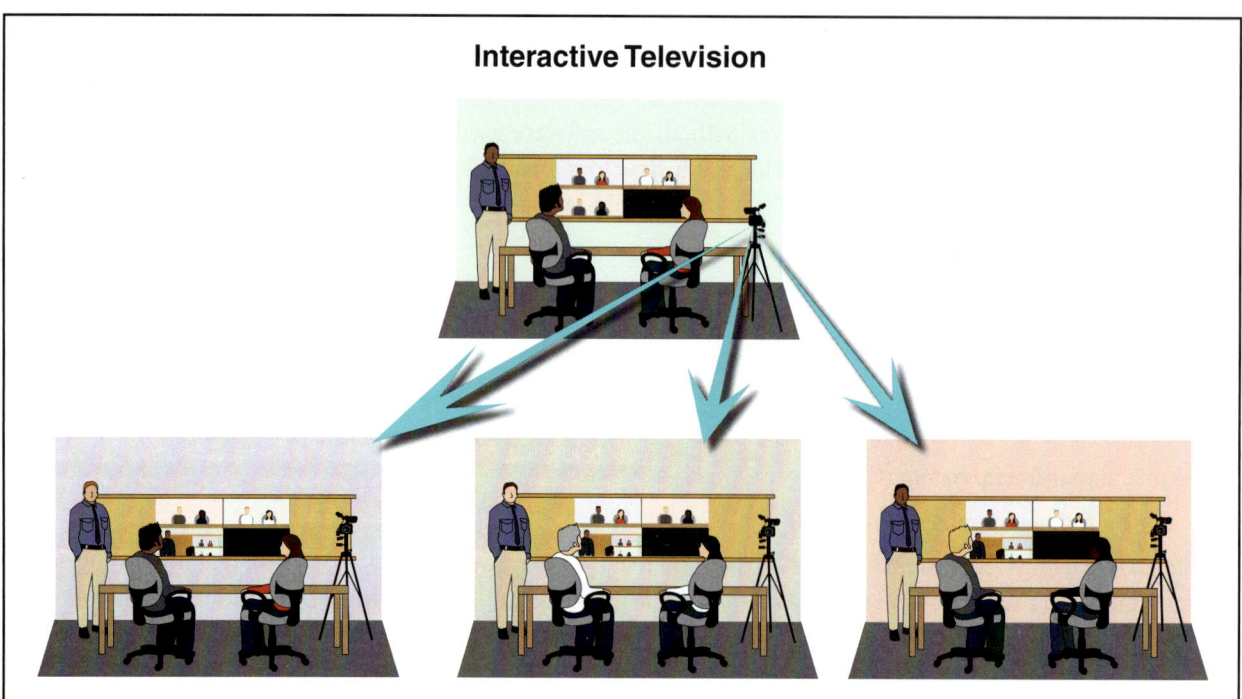

Figure 6.19 Interactive television allows students in remote areas to communicate with each other in real time.

determined by the instructor or chosen by the student. An instructor is not involved in the delivery of the training, although one may act as a facilitator and monitor to ensure that students complete the work correctly. As a result, the effectiveness of this distance learning method is heavily dependent on the student's level of motivation.

In self-directed learning, the instructor and the student schedule several meetings to monitor the progress of the independent study. Instructors are available to answer questions, evaluate learning achievements, and guide the student, but the learning process is completely the student's responsibility.

Self-directed learning often motivates students to discover information and resources beyond lesson requirements. However, instructors should be aware that not all types of training programs, particularly basic-level skill programs, are suited to this type of instruction.

Individualized Instruction

Another variation on instructor-led training, **individualized instruction**, has instructors adapt and/or adjust their teaching methods to address the needs of each student in the class and focus on his or her specific learner characteristics and learning styles. The instructor manages learning resources, guides students, and interacts with them but is not the sole or primary resource for learning as in the traditional instructor-led classroom. For example, students work with other students to meet certain requirements and share experiences. They also work with an instructor or mentor who ensures that they meet lesson objectives or individual goals.

Individualized instruction may also utilize some features of a self-directed learning plan and CBT. This approach may enable students who would not succeed in a typical learning environment to successfully achieve learning objectives.

Individualized instruction is based on the following:

- **A student's needs and learner characteristics** — The method of delivery for individualized instruction is directed at an individual student, rather than a group or class. The instructor pulls from a variety of resources to better meet the learning needs of each individual student.

- **Learning objectives or competencies required by the occupation** — Individualized instruction usually occurs within the context of a course where the learning objectives are competency-based and will be evaluated.

- **Instructional strategies and media that meet the needs of the student** — Individualized instruction is flexible in terms of the time students may take to learn the objectives and reach goals.

During individualized instruction, the instructor meets frequently with a student to evaluate progress, prescribe new learning objectives or different learning methods, and provide encouragement.

Many organizations use individualized instruction techniques in their training programs, including:

- **Learning activity packets** — Instructors prepare packets with sequenced activities and reading assignments for student use. This method is the most familiar to students and instructors.
- **Tutorial instruction** — The student receives one-on-one instructional help from the instructor or from another student, whether in the same or a slightly advanced class.
- **Programmed learning** — A systematic process of introducing information in small, sequential steps followed by questions that reinforce learning. This method may use a printed workbook or some kind of technology, such as a CBT system.

Chapter Review

Answer the following questions to review the information provided in this chapter.

1. What are some common presentation techniques that most effectively communicate information to students?
2. What are the four steps of the four-step method of instruction?
3. What are examples of instructional delivery methods?
4. What are examples of structured exercises?
5. What is competency-based learning?
6. What are examples of teaching strategies that encourage active learning?

Discussion Questions

The following questions are intended to generate discussion, expand your understanding of the chapter text, and allow you to think critically about what you have learned. Answers to these questions may vary.

1. What types of presentation techniques do adults prefer?
2. What are the purposes of the four steps of the four-step method of instruction?
3. What are your preferred instructional delivery methods?
4. Which structured exercises have provided you with better understanding of a topic?
5. Why is competency-based learning used in the fire and emergency services?
6. What teaching strategy have you experienced that helped you learn the topic at hand?

Key Terms

Agenda-Based Process — Classroom discussion format in which an agenda of topics or key points is provided to students for them to research, report on, and discuss as a group.

Associative Phase — Stage of motor learning in which the learner develops and strengthens the connections between steps in a skill, increasing fluidity of the overall skill through practice and performance.

Autonomous Phase — Final stage of motor learning in which the learner has demonstrated a mostly automatic progression through the steps of a skill, without having to rely on heavy cognitive application.

Blog — Abbreviation for web log; refers to a list of journal entries or articles posted by a single author or group of authors. Includes comment sections where readers can engage in conversation about entries.

Case Study — Description of a real or hypothetical problem that an organization or an individual has dealt with and may face in the future.

Cognitive Phase — Primary stage of motor learning in which the learner acquires the general understanding of the skill to be performed.

Competency-Based Learning (CBL) — Training that emphasizes knowledge and skills that are required on the job. Course objectives involve specific, criteria-based competence in performing tasks or understanding concepts that learners will use in their daily work. *Also known as* Criterion-Referenced *and* Performance-Based Learning.

Computer-Based Training (CBT) — A variety of self-study in which the student completes work on a computer with minimal communication with an instructor. *Also known as* E-learning, Blended E-learning, *or* Online Instruction.

Distance Learning — Generic term for instruction that occurs when the student is remote from the instructor, and when a medium such as the Internet, Interactive Television, or mail service is used to maintain communication between the two and to submit assignments.

File Sharing — Practice of making files or documents on one computer or server available to the general public, or to a selected group of individuals who are given access to the files. Allows users at remote locations to have access to the same materials without the need to put those materials on CD-ROM, memory drives, or other media.

Four-Step Method of Instruction — Teaching method based upon four steps: preparation, presentation, application, and evaluation. May be preceded by a pretest.

Individualized Instruction — Adapting teaching methods to suit individual students' specific learning styles, so that students will be better able to achieve lesson objectives.

Learning Management System (LMS) — Software application used for the administration and delivery of educational curricula, training resources, and evaluative tools.

Mastery Learning — Element of criterion-referenced or competency-based learning; outcomes of learning are expressed in minimum levels of performance for each competency.

Nominal Group Process — Classroom discussion format that requires students to follow a decision-making process similar to the processes that they will encounter in their professional duties.

Self-Directed Learning — Method of instruction in which individual students work at their own pace to accomplish course objectives. Course objectives may be determined by the instructor or chosen by the student, but course content is always determined by the instructor. *Also known as* Independent Learning.

Social Networking — Websites that allow users to be part of a virtual community. Users can communicate through private messages or real-time chat, and share photos, video, and audio.

Teleconferencing — Telephone service that allows multiple individuals at remote locations to have an audio-only meeting.

Web Conferencing — Meeting service that combines teleconferencing with an Internet-based sharing service, enabling users to communicate in real time while viewing and interacting with a computer-based presentation.

Wiki — Website that allows users to update, edit, or comment on the original content using their own Internet browser; allows for the rapid creation and deployment of websites and collaborative work on documents.

6-1
Give a prepared classroom (cognitive) lesson. [NFPA 1041 4.4.3, 4.4.4, 4.4.5]

Task Steps

NOTE to Instructor I candidate: For this skill you will be evaluated on the following:

   a. Voice is clear, appropriately pitched, and well controlled when communicating.
   b. Speech is reasonably free of language errors.
   c. Style is reasonably free of distracting behaviors or mannerisms.

Step 1: Begin the preparation step in the four-step method of instruction.
   a. Instructor I candidate introduces him- or herself to class participants.
   b. Address location of amenities (restrooms, water fountains, etc.) and address any safety issues (i.e., the location of fire exits and other places of safety).
   c. Introduce lesson objectives.
   d. Introduce subject matter.
   e. Explain why classroom material is important to class participants.
   f. Explain how material(s) will be used **(Figure 6.20)**.
   g. Establish rapport with class participants.

Step 2: Begin the presentation step of the four-step method of instruction.
   a. Use audiovisual equipment, as defined in the lesson plan, to aid in meeting learning objectives.
   b. Deliver the lesson outline **(Figure 6.21)**.
   c. If applicable, transition smoothly within and between different types of instruction (audiovisual, demonstration, discussion, etc.).
   d. Guide students toward meeting lesson objectives.
   e. Adjust teaching methods/equipment/materials to differences in class participants' learner characteristics, abilities, cultures, and behavior.
   f. If applicable, appropriately address disruptive behaviors.
   g. If applicable, ensure class continuity is maintained.
   h. Summarize the key points or objectives of the presentation.

Figure 6.20

Step 3: Begin the application step of the four-step method of instruction.
   a. Ensure that class participants are given the opportunity to apply concepts through discussions, exercises, or demonstrations, individually or as groups.
   b. Provide basic coaching and motivational techniques throughout instruction **(Figure 6.22)**.
   c. Correct disruptive behaviors.

Step 4: Evaluation/Closure step of the four-step method of instruction.
   a. Ensure students have learned the main idea of the lesson.
   b. The closure should be drawn from the students by asking them questions, asking them to summarize steps, to do another example, to apply information in a new situation or draw conclusions, take a written, oral, or practical examination to determine whether students have achieved the lesson objectives.
   c. Optional: Briefly state what the topic of the next lesson will be.

Figure 6.21

Figure 6.22

SKILL SHEETS 6-2

Give a prepared practical (psychomotor) lesson. [NFPA 1041 4.4.3, 4.4.4, 4.4.5]

Task Steps

NOTE to Instructor 1 candidate: For this skill you will be evaluated on the following:

   a. Voice is clear, appropriately pitched, and well controlled when communicating.

   b. Speech is reasonably free of language errors.

   c. Style is reasonably free of distracting behaviors or mannerisms.

Step 1: Begin the preparation step in the four-step method of instruction.

   a. Instructor 1 candidate introduces him- or herself to class participants.

   b. Address location of amenities (restrooms, water fountains, etc.) and address any safety issues (i.e., the location of fire exits and other places of safety).

   c. Introduce lesson objectives.

   d. Introduce the subject matter.

   e. Explain why the skill is important, how it relates to other skills, how many people are required to perform it, and when it should be performed **(Figure 6.23)**.

   f. Establish rapport with class participants.

Step 2: Begin the presentation step of the four-step method of instruction.

   a. Use audiovisual equipment, as necessary, to aid in meeting learning objectives.

   b. Present new concepts according to prepared lesson plan.

   c. If applicable, transition smoothly within and between different types of instruction (audiovisual, demonstration, discussion, etc.).

   d. Demonstrate skill in real time.

   e. Demonstrate skill slowly, describing each step **(Figure 6.24)**.

   f. Adjust teaching methods/equipment/materials to differences in class participants' learner characteristics, abilities, cultures, and behavior.

   g. If applicable, appropriately address disruptive behaviors.

   h. If applicable, ensure class continuity is maintained.

Step 3: Begin the application step of the four-step method of instruction.

   a. Provide students an opportunity to perform under supervision.

   b. Coach students. Check and correct any errors.

   c. Correct disruptive behaviors.

Step 4: Evaluation/Closure step of the four-step method of instruction.

   a. Ensure students have learned the main idea of the lesson.

   b. The closure should be drawn from the students by asking them questions, asking them to summarize steps, to do another example, to apply information in a new situation or draw conclusions, take a written, oral, or practical examination to determine whether students have achieved the lesson objectives.

   c. Optional: Briefly state what the topic of the next lesson will be.

Figure 6.23

Figure 6.24

138 Chapter 6 • Classroom Instruction

Student Interaction

Chapter 7
SECTION A INSTRUCTOR I

Chapter Contents

Individual Student Needs................ 141	**Strategies for Dealing with**
Diversity 141	**Classroom Issues** 151
Low Literacy Levels 142	Nondisruptive, Nonparticipating
Special Needs 142	Students 152
High Learning Abilities 146	Disruptive, Nonparticipating
Students' Rights 146	Students 153
Student Behavior Management 147	Reasons for Disruptive Behavior 154
Reviewing Policies 147	**Taking Formal Disciplinary Action** 156
Counseling 148	**Chapter Review** 158
Coaching........................ 149	**Discussion Questions**................ 158
Motivating and Encouraging Students.. 150	**Chapter 7 End Notes**................ 158
Providing Peer Assistance 150	**Key Terms**........................ 158
Mentoring....................... 151	

JPRs addressed in this chapter

This chapter provides information that addresses the following job performance requirements of NFPA 1041, *Standard for Fire Service Instructor Professional Qualifications*, 2019 Edition.

- 4.3.2
- 4.4.3
- 4.4.4
- 4.5.2

Learning Objectives

1. Describe ways instructors can address individual student needs in fire and emergency services training. [4.3.2, 4.4.3, 4.4.4]

2. Describe students' rights in an educational setting. [4.5.2]

3. Discuss techniques instructors use to manage student behavior. [4.4.4]

4. Explain strategies used by instructors in dealing with classroom behavior issues. [4.4.4]

5. Describe ways instructors can take formal disciplinary action. [4.4.4]

Chapter 7
Student Interaction

One of the most important skills that the fire and emergency services instructor must possess is the ability to interact with students. This chapter provides information on student interaction and appropriate actions an instructor may take, inside and outside the classroom or training grounds in terms of:

- Individual student needs
- Students' rights
- Student behavior management
- Reasons for disruptive behavior
- Taking formal disciplinary action

Individual Student Needs

As discussed in this manual, some students will be quick to grasp a concept or learn a skill, while others may need additional resources to remain current with the class schedule. Independent of learner characteristics or learning style, some students show behaviors and learning backgrounds that require more attention. Instructors should learn to recognize the behaviors of individuals who will need additional assistance, ideally before a class session is disrupted.

Instructors may encounter students with various learning backgrounds and learning abilities, including:

- Diverse background/history
- Low literacy levels
- Learning disabilities
- Students requiring adaptations
- High learning abilities

Diversity

Over time, research has shown the value in increasing diversity in many environments. Each unique perspective may challenge the assumptions of the others and strengthen best practices **(Figure 7.1)**. In the fire and emergency services, as in other workplaces, diversity may be based on gender, age, ethnicity, sexual orientation, and social status. Instructors must be careful to promote a welcoming environment for all students.

Figure 7.1 Diversity can strengthen best practices in a learning environment.

In contrast to the positive benefits of diversity, differences between people of different cultural backgrounds may disrupt communication. Cultural and background differences may stem from a number of variables including:

- Customs
- Behaviors
- Attitudes
- Values
- Life circumstances

In addition to cultural differences, a class member who is not a native speaker of the primary language used in the classroom may have more difficulty parsing or relating the terms used throughout the class. These complications are compounded as multiple backgrounds and languages are presented in the same classroom.

Low Literacy Levels

Literacy, the ability to read, write, and generally function in a given language, is a critical skill in most workplaces, including the fire and emergency services. To some degree, skills may be taught during hands-on classes, but most classwork assignments and many evaluations will have a written component. The U.S. *Workforce Investment Act of 1998* indicates that adults should reach an 8th-grade reading level before entering the workforce. Students who do not meet that benchmark, whether because they are not native speakers of a language or some other reason, may have difficulty taking notes and keeping up with the material as presented. Should the gap in a student's literacy level be significant, an instructor may help the student research and access any available assistance, including referrals to adult education programs, employee assistance programs (EAPs), or personal tutors.

If students demonstrate low literacy levels, they may be self-conscious about gaps in knowledge and/or abilities. They may become disengaged in classroom discussion and display disruptive or unsafe behavior to disguise their low literacy or other potential vulnerabilities. An instructor must be careful not to imply that illiteracy is a factor of the student's intelligence and instead work with the student to find resources that will help the student succeed.

Some students may complete twelve years of schooling while functionally illiterate, and they may not know they have a gap in knowledge. In this case, it may be sufficient for an instructor to make simple modifications to resources. To aid students with low literacy, the instructor should make sure that locally produced documents have the following features:

- Short sentences and paragraphs
- Double-spaced lines
- Directional headings and wide margins
- Font size and type that is large and clear enough to be read easily
- Vocabulary should be simple, and all uncommon terms should be easily found in a glossary

When giving exams, the instructor should always make sure that directions are simple. Instructors may include pictures, tables, graphs, or charts to illustrate concepts and divide text.

Special Needs

Students with any kind of special need are at a disadvantage when compared with students without a special need. Instructors will usually be able to identify students who are struggling with the requirements of a class soon after beginning. For example, a student may not appear to comprehend a class session or assignment regardless of instructions that are understood by the rest of the class. Instructors should communicate at the beginning of a course or lesson that anyone seeking reasonable accommodations should contact them privately. The private meeting will allow the instructor to identify appropriate accommodations. This availability for consultation, and indication of the possibility for accommodations should also be included in the course syllabus.

Students with limited learning disabilities may be able to perform class requirements at or above the level of their cohort with some accommodations in specific areas. Students with specific learning disabilities may not be able to perform at the rate of their classmates in certain subjects or tasks. These students will be able to learn and perform at the same level but at a slower pace.

Learning disabilities are not an indicator of intelligence or motivation. They are, in contrast, a wide variety of neurological-based processing difficulties. As learning disabilities are increasingly researched and managed, more people who experience difficulties in a learning environment are evaluated. Adults with a learning disability may not know why they have trouble in some aspects of learning, in part because available evaluations were not available when they were children.

Officially, the U.S. *Americans with Disabilities Act (ADA) of 1990* divides learning disabilities into three major categories:

- **Speech and language disorders** — Difficulty producing speech sounds, using spoken language, or understanding what other people say
- **Academic skills disorders** — Difficulty reading (dyslexia), writing (dysgraphia), and calculating (dyscalculia)
- **Miscellaneous learning disabilities** — Difficulty performing fine motor skills (dyspraxia), learning nonverbal skills, and other difficulties

Students may have average to high intelligence but perform poorly on tests due to their specific learning disability.

In a legal liability context, students seeking accommodations are required by federal law to show documentation proving they have a specific learning disability. To acquire that documentation, students need to be diagnosed by an appropriate professional. An organization's human resources department may be able to provide testing and referral assistance for students. Government agencies and large private organizations that provide training may offer evaluation and assistance for employees with learning disabilities or refer them to appropriate agencies.

Some accommodations may require documentation such as a written diagnosis. Accommodations that may be available at the discretion of the instructor include:

- Tutoring
- Developing individualized instruction
- Providing additional feedback on progress
- Allowing more time on tests
- The option to take tests in a place free from distractions
- Providing a reader for, or audio recording of, assigned reading **(Figure 7.2)**
- Assigning one student to take notes for another

The instructor is encouraged to read a variety of resources on respectfully working with and accommodating students with disabilities. In the U.S., persons who have learning disabilities are protected from discrimination by three laws: Individuals with Disabilities Education Act (IDEA), Rehabilitation Act, and ADA. In particular, when an instructor knows that a student with a specific learning disability will be in his or her class, the instructor should focus on that condition. See **Table 7.1, p. 144-145** for specific disability categories as defined under IDEA.

Figure 7.2 Accommodations for students can include audio recordings of lessons or assigned readings.

Table 7.1
Categories of Disability Under IDEA Law

Disability Category	Definition
Autism	*Autism* means a developmental disability significantly affecting verbal and nonverbal communication and social interaction, generally evident before age three, that adversely affects a child's educational performance. Other characteristics often associated with autism are engagement in repetitive activities and stereotyped movements, resistance to environmental change or change in daily routines, and unusual responses to sensory experiences. Autism does not apply if a child's educational performance is adversely affected primarily because the child has an emotional disturbance, as defined below. A child who manifests the characteristics of autism after age three could be identified as having autism if the above criteria are satisfied.
Deaf-Blindness	*Deaf-blindness* means concomitant hearing and visual impairments, the combination of which causes such severe communication and other developmental and educational needs that they cannot be accommodated in special education programs solely for children with deafness or children with blindness.
Deafness	*Deafness* means a hearing impairment that is so severe that the child is impaired in processing linguistic information through hearing, with or without amplification that adversely affects a child's educational performance.
Developmental Delay	*Developmental delay* means, for a child under 3 years of age (Part C of IDEA) and for a child aged 3 through 9 (Part B of IDEA), a delay in one or more of the following areas: physical development, cognitive development, communication development, social or emotional development, or adaptive development as measured by appropriate diagnostic instruments and procedures.
Emotional Disturbance	*Emotional disturbance* means a condition exhibiting one or more of the following characteristics over a long period of time and to a marked degree that adversely affects a child's educational performance: a. An inability to learn that cannot be explained by intellectual, sensory, or health factors. b. An inability to build or maintain satisfactory interpersonal relationships with peers and teachers. c. Inappropriate types of behavior or feelings under normal circumstances. d. A general pervasive mood of unhappiness or depression. e. A tendency to develop physical symptoms or fears associated with personal or school problems. Emotional disturbance includes schizophrenia. The term does not apply to children who are socially maladjusted, unless it is determined that they have an emotional disturbance.
Hearing Impairment	*Hearing impairment* means an impairment in hearing, whether permanent or fluctuating, that adversely affects a child's educational performance but that is not included under the definition of deafness in this section.
Intellectual Disability	*Intellectual disability* means significantly subaverage general intellectual functioning, existing concurrently with deficits in adaptive behavior and manifested during the developmental period, that adversely affects a child's educational performance.
Multiple Disabilities	*Multiple disabilities* means concomitant impairments (such as mental retardation-blindness or mental retardation-orthopedic impairment), the combination of which causes such severe educational needs that they cannot be accommodated in special education programs solely for one of the impairments. Multiple disabilities does not include deaf-blindness.

Disability Category	Definition
Orthopedic Impairment	*Orthopedic impairment* means a severe orthopedic impairment that adversely affects a child's educational performance. The term includes impairments caused by a congenital anomaly, impairments caused by disease (e.g., poliomyelitis, bone tuberculosis), and impairments from other causes (e.g., cerebral palsy, amputations, and fractures or burns that cause contractures).
Other Health Impairment	*Other health impairment* means having limited strength, vitality, or alertness, including a heightened alertness to environmental stimuli, that results in limited alertness with respect to the educational environment, that—a. Is due to chronic or acute health problems such as asthma, attention deficit disorder or attention deficit hyperactivity disorder, diabetes, epilepsy, a heart condition, hemophilia, lead poisoning, leukemia, nephritis, rheumatic fever, sickle cell anemia, and Tourette syndrome; and b. Adversely affects a child's educational performance.
Specific Learning Disability	*Specific learning disability* means a disorder in one or more of the basic psychological processes involved in understanding or in using language, spoken or written, that may manifest itself in the imperfect ability to listen, think, speak, read, write, spell, or to do mathematical calculations, including conditions such as perceptual disabilities, brain injury, minimal brain dysfunction, dyslexia, and developmental aphasia. Specific learning disability does not include learning problems that are primarily the result of visual, hearing, or motor disabilities, of mental retardation, of emotional disturbance, or of environmental, cultural, or economic disadvantage.
Speech or Language Impairment	*Speech or language impairment* means a communication disorder, such as stuttering, impaired articulation, a language impairment, or a voice impairment, that adversely affects a child's educational performance.
Traumatic Brain Injury	*Traumatic brain injury* means an acquired injury to the brain caused by an external physical force, resulting in total or partial functional disability or psychosocial impairment, or both, that adversely affects a child's educational performance. Traumatic brain injury applies to open or closed head injuries resulting in impairments in one or more areas, such as cognition; language; memory; attention; reasoning; abstract thinking; judgment; problem-solving; sensory, perceptual, and motor abilities; psychosocial behavior; physical functions; information processing; and speech. Traumatic brain injury does not apply to brain injuries that are congenital or degenerative, or to brain injuries induced by birth trauma.
Visual Impairment Including Blindness	*Visual impairment* including blindness means an impairment in vision that, even with correction, adversely affects a child's educational performance. The term includes both partial sight and blindness.

"Individuals with Disabilities Education Improvement Act of 2004," (IDEA) Public Law 108-446 (December 3, 2004) Implementing Regulations for IDEA, 34 CFR Part 300 "Rosa's Law" (October 5, 2010)

The instructor may need to arrange the following accommodations for students who are struggling to learn new concepts or skills:

- Private conferences
- Special assignments
- Individual instruction

An instructor may also reevaluate and revise the subject matter or instructional methods so that all learners are able to meet the course objectives and requirements. A student may have difficulty grasping information and skills in one lesson or category yet may excel in other areas. Therefore, instructors should develop assignments and lesson plans that can be tailored to individual learning styles and student needs.

Some students can become conscious of their difficulties as other students work faster with fewer mistakes. Instructors must be careful to encourage all students to treat each other equally. Instructors should always provide positive feedback as students accomplish course objectives.

High Learning Abilities

Some students appear to learn without trying. These students will usually accomplish more than is expected of average students, and they may study and learn very well without much supervision.

When students perceive that they can understand and retain material more quickly and accurately than their peers, they may lose motivation and become disengaged from the learning process, or even become disruptive and unsafe to other students. They may leave the course with a low opinion of the instructor, the curriculum, or the training department as a whole.

Instructors can turn the presence of high ability learners into an asset. These students can be given more creative assignments and asked to present their findings to the class. Another approach for instructors is to task a high ability learner as a tutor or note taker for another student **(Figure 7.3)**. While assisting other individuals, adult high ability learners improve their own retention. At the same time, the student receiving the help may feel more comfortable learning from a peer rather than an instructor. Students with high learning abilities need to be engaged with assignments that meet their ability level, even if these assignments are beyond the scope of what the rest of the class is doing.

Figure 7.3 Some students may achieve greater understanding by getting help from another student instead of the instructor.

Students' Rights

Students must have equal access to fundamental rights and privileges as guaranteed by law. In an educational setting, this includes:

- Privacy of records and test scores
- Freedom to hold and express an opinion different from that of the instructor or organization
- Equal access to the learning environment
- Fair and equal treatment in class
- A learning environment free of discrimination or harassment
- A safe learning environment

Any rules or regulations that may be perceived as infringing on students' substantive rights should be reviewed, validated, and communicated to students at the beginning of instruction. These rules and regulations must be consistently applied throughout the program.

In some cases, it is necessary to protect the physical safety of students during the learning process. A reasonable regulation for student safety may be seen as an infringement of student rights. For example, an instructor may need to ask students to:

- Cover religious caps with safety hard hats.
- Remove cultural garbs to accommodate full turnout gear.
- Wear protective shoes in place of bare feet or other foot coverings.
- Remove facial piercings and/or facial hair to accommodate protective masks or head coverings.

When regulations appear to be unreasonable, but they still must be enforced, instructors should be prepared to explain the regulation to the students.

NOTE: For information regarding the Family Educational Rights and Privacy Act (FERPA), refer to Chapter 1, The Instructor as a Professional, Chapter 14, Supervisory and Administrative Duties, and Chapter 18, Training Program Administration.

Student Behavior Management

Managing student behavior is one of an instructor's most important skills. Instructors must be able to prevent disruptive students from interfering with the class, or they risk losing the other students' respect and failing to accomplish the learning objectives. As mentioned in the above section on individual student needs, to keep students engaged in the learning process, instructors may need to address different learner characteristics and learning styles and adjust the lesson accordingly. However, when disruptive students are beyond the instructor's management skills, formal disciplinary action may be necessary to remove them from the course or use outside influences to control their behavior. In any case, the ultimate goal of student behavior management is the steady, safe participation by all students in a learning environment.

With adults, peer pressure can be effective to manage behavior, but it is ultimately the instructor's responsibility to deal with disruptive students. Even when peer pressure is partially effective, an instructor may speak to disruptive students in private or take stronger disciplinary measures.

Most organizations have policies for managing issues involving student problems and performance; however, the best action is to prevent problems before they occur. The next few sections provide information on the following techniques for correcting or changing student behavior:

- Reviewing policies
- Counseling
- Coaching
- Motivating and encouraging students
- Providing peer assistance
- Mentoring

Reviewing Policies

One of the first priorities for an instructor is to review the rules, regulations, and policies of the training division with students. This strategy promotes, supports, and enforces these policies so that students follow and respect them.

Instructors and students who travel to learning sites must become familiar with and follow each facility's rules and regulations **(Figure 7.4)**. Facility and training ground rules generally address:

- Safety
- Facility layout
- Attendance and tardiness
- Responding to emergencies from class
- Level of expected class participation
- Methods of evaluation
- Use of personal electronic devices
- Assignment due dates

Figure 7.4 Instructors and students must follow all rules, regulations, and policies of the training facility.

- Class cancellations due to weather or other causes
- Student parking
- Dress and grooming
- Breaks or rest periods
- Use and cleanup requirements
- Off-limit areas
- Smoking regulations (specifically smoking areas)

NOTE: The instructor should brief students on the rules of the facility early in the course or class.

Counseling

Some students engage in disruptive behavior when they have difficulty meeting learning objectives or adjusting to training. Such students may benefit from **counseling**.

For the purpose of this manual, counseling refers to the act of helping a student to adjust, redirecting a student's behavior, or eliminating barriers to learning. Counseling actions may include:

- Suggesting resources and concepts that may reframe a question
- Discussing options available based on current circumstances
- Administering or arranging for tests that help identify problem areas
- Providing vocational assistance based on an individual student's aptitude and interests

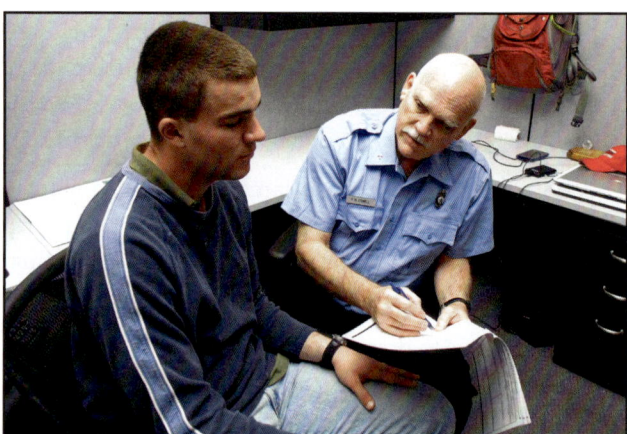

Figure 7.5 Instructor-student counseling sessions can focus on a number of student issues, including class anxiety.

Counseling sessions between instructors and students should focus on issues relevant to the students' progress in class, which may include a student's social anxieties, or personal concerns that prevent participation in class activities or course requirements **(Figure 7.5)**. Unqualified instructors should never assume the role of therapist when a student appears to have a psychological or emotional problem. In that situation, the instructor should follow organizational procedures, such as notifying the student's or the instructor's supervisor. Students who need professional counseling can typically access these services through their employee assistance programs.

CAUTION: Instructors and staff members should seek additional help if a student's concerns appear to be more complicated than they can address in their formal capacity or level of training.

Counseling sessions between students and instructors must take place in a private and controllable area. The instructor should be careful to emphasize that the counseling session is intended to resolve problems, not to force any specific resolution. When the instructor is sincerely interested in the student's well-being, and considers the individual as a potential colleague, the student will more easily recognize that the purpose of the conference is to help resolve problems. Students may not willingly offer information about troublesome situations; instructors must encourage students to express and explain their feelings.

The sense of partnership created during successful counseling makes students more likely to listen to the instructor's advice. Instructors must learn and practice appropriate counseling and positive reinforcement methods that motivate students to perform properly.

Training organizations usually have a formal process for documenting behavioral issues, and it typically involves completing some type of counseling form. New instructors must be aware of the organization's process, and they need to realize that it is more than the simple and immediate measures they can implement in class. Some organizations may be required to maintain records on counseling for accreditation purposes.

Instructors must consult with the organization and follow its policies on counseling students and completing the counseling form. The general guidelines for an instructor are:

- Meet and talk with the student.
- Use facts to precisely describe the individual's behavior.
- State student expectations and record them on the counseling form.
- Discuss the issue and possible solutions with the student.
- Explain what actions will be taken if the student does not comply with the objectives and solutions.
- Have the student sign a form to acknowledge counseling.
- Give a copy of the counseling form to the student and retain the original.
- Forward a copy of the form to the training institution.
- Forward a copy of the form to the student's supervisor, chief, or manager.

If counseling fails, instructors must take steps to alleviate the situation. Further conversations may help the situation, in which case, the instructor must again conduct a meeting privately, calmly, and with acknowledgement for good work and focus on constructive steps that can be taken. Demonstrating a sincere interest in students is more effective than making them feel inferior or inadequate.

Coaching

Disruptive behavior often results from frustration or confusion, and instructors can help students overcome these behaviors through **coaching** techniques. For instructors, coaching is the process of giving motivational correction, positive reinforcement, and constructive feedback to students to maintain and improve their performance.

To be effective, feedback needs to be positive, immediate, direct, and frequent. It can either be as simple as telling the student how well a task has been completed or involve a formal counseling session when a change in negative or inappropriate behavior is required. Private industry generally subscribes to a formal coaching model that contains the following four steps, modified for an instructional setting:

Step 1: Describe the current level of performance — Describe levels in a positive manner. Specify behaviors that could be improved using specific examples. Do not state that a student is doing something wrong; simply say that it can be done better, safer, or more efficiently.

Step 2: Describe the desired level of performance — State exactly what is required for the student to succeed; provide clear direction for the student to meet the goal.

Step 3: Obtain a commitment for change — Ensure that the student commits to the desired level of performance. In some cases, this commitment is considered a binding contract and becomes part of a student's permanent record.

Step 4: Follow up on the commitment — Observe the student to determine whether his or her performance has improved, or schedule a follow-up meeting to discuss progress. If the student does not perform to the desired level, more coaching may be necessary. If this is unsuccessful, the next step would be formal, professional counseling.

Feedback from instructors must be objective and precise in its description of desirable and undesirable behavior. Instructors must not fall into the trap of just providing critical phrases, even if they are intended to motivate a student to try again. That type of feedback does not inform the students of their error or explain how to correct it. Good feedback hints at the right answer, such as: *"That was good, but you've forgotten what the rope needs to have on the running end before you tighten it."*

Asking students questions allows them to stop, think, and recall the information they are attempting to apply. Some students will take longer than others to perform this process, but they will all remember better because they had to recall it themselves. On occasions when students do not recall the information, the instructor can carefully review it with them.

Motivating and Encouraging Students

Most students arrive at a training session with the desire to complete the course or curriculum successfully. This self-motivation may arise from their desire to do the following:

- Keep a job
- Get a promotion or raise
- Gain recognition
- Feel important, competent, or successful
- Join a group (organization or club)

Fear of failure is a psychological barrier that may prevent people from taking risks. If students believe that an instructor genuinely wants them to succeed and will give them the tools to do so, they are more likely to overcome a fear of failure **(Figure 7.6)**.

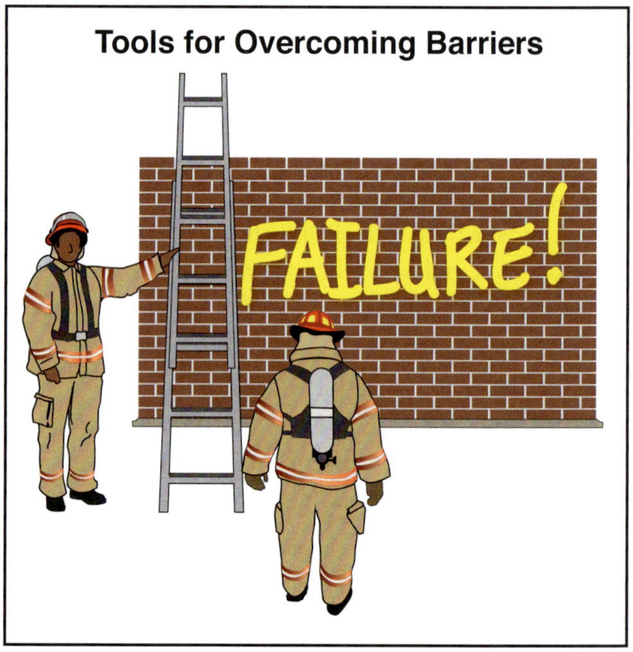

Instructors can motivate students and encourage their successes by offering the following instructional enhancements:

- **Provide quality instruction that helps students who try to learn** — Make the first experience with a subject or topic safe, successful, and interesting. First impressions have a lasting effect on learning.
- **Provide evidence that student efforts make a difference** — Stress the importance of the amount and quality of effort needed to learn tasks before students begin. Emphasis on the effort accomplishes the following objectives:
  — Establishes student responsibility
  — Reduces feelings of helplessness
  — Increases perseverance
  — Generates feelings of pride and accomplishment
- **Establish clear expectations** — Make the learning objectives and evaluation criteria clear. When students know what to expect and how they are progressing, they are more likely to succeed. In preparation for testing, they need to know exactly what they are to learn and how well they are performing.
- **Provide continuous feedback about student progress** — Instructors who provide learning and testing criteria, continuous and constructive feedback, and appropriate coaching and encouragement should have confident and successful students.

Figure 7.6 Instructors should strive to give their students all the necessary tools for overcoming psychological barriers, such as fear of failure.

Providing Peer Assistance

Peer assistance involves having students help each other in the learning process. For example, some students may shy away from demonstrations of skills in front of an instructor until they feel confident in their abilities. These students feel more comfortable practicing with a peer.

Students who have grasped a knowledge or skill and can explain it to others make good peer assistants. It helps if they have background experiences that enable them to explain how classroom activities can be applied on the job.

Mentoring

In contrast to a peer assistance arrangement, mentoring situations occur outside the classroom, usually in the job environment. Many instructors have acted as mentors. Instructors often guide the actions of new employees, just as they guide the actions of new students in the learning environment or training evolution. Instructors, or the organization, may assign a mentor when they know someone who can guide student actions in real experiences on the job. Mentors must be selected for their work experience, subject matter interest, patience, good opinion of the organization, and communication abilities.

The primary purpose of mentoring is to prepare students for advancement within the organization with the direction of a positive role model. Mentoring programs enhance management skills, improve productivity, and encourage diversity. Mentors assist their students (protégés or mentees) in the following ways:

- Serving as role models
- Providing guidance in career planning
- Assisting in gaining specialized training
- Providing outside resources
- Providing challenging work assignments
- Monitoring students' achievements

The concept of mentoring begins with the direct supervisor. The immediate supervisor or instructor is customarily designated as the primary mentor (coach, counselor, guide, role model, etc.) for each student. In military functions, an intermediary or secondary mentor may not be assigned, but a student may seek counseling and professional development advice from other sources.

Some nonmilitary organizations will require a mentorship program with new student or new hire orientation. Mentors may be given an opportunity to volunteer, or they may be asked privately if they would be willing to serve in this role. Regardless how an organization sets up the mentor-student relationship, all participants must understand the time and parameters for each of the roles.

Strategies for Dealing with Classroom Issues

When it is necessary to regain the attention of disruptive students who are talking among themselves off topic, instructors may call on one of the individuals. While trying to not embarrass the person, the instructor summarizes the topic and asks a simple opinion question that directs attention back to the lesson. Doing this more than once is not advised.

Some students feel that they already have superior knowledge to instructors and do not need to be in class. These students may attempt to goad the instructor into revealing inadequacy or ignorance of their subject matter. Such students may also desire to display their knowledge to their peers. The following list offers suggestions to deal with these types of students:

- Show confidence in your role as content expert. Remember that these students would not be in the class if they were established as the experts.
- If you do not know the answer to a question, admit it, and bring the answer back to the class when you can. Being responsive and accurate is more important than being perfect.
- Respond to the student, but continue to manage class time, and consider the interests of others when engaged in a dialogue with an individual.
- Always smile. A smile is disarming and indicates willingness to listen and discuss.

- Do not embarrass students when they expose their limitations. In doing so, you can lose their respect and that of the rest of the class.
- Avoid getting into a battle of wills with students. Allowing conflicts to take an antagonistic tone robs the class of their information and disrupts lessons.
- Let participants know that their questions have merit.

When a student asks a challenging question, chances are that other students have the same one and are too shy or embarrassed to ask. An instructor can redirect the question into an open discussion to get a wider perspective on the topic. Because of the positive reinforcement, everyone gets involved, all benefit from the discussion, and everyone grows comfortable participating.

A disruptive student may be seeking attention, but instructors must be careful not to reward this behavior by paying too much attention to the actions. Some behaviors will simply stop; some will not. When it does not, the L-E-A-S-T method of progressive disciplinary action is suggested as follows (**Figure 7.7**):

- **Leave it alone** — Wait to see if the behavior goes away; it might be an isolated occurrence.
- **Eye contact** — Look at the student long enough to make eye contact. Eye contact tells the student *I see what you are doing* and implies *Now stop it*.
- **Action** — Take action when the behavior continues. Actions may be a simple comment to the entire class on the importance of being attentive or a question directed to the disrupting student. This can remind the class that the instructor is aware of student behavior.
- **Stop the class** — Stop the class, preferably by dismissing the other students on a break, and ask the disruptive student to stay to discuss the problem.
- **Terminate the student** — Expel the student from the class if discussing the problem fails to solve it. Take and document appropriate disciplinary measures.

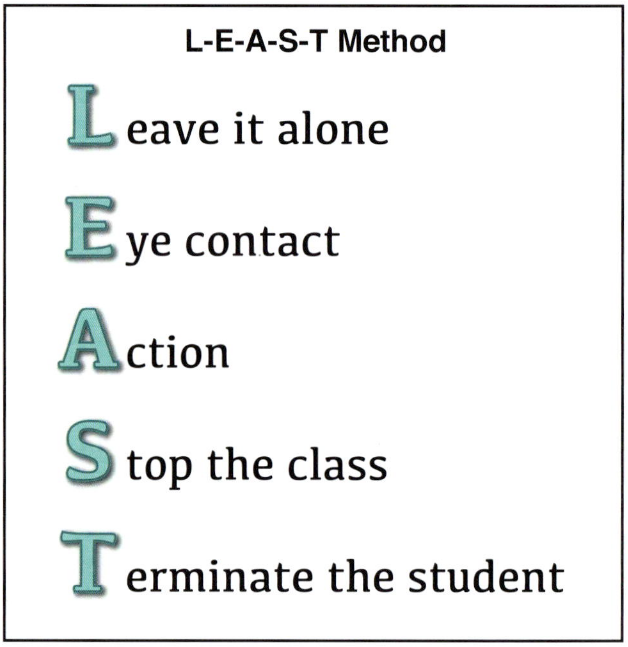

L-E-A-S-T Method

Leave it alone

Eye contact

Action

Stop the class

Terminate the student

Figure 7.7 When a disruptive student seeks attention, the L-E-A-S-T method can be a valuable approach to discipline.

An advantage of being in a paramilitary organization such as the law enforcement, fire, and emergency services is that the instructor generally has the authority of an officer in the classroom. Instructors should be prepared to use this authority in order to maintain classroom control, but also plan redirects to avoid those circumstances.

Nondisruptive, Nonparticipating Students

Nondisruptive and nonparticipating students may appear to be daydreaming, distracted, or uninterested. In courses where students attend instruction for long periods of time, instructors may use several methods to engage them.

Instructors should ask students casually about their attitudes toward the class, or any distractions they may have. Simple comfort questions during breaks may encourage students to readily provide answers because they appreciate the interest and attention. Depending on the circumstance, the instructor may remind students that they have a responsibility to participate. These students will normally make every attempt to fulfill the class requirements and please an instructor who shows interest in them.

Shy or Timid

The shy or timid individual may be hesitant or at a loss for words when expected to respond or participate, and may actually be intimidated to respond aloud or participate during class discussions. When an instructor recognizes this, he or she should avoid calling on these students until they have become comfortable with the class and the instructor. To help facilitate this, instructors should encourage shy students to participate in informal or small-group discussions. Success there will encourage them to be more outgoing in formal class discussions or when giving a presentation.

The instructor may also talk with these students during breaks to help them become comfortable with the learning process, the instructor's teaching style, and course expectations. Instructors who use these simple methods of making students feel comfortable will help them overcome shyness and encourage them to participate in the class.

Bored or Uninterested

Many quiet students may be above average in ability, but they may drift mentally due to many circumstances, including a perceived lack of opportunity to interrupt a lecture for a question. Instructors can redirect attention by asking direct questions or beginning activities that require student participation. Instructors should be alert for the following signs of daydreaming and boredom:

- Glazed looks
- Gazing around the room
- Doodling
- Thumbing through materials not related to the subject

Students who are uninterested often display little energy and attention. When this behavior continues through multiple types of instructor interventions, instructors may check with their supervisors or other instructors to determine whether the students have exhibited personal or other problems. Instructors will want to gather as much information as possible to plan strategies for working with these students. A supervisor or other instructor may indicate that a student experienced the following situations:

- Required counseling or tutoring in other courses
- Attempted to take a class beyond the level of readiness
- Coped with health, emotional, family, or learning difficulties that interfered with classroom performance

Disruptive, Nonparticipating Students

In contrast to students who are quietly uninterested in a class session, disruptive, nonparticipating students act in a way that is noticeably inappropriate for a learning environment. These students may distract the class by being overly talkative, showing a lack of respect for the instructor or the course, or trying simply to draw attention to themselves. In addition to stalling class progress, this behavior may create a safety hazard.

Instructors should not allow these students to control the class. If possible, the instructor should attempt to redirect students' energy toward constructive participation. But if tactful attempts to change the students' behavior fail, instructors should take another action, such as asking them to leave the session. Instructors should clearly inform disruptive students that cooperation and constructive participation is expected upon their return. Continually disruptive students may be reported to supervisors or commanding officers.

Talkative Students

Talkative students may monopolize discussions and prevent others from participating without intending harm. These students may have friends or acquaintances in the class and see the time together as a chance to multitask. Alternately, a group may sit together and entertain each other during discussion or lecture intervals. In these situations, instructors must recapture the attention of the group. Strategies may include making a private appeal to one or more students in the disruptive group or making pointed recommendations to the group to continue their discussions outside of class time.

Attention-Seeking Students

Students who thrive on attention may use a group situation to perform and gain attention for themselves. Sometimes their performances help put other students at ease and get others to participate, but these students may need coaching to know when to stop and share the focus. If the attention-seeking student tries to respond to all the instructor's questions, the instructor should encourage other students to respond and remind the student that his or her classmates need an opportunity to answer as well. As a redirect, the instructor may be able to recruit the attention-seeking student to stand-by for opportunities to help clarify a point when a different student offers an impartial answer.

When the attention-seeking student does not demonstrate willingness to work with an instructor's redirects, the instructor may offer an ultimatum, indicating the repercussions for repeated instances of disruptive behavior. The timeliness and manner in which an instructor delivers such an ultimatum determines its effectiveness. When students understand that safe behaviors and cooperation are important, expected, and enforced, instructors will have few if any problems (**Figure 7.8**). When problems persist, the instructor should always follow the organization's disciplinary policies and procedures.

Instructors should never tolerate inappropriate behavior that creates unsafe conditions for students. Examples of safety violations related to an attention-seeking student's behavior may serve as a strong object lesson for the entire class. Instructors must be careful not to use these examples as a reprimand to the disruptive student, but as a teaching opportunity for the class.

Figure 7.8 An instructor should set expectations for safe behavior and follow through with appropriate enforcement.

Reasons for Disruptive Behavior

Regardless of the cause, disruptive behavior distracts from the learning environment, may result in an unsafe practice, and wastes valuable time. The sections that follow describe instructor-and student-caused disruptions, with the aim of preventing and managing those behaviors.

Instructor-Caused Disruptive Behavior

Some disruptive behavior results from the actions of the instructor (**Table 7.2**). For example, instructors who are intimidating, overly controlling, unprepared, or not in control of their allotted time may cause students to react in a way that is disruptive.

Table 7.2
Unfavorable Student Reactions to Instructor Actions

Instructor Actions	Student Reactions
Intimidates	Feels insecure
Shows impatience	Shows fear
Attempts to overcontrol	Rebels
Rambles without a goal	Shows no interest
Gives dull lectures	Feels unstimulated or bored
Runs overtime	Becomes fidgety

Some instructors project authority in a way that some students may find intimidating. While an instructor must demonstrate that he or she is a master of the information presented in the classroom, this authority must never be used to intimidate a student in any way. In contrast, the instructor should practice demonstrating respect and patience to reinforce the image that the instructor is concerned with the student's learning experience. Patience is especially important with students who require additional time to develop an answer, whether because they have difficulty in formulating the words or whether because they may have too many ideas to express simply on the first try. When an instructor looks for ways to improve a student's learning experience, rapport can be built in the process.

Overseeing some types of learning environments, particularly skills training, can be frustrating to an instructor who wants to help students succeed. Sometimes, the instructor will try to create a rigid environment to push students into learning in a way that makes sense to the instructor. This can result in the learning experience becoming too structured for some student learning styles. Instructors should not rely on strict rules that may become barriers to learning. Instead, the instructor should prepare to adjust to the needs of students, the changing environment, and the subject matter.

Students can often easily recognize an unprepared instructor. This lack of preparedness indicates to a student that the instructor does not respect the students, topic, and learning experience. When an instructor prepares for a class, elements such as breaks in the lecture, anecdotes, and increased chances for student participation are built into the instruction. Instructor preparation and practice increase the ability for all participants to enjoy learning.

Finally, an instructor must maintain awareness and control of the timing of the class. Preparation ahead of a class session should include breaks and time for interruptions. When these factors are not considered, a class session may go beyond its time limit. Students may accommodate a late class dismissal on a rare occasion, especially when they can plan, or know why it was necessary. However, students can quickly come to resent instructors who perpetually exceed the time limit. This may also indicate to students that they may break rules. As a student's resentment about a course increases, the likelihood that the student will behave disruptively will also increase.

Student-Caused Disruptive Behavior

Students may unintentionally cause disruptions. Regardless of intent, student-caused disruptive behavior may take the following forms:

- Arriving late
- Speaking at inappropriate times
- Talking with others during class session times about unrelated topics
- Sleeping in class, especially snoring

- Engaging in horseplay or attention-seeking behaviors
- Interrupting others
- Sidetracking discussions
- Acting blatantly disrespectful of the instructor and/or other students

Students may not intend to present disruptive behaviors. Instructors should try to determine why some students disrupt the class. Knowing where these behaviors come from will help an instructor tailor an appropriate and effective response. Causes of disruptive behaviors may include:

- Distractions in a student's personal life, such as long work hours that erode sleep and study time
- Prior negative experiences in training and classroom environments
- Boredom or resentment of class requirements
- Demands to balance competing interests and commitments

Some students may have had negative experiences in a classroom and may not know what behaviors are expected in the fire and emergency services training environment. Other students may feel coerced into taking a course, and they may resent the time it takes away from other responsibilities. Still other students may try to demonstrate that they are knowledgeable on a subject at the expense of the instructor or another student if they feel insecure in the learning environment. Finally, students may attempt to use disruptive behavior to gain attention, particularly if they feel a classroom related-need is being ignored. In each of these situations, the instructor may have to rely on tactics that have been developed by the profession. The instructor's supervisor may be asked to help address problematic behaviors.

Some students may understand why they need a class but be uninterested in the subject matter. Instructors may find activities that are particularly helpful in making dry content interesting and engaging. Alternatively, an instructor may prepare an introduction for a section that warns students that a particular block of instruction is going to be dry, and will get more interesting at a stated upcoming time.

When an instructor needs more direction in matching a student's needs with the teaching style, he or she may spend a portion of a class period to ask students to express their internal and external motivations as well as their professional needs. In some cases, the instructor may be able to help students to determine the relationship between their needs and the course content for themselves. When students feel that their opinion is valued, and they have a venue to express needs in a productive manner, they may self-regulate their behaviors.

Taking Formal Disciplinary Action

When an instructor is unable to help a student accept or settle into a routine, disciplinary action becomes necessary. Instructors should be familiar with organizational policies and follow them closely. The student's behavior should be thoroughly documented in order to justify disciplinary actions **(Figure 7.9)**. This may include collecting data such as:

- Record the date and behavior whenever class must be interrupted to manage a specific student's behavior.
- Keep documentation in a secure location.
- Draft a memo to the student, detailing the disruptive behavior. This memo should state that you are seeking formal disciplinary action based upon the student's classroom behavior. Also, you may make a suggestion for the action you recommend be taken.
- Send a copy of this memo to the chief/manager and training officer.
- Await formal action by the training division.

Training organization regulations may provide instructors with the option of dismissing an individual from a class. If so, a paper trail of notes, memos, and counseling forms is essential to prove that the action was taken with justification. The documentation must show progressive reprimands or disciplinary actions that may end with or result in an individual being removed from a class.

Documenting Discipline:
Solving Problems in Progressive Steps

The following steps provide progressive actions to take when employee or student discipline is necessary:

Facts: State the problem and focus on specific behavior. Document all behavior and conversations:

1. ***Keep an incident diary/calendar or file.*** Begin to make notations of incidents when it has become necessary to speak to an individual about undesired actions. Note undesirable behaviors, frequency, discussions about behaviors.

2. ***When making notes, state facts (problem or issue).*** Write down what the individual did or did not do and when based on expectations; state the conversation with the individual.

3. ***Use these facts to write a disciplinary action memo.*** Include who, what happened, where, when, and what will be done. Follow organization policy on filing or distributing the memo to appropriate personnel and to the individual.

4. ***Include in the memo a specific objective, solution, and action that will be taken.*** Show your attempt to help the employee/student succeed (see below).

Objectives: Set objectives that will communicate a specific expectation:

1. ***If undesired behavior continues, meet formally with the individual.*** Discuss the problem, refer to previous discussions and file notes, and set and state expectations for the individual to meet in order to change behavior. Be specific.

2. ***Follow up the meeting with a written memo.*** State in writing the problem and the expectations discussed in the meeting. Follow organization policy on filing or distributing the memo to appropriate personnel and to the individual.

Solutions: Determine solutions that will help the individual reach the objective:

1. ***In the meeting, offer solutions or plans that will help solve or correct the problem.*** Suggestions may include (1) attending a class or training seminar or working with a tutor and (2) getting assistance or coaching from a specific individual in setting priorities, completing assignments, getting information or resources, and checking work or progress.

2. ***Give solutions orally to the individual and follow up with a written memo.*** In the memo state facts, objectives, solutions, actions, and consequences (such as suspension, pay reduction, transfer, or termination). Follow organization policy on sending copies to the individual's immediate supervisor, personnel file, or other references.

3. ***Always follow oral warnings with a written warning.*** A written follow-up to a discussion verifies that the discussion occurred and clearly identifies steps for the individual to follow.

Actions: State what actions will be taken if objectives are not met, and follow up with written memo.

Implement actions as necessary and as stated in the memo.

Adapted from: *Documenting Discipline*, American Media, Inc., West Des Moines, Iowa 50265

Figure 7.9 Instructors must perform disciplinary action in a methodical and professional manner.

Legal and ethical issues may require or dictate that instructors have supervisory authority while conducting classes. This level of implied authority becomes critical when facing disruptive behavior and trying to maintain an appropriate classroom atmosphere. Instructors must understand and not act beyond their levels of responsibility and authority as directed by local policy.

Chapter Review

Answer the following questions to review the information provided in this chapter.

1. How do instructors adjust their instructional delivery to address individual student needs?
2. What are students' rights in an educational setting?
3. What are some techniques instructors can use to manage student behavior?
4. What are some common learning environment behavior issues?
5. How can instructors take formal disciplinary action with their students?

Discussion Questions

The following questions are intended to generate discussion, expand your understanding of the chapter text, and allow you to think critically about what you have learned. Answers to these questions may vary.

1. How have instructors adjusted their instructional delivery methods to help you in other courses you have taken?
2. Why are students' rights important in an educational setting?
3. What types of techniques have you seen used by instructors to manage student behavior?
4. What instructor strategies do you feel would be successful for dealing with behavior issues in a learning environment?
5. What would constitute formal disciplinary action of students by instructors?

Chapter 7 End Notes

Table 7.1 Citation

Taymans, J. M., National Institute for Literacy, Learning to Achieve: A Professional's Guide to Educating Adults with Learning Disabilities, Washington, DC 20006.

Key Terms

Coaching — Process in which instructors direct the skills performance of individuals by observing, evaluating, and making suggestions for improvement.

Counseling — Advising learners or program participants on their educational progress, career opportunities, personal anxieties, or sudden crises in their lives.

Chapter 8

Skills-Based Training Beyond the Classroom

SECTION A — INSTRUCTOR I

Chapter Contents

Resources: Safety Guidelines, Regulations, and Information 161	Psychomotor Skills Demonstrations 167
Federal Government Agencies 161	Evolution Control 168
State/Provincial and Local Safety and Health Agencies 162	Simple Training Evolutions 168
Standards-Writing Organizations...... 163	Increased Hazard Exposure Training..... 170
Professional and Accrediting Organizations 164	Live-Fire Training 171
Instructor as Safety Role Model 165	Increased Hazard Exposure Training ... 177
Planning for Safe Training 166	**Legal Liability** 178
Verifying Instructor Skill Level......... 166	Vicarious Liability.................. 179
Inspecting and Repairing Facilities and Props 166	Foreseeability 179
Identifying Training Hazards 167	Liability Reduction 180
	Chapter Review 181
	Discussion Questions 181
	Key Terms 182

JPRs addressed in this chapter

This chapter provides information that addresses the following job performance requirements of NFPA 1041, *Standard for Fire Service Instructor Professional Qualifications*, 2019 Edition.

4.3.2   4.4.4
4.4.2   4.4.5
4.4.3

Learning Objectives

1. Discuss agencies and organizations where instructors can find the most current information about safety guidelines and regulations. [4.3.2]

2. Describe the responsibilities of the instructor as a safety role model. [4.4.2, 4.4.3, 4.4.4]

3. Identify ways that instructors plan for safe training. [4.4.2, 4.4.3, 4.4.5]

4. Describe the information given by an instructor during a psychomotor skills demonstration. [4.4.2, 4.4.3]

5. Identify the elements of evolution control. [4.4.2, 4.4.3]

6. Explain a simple training evolution. [4.3.2, 4.4.2, 4.4.3]

7. Identify types of increased hazard exposure training. [4.4.2, 4.4.3]

8. Identify types of legal liability for which instructors and/or organizations may be held accountable. [4.4.3]

Chapter 8
Skills-Based Training Beyond the Classroom

Regardless of the location of a class session, instructors must understand their responsibilities when teaching skills. The first responsibility is to maintain a safe training environment. Injuries and especially fatalities are unacceptable in training scenarios. The second responsibility is to teach skills with the students' mastery as the ultimate goal. Realistically, students may not reach mastery during one class but may reach a level of competency necessary to satisfy learning objectives. Students' future safety and the safety of their classmates depends upon instructors who teach them to perfect and maintain key skills. This chapter presents instructors with the basic information needed to safely teach skills including:

- Resources: safety guidelines, regulations, and information
- Instructor as safety role model
- Planning for safe training
- Psychomotor skills demonstrations
- Evolution control
- Simple training evolutions
- Increased hazard exposure training
- Legal liability

Resources: Safety Guidelines, Regulations, and Information

Instructors should remain current with ever-changing safety guidelines and regulations. As research continues to confirm or refine best practices, tactics that had been considered safe five years ago may now be known to have hidden hazards or inefficiencies. A wide variety of information is readily available from entities such as:

- Federal government agencies
- State/provincial and local governmental occupational safety and health agencies
- Standards-writing organizations
- Professional organizations and associations

Instructors must ensure that they receive information from credible sources. Awareness of these resources and access to their contact information are valuable assets for any instructor. These resources are further described in the following sections.

Federal Government Agencies

Numerous federal government agencies in North America are responsible for developing, regulating, and ensuring safe workplace policies. Four based in the U.S. and Canada that may be of the greatest use to instructors are:

- **National Institute for Occupational Safety and Health (NIOSH)** — U.S. agency responsible for investigating, researching, and evaluating safety and health hazards in the workplace.

- **Occupational Safety and Health Administration (OSHA)** — U.S. agency responsible for setting and enforcing workplace safety and health standards; can issue citations and fines. In federal OSHA states, OSHA enforces regulations that apply only to private and federal firefighters in non-state-plan states.
- **Canadian Centre for Occupational Health and Safety (CCOHS)** — Canadian federal government agency that provides information and policy development regarding work-related injury, illness prevention initiatives, and occupational health and safety information.
- **National Institute of Standards and Technology (NIST)** — U.S. agency that promotes standardization and measurement of various sciences and technologies. NIST often has information on fire science or testing results for fire equipment.

> **Occupational Safety Regulations**
>
> Many states have their own occupational safety regulations that state agencies enforce. State plans must meet or exceed federal requirements. Instructors are encouraged to have a current copy of their state's plan if one exists. Regulations that apply to fire and emergency services personnel may vary by state where not federally required.

Other organizations may have more specialized regulatory information that instructors may need to seek out for specific training situations. The following U.S. federal agencies can provide information on weapons of mass destruction (WMD), PPE for high-threat incidents, and bioterrorism training, among other training topics:

- **Department of Homeland Security (DHS)** — Agency tasked with preventing terrorist attacks, and minimizing damage from potential terrorist attacks and natural disasters. Other federal agencies that are under DHS include:
  - *Federal Emergency Management Agency (FEMA)* — Agency tasked with responding to, planning for, recovering from, and mitigating disasters.
  - *United States Fire Administration (USFA)* — Provides national leadership to foster a solid foundation for the fire and emergency services; emphasizes information about fire prevention, firefighter preparedness, and emergency response. The National Fire Academy (NFA) is part of the USFA.
  - *Emergency Management Institute (EMI)* — Serves as the national focal point for the development and delivery of emergency management training to emergency services responders at all levels, public and private, to minimize the impact of disasters on the American public.
- **Centers for Disease Control and Prevention (CDC)** — Agency that collects and analyzes data regarding disease and health trends.
- **Environmental Protection Agency (EPA)** — Agency that sets policy and regulations for protection of the environment.
- **Department of Transportation (DOT)** — Agency that establishes the national curriculum and regulations for EMS personnel and equipment.
- **Department of Defense (DoD)** — Agency that regulates military, civilian, and federal fire and emergency services personnel stationed on federal property.

NOTE: Agencies in both Canada and the U.S. share information and materials. Some Canadian safety regulations are based on or modeled after NFPA standards and OSHA regulations.

State/Provincial and Local Safety and Health Agencies

States/provinces and local governmental occupational safety and health agencies often include review and enforcement functions. In Canada and the U.S., states/provinces or local agencies may have to follow regulations that differ from, expand upon, or exceed national rules. They may also have to follow additional regulations that are not addressed at the federal level, such as those mandated by state/provincial environmental agencies.

Instructors should know the federal, state/provincial, and local regulations that are likely to impact their training evolutions. When in doubt about regulations, instructors should know how to contact their state/provincial and local governmental occupational safety and health agencies, state/provincial fire training academies, state/provincial fire marshal offices, or state EMS agencies to inquire about which regulations may be applicable for a particular training evolution. Instructors should not limit their inquiries to regulations for fire and emergency services. State/provincial and local agencies may also have safety regulations written for other industries or organizations that apply to fire and emergency services training. For example, regulations for construction workers on a roof may be applicable to firefighters training for vertical ventilation scenarios. Fire and emergency services trench-rescue operations may be governed by regulations for shoring and cribbing that apply to private contractors who lay underground cables and pipelines **(Figure 8.1)**.

Regulatory agencies may have consultants or educators who can review pertinent safety regulations with instructors who are planning a training curriculum or course. State/provincial and local health departments or agencies can also provide statistical data and safety programs, as well as information on diseases and their prevention.

Figure 8.1 Instructors should be aware of industry safety regulations that may apply to emergency services training.

Standards-Writing Organizations

Standards-writing organizations develop and issue operating procedures and design requirements for various industries, including the fire and emergency services. These organizations arrange for a committee, composed of industry representatives, to build an overall standard as members develop a consensus on individual topics. These standards do not become law until a government authority adopts them. The primary standards-writing organizations involved with the fire service are:

- **National Fire Protection Association (NFPA)** — Develops minimum safety standards and guidelines. Training organizations, government functions, and other agencies can adopt NFPA standards as their guidelines for safety compliance. Instructors must be aware of and familiar with NFPA standards that relate to safety. Organizations that do not adopt these standards may be viewed negatively or even experience legal ramifications if an injury occurs during training, especially if that injury could have been prevented by following an established standard. These standards guide the performance of live-fire training evolutions and other high-hazard training, and are updated on a schedule.

- **American National Standards Institute (ANSI)** — ANSI does not develop standards but facilitates their development through the consensus process. This work is coordinated via accredited member organizations to help private sector organizations voluntarily standardize their systems. ANSI-approved standards include:
  — Respiratory protection practices
  — Physical qualifications for using respiratory protection equipment
  — Fit-testing methods

- **Underwriters Laboratories Inc. (UL)** — Tests and certifies fire-extinguishing agents and equipment, among many other duties. In Canada, this organization is known as Underwriters Laboratories of Canada (ULC). Agents and equipment that meet these requirements are said to be UL/ULC listed. Certification tests and UL/ULC acceptance provide consumers with independent documentation on product-performance characteristics.

> **Some Notable NFPA Standards**
>
> Instructors should be familiar with the following NFPA standards:
>
> - **NFPA 1041,** *Standard for Fire Service Instructor Professional Qualifications* — Identifies the professional levels of competence required of fire service instructors.
> - **NFPA 1402,** *Guide to Building Fire Service Training Centers* — Lists guidelines to follow when building training facilities (including burn buildings, smoke buildings, and combination buildings) and when conducting outside drill ground activities.
> - **NFPA 1403,** *Standard on Live Fire Training Evolutions* — Lists guidelines to follow during live-fire training evolutions at acquired structures, facilities designed for live-fire training, and exterior props, including information on student prerequisites, structures and facilities, fuel materials, safety, and instructor qualifications.
> - **NFPA 1410,** *Standard on Training for Initial Emergency Scene Operations* — Contains minimum requirements for evaluating training for initial fire suppression and rescue procedures. These procedures are used by fire department personnel engaged in emergency scene operations.
> - **NFPA 1500™,** *Standard on Fire Department Occupational Safety and Health Program* — Contains guidelines for fire departments to follow to ensure the health and safety of firefighters.
> - **NFPA 1583,** *Standard on Health-Related Fitness Programs for Fire Department Members* — Outlines a complete health-related fitness program designed for fire departments, including job descriptions, rehabilitation, nutrition, and wellness components.
> - **NFPA 1584,** *Standard on the Rehabilitation Process for Members During Emergency Operations and Training Exercises* — Contains guidelines for developing rehabilitation programs that can be implemented at incident scenes or at training exercises.

Professional and Accrediting Organizations

The following fire and emergency services professional organizations are important sources of safety information and model programs:

- International Association of Fire Chiefs (IAFC)
- Canadian Association of Fire Chiefs (Association Canadienne des Chefs de Pompiers) (CAFC/ACCP)
- International Association of Fire Fighters (IAFF)
- National Volunteer Fire Council (NVFC)
- North American Fire Training Directors (NAFTD)

Fire and emergency services instructor associations and safety organizations also provide information on safety and networking opportunities. These organizations often provide information and sources to instructors who are not members, but who need information for their programs. Some of these organizations include:

- Fire Department Safety Officer Association (FDSOA)
- International Society of Fire Service Instructors (ISFSI)
- American Society of Safety Engineers (ASSE)

Two national organizations provide **accreditation** to fire and emergency certifying entities:

- International Fire Service Accreditation Congress (IFSAC)
- National Board on Fire Service Professional Qualifications (ProBoard)

Instructor as Safety Role Model

The instructor is a primary role model for safety and must take that role seriously. In addition to describing safety guidelines, instructors must demonstrate and reinforce these guidelines equally with students and staff. Following safety guidelines or plans has a significant effect on reducing injuries and fatalities in training and at emergency incidents. Instructors must be empowered to plan, change, and present courses that follow safety requirements.

> **CAUTION:** If a policy is in place and you, as an instructor, do not follow its guidelines, you may be held liable for injuries to students.

Instructors must devote appropriate time to discussing aspects of safety. Safety provisions and briefings should be planned well before the activity starts and reflected during the training session through awareness and practice.

Instructors can increase awareness and help prevent accidents in the learning environment by performing the following actions either before or during training:

- Describing applicable safety requirements or procedures to students in the following forms:
  — Providing rules and guidelines in writing.
  — Reading the written rules and guidelines aloud as the students read them silently.
  — Having students sign a statement that they have read and understood all safety rules and regulations. In some departments and organizations, this signature is mandatory.
- Creating a training and safety plan for any high-hazard drill or application, and briefing all instructors and students on the plan before training begins.
- Planning carefully for training scenarios.
- Ensuring that appropriately trained personnel assist in supervising scenarios.
- Inspecting and repairing tools, equipment, props, and apparatus before starting training sessions.
- Assigning a safety officer to each training scenario based upon a prepared training and safety plan (**Figure 8.2**).
- Assigning additional personnel (more than the minimum) for safety positions depending upon the scale of the training evolution and the severity of danger.
- Describing the proper safeguards and equipment used for preventing accidents.
- Describing possible hazards and explaining the necessary precautions (**Figure 8.3**).
- Briefing students on relevant techniques, procedures, tools, facility characteristics, and appropriate safety rules before starting the evolution.
- Reviewing emergency procedures, emergency evacuation plans, and verbal or alarm alerts with students before the evolution.

Figure 8.2 To help prevent accidents and increase safety awareness, a safety officer may be assigned to oversee training.

Figure 8.3 Instructors should inform students about likely hazards before practicing skills. *Courtesy of U.S. Air Force.*

- Modeling and reinforcing safety policies and procedures by personally adhering to them.
- Watching for behaviors that can contribute to unsafe conditions, such as improper attitude, complacency, or lack of knowledge or skill.

When planning practical training evolutions, instructors must identify and eliminate potential hazards. Instructors must also plan to address the necessary precautions to prevent injury while training, train students to recognize job hazards, and teach them how to control or eliminate these hazards. These precautions help minimize the level of risk and prevent injuries.

As stressed above, when planning high-hazard training evolutions, the instructor should appoint another qualified individual as the safety officer for the training evolution. The safety officer must be familiar with the logistics involved in the evolution including the purpose and scope of the training, and what may happen at any given time.

Fire and emergency services programs often train nonemergency services employees. Although these students may not be required to perform all aspects of the curriculum on their jobs, instructors must enforce all safety procedures as these employees participate in the course. Regardless whether all training procedures apply to these students, they must still follow safety guidelines.

NOTE: Establish and use training and safety plans during all types of emergency incidents and high-hazard training evolutions. Also include use of a personnel accountability system.

Planning for Safe Training

Training ground instruction requires thorough planning. The sections that follow discuss important actions regarding planning safe training.

Verifying Instructor Skill Level

Even though verification of instructor skill levels ultimately falls on the Instructor II and III levels, instructors should be honest with themselves about their own ability levels and knowledge. An instructor who is certified to a proficiency level in a skill may find that he or she needs to relearn the skill if it has not been actively used in some time. In addition, some skills change over time. When an instructor has the appropriate certifications but is not confident in his or her competence in teaching the skills needed for a class, the instructor should seek out courses to gain the current skills or recommend a better qualified instructor to teach the course.

In contrast, an instructor may be capable of teaching the skills at the appropriate level, but may not have collected the appropriate certifications to teach a topic. In this case, an instructor should work with supervisors to update their certifications whenever possible.

Training cannot continue without an available instructor who has the appropriate skills and qualifications. Instructors should be careful to work with their organization to take advantage of continuing education opportunities that will benefit their ability to offer timely, accurate, and safe training.

Inspecting and Repairing Facilities and Props

Equipment evaluation and inspections ensure that class props and all other instructional technology tools are available and in working condition before a class session begins. If inspections show that equipment is unsafe or damaged, the instructor should report this to the AHJ. Depending on resources available, the training session may be rescheduled, adjusted to exclude the type of equipment, or moved to another location with the specified equipment in functional repair.

During entry-level recruit classes, some inspection, maintenance, and repair of facilities and props can be assigned to class members as a training activity. Care must be taken to not assign tasks that require certified or authorized personnel, such as evaluating the implications of beam alignments, or apparent cracks in structural or aesthetic elements. The AHJ should establish an inspection time schedule based on industry practice, manufacturer's recommendations, and local needs. Generally, inspections of tools, facilities, apparatus, equipment, and any other instructional technology tools should occur on the same schedule as in the stations **(Figure 8.4)**.

Keep thorough records of all maintenance, repairs, inspections, and replacements **(Figure 8.5)**. The records provide a basis for developing an accurate operating budget, justifying repairs or replacements, and assessing equipment's overall value. See Chapter 10, Records, Reports, and Scheduling, for additional information. In all cases, maintenance and repairs beyond the capabilities of students or instructors should be delegated to certified or authorized repair personnel.

Identifying Training Hazards

Instructors should be familiar and comfortable with prepared lesson plans, the training facility, any training props, and the skills and hazards involved in the evolution to be conducted. Students should be informed during the training briefing of hazards they will face as part of training. Instructors will need to modify the lesson plan to include appropriate safety instructions.

Psychomotor Skills Demonstrations

Skills demonstrations begin with giving the following information:

- An explanation of the skill
- Discussion of why it is important
- How it relates to other skills
- How many people are required to perform it
- When it should be performed

After this general overview, the instructor should perform the skill at normal speed, then perform it again slowly while explaining each step. Next, the instructor should encourage students to ask questions. The instructor should repeat the slow-speed demonstration until students are able to verbalize the steps of the skill. At this point students are ready to practice the skill themselves. The slow-speed demonstration is the transition between the presentation step and the application step.

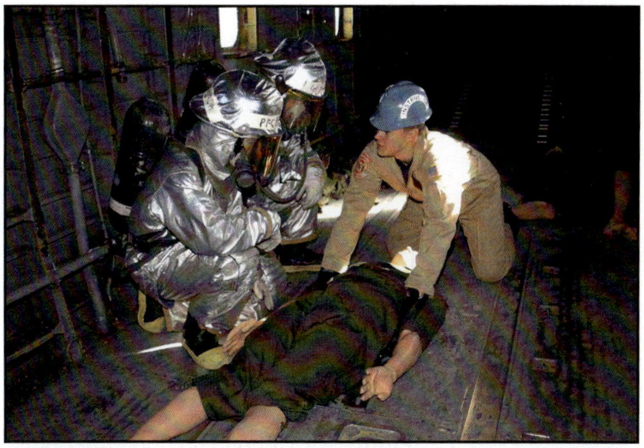

The instructor should guide and coach the students as they practice, then have the students critique and coach each other as they practice in small groups **(Figure 8.6)**. Once or twice during a practice session, instructors may need to demonstrate the entire skill. Instructors must show the skill steps correctly and in sequence. The end of the session is a particularly good time to do this. Students should be encouraged to practice skills during rest breaks, during free time, and at the beginning of the next training session. When students have perfected the skill, they are ready for evaluation.

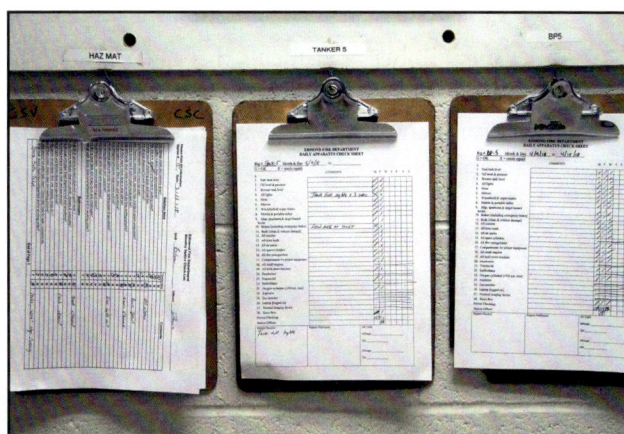

Figure 8.4 Inspect training equipment on the same schedule as other fire station equipment.

Figure 8.5 Thorough record-keeping is a necessity for all equipment and apparatus.

Figure 8.6 Students can reinforce skills by observing each other's work in small teams.

Planning, preparation, and practice are essential to a successful skills demonstration. The instructor who does not appear proficient at a skill will lose credibility with students and waste valuable training time by having to repeat or correct skill steps.

Evolution Control

Practical training evolutions must be controlled, regardless of their size and complexity. Controlling an evolution involves the following elements:

- **Supervising** — Instructors actively critique individual students' activities to make sure they practice skills safely and correctly.
- **Monitoring** — Instructors observe the overall progress of the evolution to make sure students work together to accomplish the learning objectives.
- **Teaching** — When appropriate, instructors present new or related information during the evolution.
- **Managing** — Instructors apply the elements of the NIMS-ICS to control and coordinate the evolution to the same management and communications requirements as an actual emergency situation.

The number of elements that must be maintained during a controlled evolution will require a relatively large number of instructors. A ratio of three to seven students to one instructor is optimal in most hands-on training environments. In high hazard training evolutions, the ratio may require more instructors or fewer students. For example, NFPA 1403 mandates a maximum student to instructor ratio of five to one **(Figure 8.7)**.

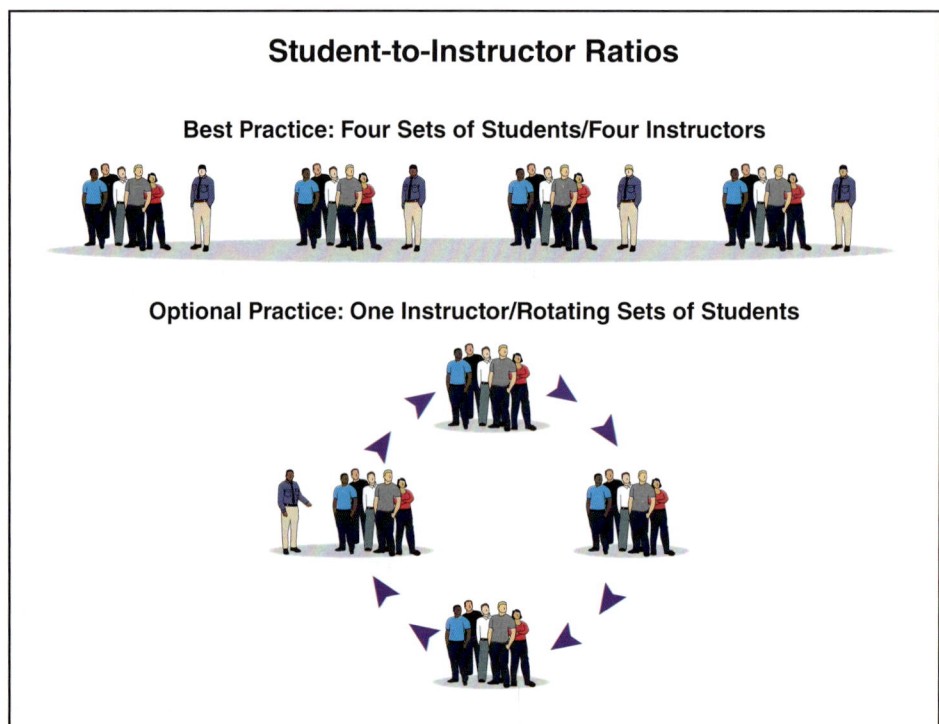

Figure 8.7 Depending upon the number of available instructors, adjustments may be needed to maintain an optimal student-to-instructor ratio.

Simple Training Evolutions

Simple training evolutions share some training techniques in common with psychomotor skills training. The main difference is the number of students involved and the training location. Psychomotor skills training is often conducted in a classroom or lab, with the potential for a large number of students learning at the same time.

Figure 8.8 Lifting and placing a ground ladder, using a portable fire extinguisher, and unloading a hoseline are all examples of training evolutions that require students to perform only a few tasks.

Simple training evolutions involve small numbers of students performing a single skill that requires only a few tasks. Training evolutions typically take place on a fireground or other outdoor learning environment regardless of their size or complexity. Examples include **(Figure 8.8)**:

- Lifting and setting a ground ladder
- Using a portable fire extinguisher
- Lifting and moving patients
- Forcing entry through a door
- Taking and recording patients' vital signs
- Deploying and advancing an attack hoseline
- Driving and parking a fire apparatus or ambulance

To conduct team evolutions, the instructor should inspect and familiarize him- or herself with the tools and equipment required to complete the learning objectives.

Training evolutions may involve small groups of one to five active students and require only one or two instructors for supervision. Students repeat the evolution until they are able to perform it without error. When more than one student is involved, they rotate positions so all students have the opportunity to experience and practice each part of the skill.

The instructor begins an evolution by performing the following actions:

- Explaining the purpose of the evolution. This should include an indication of how the evolution relates to the learning objectives or outcomes.
- Emphasizing the safety requirements for the evolution, including identification of training hazards.
- Demonstrating the evolution. When this action involves more than one student, a group of experienced responders may be tasked to show individual students the steps in whole or in part.
- Relating the evolution to the classroom lecture.

- Monitoring student performance. If students perform poorly, make an error, or violate safety protocols, the instructor or designated staff/assistant should stop them immediately and correct their behavior. Immediate correction forces students to recognize the problem and adjust their behavior.

NOTE: Remember to use positive reinforcement when applicable.

Training evolutions should be performed as though students were responding to an actual emergency incident including typical staffing levels and appropriate PPE **(Figure 8.9)**. All applicable policies and procedures must be followed, including using the NIMS-ICS.

Figure 8.9 Students should train with the same equipment they would use during an actual incident.

Increased Hazard Exposure Training

NFPA 1041 includes the following examples of increased hazard exposure training scenarios:

- Live-fire exercises
- Hazardous materials
- Above and below grade rescue
- Evolutions that require the use of power tools

Increased hazard exposure training is dangerous even under the very controlled conditions found at permanent training facilities. NFPA 1041 therefore requires that a Level II Instructor supervise both students and Level I Instructors during this type of training.

An instructor tasked with leading increased hazard exposure evolutions must be prepared and qualified. The instructor should learn the qualifications that his or her state/province and municipality requires for these types of training evolutions and ensure that the qualifications are met before continuing with the training. If the instructor has any doubts about his or her qualifications or experience, the training should be postponed until the instructor has gained the proper qualifications and become more knowledgeable about the standards and regulations that apply.

WARNING: An instructor can be held liable for injuries and fatalities on a fire ground.

Live-Fire Training

Live-fire training is an important part of firefighter training. Entry-level personnel learn new skills and experienced personnel develop their existing skills. The decrease in actual fire fighting incidents, the increase of internal hazards in fires involving modern lightweight and/or synthetic materials, and the assignment of entry-level personnel to EMS duties have resulted in fewer opportunities for personnel to experience live-fire situations. Therefore, live-fire training evolutions may be the only experience some personnel receive. Despite the decrease in the prevalence of structure fires and the increase in their hazardous conditions, live-fire training still serves the purpose of creating a cadre of responders who can read and react appropriately to fire conditions **(Figures 8.10 a and b)**.

The following sections explore categories of live-fire training environments:

- Safety during live-fire training evolutions
- Acquired structures
- Purpose-built structures
- Exterior and wildland fires

Figures 8.10 a and b Training for fighting exterior fires may take place at a training facility (a) or at a remote site (b).

Safety during Live-Fire Training Evolutions

To ensure the highest level of safety, all live-fire training evolutions must meet the requirements of the appropriate sections of the most current edition of NFPA 1403, *Standard on Live Fire Training Evolutions*. This standard emphasizes strict safety practices for structures selected for live-fire evolutions.

While this standard is not legally mandated in all jurisdictions, noncompliance with the standard has resulted in injuries and fatalities for which instructors were held criminally and civilly responsible. In November 2004, the Centers for Disease Control and Prevention (CDC) from the U.S. National Institute for Occupational Safety and Health (NIOSH) issued a Workplace Solutions document, titled "Preventing Deaths and Injuries to Fire Fighters during Live-Fire Training in Acquired Structures," that outlines the precautions to take when engaging in live-fire training. Any instructor leading live-fire training should make every effort to prepare and instruct based upon the NFPA and CDC guidance.

NIST Study on Fuel Pallet Size in Training Evolutions

In 2007, NIST published a scientific study of two firefighter training deaths at live-fire incidents. The first was in an acquired structure; the second was in a purpose-built burn building at a training facility. In both cases, the supervising instructors increased the fire loads beyond safe levels, despite having started the scenario with fire pallets that were within acceptable limits.

In the case of the acquired structure, the instructor in charge added a polyurethane couch mattress to the fuel load after the original fuel load produced smoke but very little fire. After allowing the mattress to burn in a closed room to develop the smoke conditions, a search and rescue team entered the building, followed by two hose crews. But the structure was filled with smoke and the hose crews could not locate the search and rescue team. After they had been in the structure for three and half minutes, the outside vent team broke a window in the burn room. The resulting flashover led to the deployment of the standby rapid intervention team (RIT), and then to the evacuation of all personnel. When the fire was extinguished, the bodies of the two search and rescue members were found in the fire room.

In the case of the purpose-built burn building, the instructor was killed when his self-contained breathing apparatus (SCBA) facepiece failed. The training fire had been burning all day, with the instructor adding fuel periodically to ensure that the fire was burning adequately for the next training evolution. But the structure had no windows and was built to contain heat, so after five evolutions, the heat trapped inside the structure had risen to dangerous levels. When the instructor entered to add fuel for the sixth evolution, his PPE could not withstand the high temperatures.

NIST's ongoing research into these incidents has included experiments using fuel pallets of varying sizes and with varying materials to better understand what fuel materials and pallet sizes are the safest for use in training scenarios. As research continues, instructors are cautioned to always follow the proper protocols for fuel loads and pallet sizes.

Further information on this study can be found in the following NIST document: *Fatal Training Fires: Fire Analysis for the Fire Service* by Daniel Madrzykowski.

Acquired Structures

Instructors may be approached by a homeowner seeking to donate a building for use in training evolutions. Instructors and training administrators must take great care when designing or leading training that uses **acquired structures** (Figure 8.11). Firefighter injuries and fatalities during live-fire training inside of acquired structures have contributed to many of the training-related casualty statistics in recent years.

Figure 8.11 Acquired structures used for live-fire training must adhere to all NFPA 1403 safety requirements. Exterior siding on the structure pictured here has been removed for asbestos abatement.

Factors that may influence whether acquired structures may be used in internal live-fire evolutions include:

- Acquired structures are usually in deteriorating condition and may have been scheduled for demolition.
- Environmental laws may prohibit the burning of a structure because of its location.
- The structure may also have been designated as a historical landmark.
- Acquired structures may have lightweight construction elements that cannot withstand fully involved fire conditions.

After a structure is acquired, an instructor may be assigned to oversee live-fire evolutions. When a jurisdiction intends to use an acquired structure for live-fire training, safety protocols must be followed.

NOTE: Refer to NFPA 1403's Live Fire Evaluation Sample Checklist (Annex B) for information on conducting safe, legal training at an acquired structure.

The building must undergo a complete and thorough inspection, and may require permits. If the instructor is not qualified to perform this inspection, a trained inspector should be brought in to perform this task. The authority having jurisdiction must apply for and obtain any necessary permits to conduct live-fire training evolutions.

Acquired structures that are used for live-fire training must meet the safety requirements of NFPA 1403, which may require that improvements are made to the structure before training. An instructor should consider the cost of making the acquired structure safe enough for the proposed type of training.

WARNING: Any structure that cannot be made safe cannot be used for interior structural fire fighting training.

When a structure cannot be brought to a state of repair suitable for internal live-fire training evolutions, the structure may alternatively be designated as a training site for lower-hazard evolutions. Depending on the local conditions, AHJ, and structure suitability, training evolutions may include:

- Evaluation of building construction
- Ladder evolutions
- Search and rescue
- Forcible entry
- External live-fire
- Exposure protection

Once a site has been selected, the structure has been acquired, inspected, and repaired, and all necessary permits for the live-fire exercise have been obtained, an instructor then considers the type of training best suited for the specific structure. The instructor must be careful to set goals and learning objectives that can be achieved given the limits imposed by the particular structure. The following factors establish the limits of the types of training that can be performed at the structure:

- Total number of students
- Instructor to student ratio
- Instructors' qualifications
- Student prerequisites
- Safety considerations
- Equipment availability
- Location and access

Advanced planning is critical to the success of any practical training evolution involving acquired structures. When planning for a live-fire exercise in an acquired structure, instructors must consider a variety of issues or factors that can have an effect on the training evolution. No live-fire training should be attempted until the following issues are considered and plans made to address them:

- Extreme weather
- Structural condition
- Training possibilities
- Permits and inspections
- Notice and documentation
- Fuel materials
- Water supply

Extreme weather. While climate conditions affect every training evolution, their effect on acquired structures can be extreme. Wind speed and direction, humidity, temperature, and time of day all affect the training environment. Weather extremes are common in many locations, from below-freezing cold to high temperatures and humidity. Hazards such as lightning, hail, and high winds may also present additional dangers when severe storms approach.

Consider the condition, location, access, and exposures of the structure and its surrounding area when evaluating how weather may affect it and any planned evolutions. For more information on the forces that weather can exert on older construction, refer to the IFSTA manual, **Building Construction Related to the Fire Service**.

Structural condition. Acquired structures must comply with NFPA 1403, specifically the Structures and Facilities section of its Chapter 4. Before conducting live-fire training, the structure must be inspected to determine whether repairs are necessary to comply with the standard.

> **Asbestos at Acquired Structures**
>
> When a certified inspector has detected the presence of asbestos, the building owner is required to use certified personnel to remove and dispose of the materials. Safety officers and instructors should be familiar with federal, state/provincial, and local regulations for asbestos abatement and hazardous waste disposal. By complying with these laws, both instructors and organizations avoid the dangers of liability created by exposure to and removal of asbestos.

Training possibilities. An instructor must first determine whether it is even possible to use a structure for live-fire training. Environmental laws, for example, may prohibit the burning of a structure because of its location. The structure may also have been designated as a historical landmark. If live-fire training is not possible, the structure might be suitable for forcible-entry training, ladder evolutions, search and rescue operations, or a class on building construction.

Permits and inspections. The instructor must have written permission from the rightful owner of the property in order to burn a structure. The instructor must also apply for and receive the appropriate permits from the local jurisdiction. Permits may be required from the fire and life safety division, the building inspections office, state/provincial environmental authority, or the local residential code enforcement office. Having these permits on display is an important part of public relations, demonstrating to citizens that the fire and emergency services organization adheres to the same restrictions as the rest of the community.

When using an acquired structure, instructors conducting live-fire or other training evolutions must consider the effect on the surrounding neighborhood. Refer to the IFSTA **Fire and Emergency Services Instructor, Ninth Edition** Curriculum Suite for sample Structural Live-Fire Training Forms that may help instructors prepare letters and public announcements.

Notice and documentation. In order to meet his or her responsibilities as an instructor, the instructor should perform the following actions:

- Distribute a notice (letter or brochure) to each resident living within a reasonable distance of the structure, informing them of the date and time of the training event, a description of the training activity, and its effect (such as street closures) on the surrounding area.
- Plan the placement of hoselines and apparatus carefully, and consider how they may least restrict access to the neighborhood.
- Notify the water department when hydrants are involved.
- When an acquired structure is located in an area with limited water flow, flush water mains so that rust and sediment do not cause problems for pumping operations or surrounding households.
- Prepare water supply and flow analyses. Instructors must know the required fire flow for the structure, including safety margins.

- Video or photograph surrounding structures, vehicles, and grounds. If neighboring property is damaged by the training evolution, these videotapes and photographs help document the conditions present before the training activity began. Video documentation may be important if legal claims arise later.
- Hold a briefing with all participants to explain the training evolutions. Take all participants on a walk-through inspection to familiarize them with the structure's layout and exits.

Fuel materials. Misuse of fuels is a contributing factor in many live-fire training incidents. Although all live fires in acquired structures are dangerous, following the NFPA 1403 requirements regarding fuel will lessen the possibility that students or instructors will be injured. There may also be additional requirements for fuel usage in the state's OSHA regulations for live-fire training that the instructor should adhere to. Safety is always the primary concern during live-fire training exercises. It is generally safe to use propane torches or fireplace lighters to ignite the fuel and then remove the igniting device from the structure once the fire has started. Instructors must adhere to the following requirements:

- All fuels must have burning characteristics that are known and controllable.
- Never use unidentified materials, such as debris found around or in a structure. Remove all such materials before training takes place (**Figure 8.12**).
- Never use pressure-treated materials or any materials containing pesticides or harmful chemicals.
- Use only enough fuel to create a fire of the desired size. Never use more than this amount.
- Control the structure's fuel load to prevent flashover or backdraft conditions.
- Never use flammable or combustible liquids (accelerants) in live-fire training at acquired structures.
- Assess all factors in the fire room that could affect the growth, development, and spread of the fire, in order to predict fire behavior.
- Remove all highly combustible materials from the structure, including carpets, floor coverings, foam mattresses, and furniture.

Figure 8.12 Debris found in or around an acquired structure should be removed before training.

- Document and check against the materials and construction requirements allowed in a live-fire training evolution per NFPA 1403:
  — Furnishings
  — Wall and floor coverings and ceiling materials including varnish and paint on cabinets and other built-in furnishing.
  — Type of construction of the structure (type of roof and combustible void spaces)
- Document the dimensions of the fire room.
- Ensure that a safety officer supervises while a designated ignition officer ignites the fire.

Water supply. Water supply requirements for fire fighting in rural and suburban areas are provided in NFPA 1142. These requirements should be applied to fire attack during live-fire training evolutions. There must be reliable water sources for the duration of any live-fire evolution. In this sense, water supply operations during a live-fire exercise represent the equivalent water supplies needed at actual fires.

Many acquired structures are in locations that are not convenient to hydrants or other water sources. These locations provide an excellent opportunity for training evolutions on water shuttle, portable dump tank, large-diameter hose deployment, drafting, and relay pumping operations. After completing training about how to get water to the location, the live-fire exercise can continue at the acquired structure.

Purpose-Built Structures

Purpose-built structures are usually found at permanent training facilities (**Figure 8.13**). These structures typically have temperature sensors to monitor the rate of temperature rise within the unit. Fuel sources may be LPG, natural gas, or Class A materials such as untreated lumber or straw.

All interior structural fire training evolutions must meet the requirements of NFPA 1403. The safety requirements discussed earlier in this chapter from NFPA 1403 that apply to acquired structures also apply to purpose-built structures with the following exceptions:

Figure 8.13 Purpose-built training structures are found on many fire training grounds.

- Multiple fires can be lit in purpose-built structures.
- A single water supply may be used.

Take extreme caution that complacency does not occur in terms of recognizing the potential for collapse of the building, the structural condition, the heat in the building, and the ability of fire to spread in ways we cannot predict. The instructor must combat a complacent attitude and reinforce the need for situational awareness.

Exterior and Wildland Fires

Exterior fire-suppression training may include evolutions that simulate low fire load fires, transportation fires, flammable/combustible liquids fires, and **wildland fires**. This training can be conducted at a permanent training facility or a remote site, with one or more fire companies or training agencies. Depending upon the location of the training and the type of fuel used (such as Class B fuels), there may be environmental requirements that should be met. Instructors should check with the AHJ that enforces the federal environmental regulations. In addition, many exterior fires such as those with exterior props or Class B materials may be governed by requirements in NFPA 1403.

Figure 8.14 Small prop fires are versatile in outdoor training evolutions.

Small prop fires. These fires are used to train entry-level students and industrial fire brigade members how to use portable handheld extinguishers to control all classes of fires (**Figure 8.14**). Training evolutions usually take place outdoors, in an area where the spread of fire is limited or nonexistent. They involve small quantities of fuel, which typically consists of Class A materials (in the form of shipping pallets) or Class B materials contained in small burn pans. Similar LPG- or natural gas-supplied burn pans are also available.

Instructors demonstrate the appropriate procedure based on the type of fuel and extinguisher, and students then repeat the procedure, practicing it until they are proficient. In all cases, a backup extinguishing system must be present, usually in the form of an experienced crew with a charged attack hoseline.

Medium to large prop fires. These training fires use permanent training props that are contained in burn pits. They are typically fueled by Class B materials such as LPG, natural gas, or other flammable/combustible liquids. Valves located outside the pit and supervised by a fuel-control officer are used to turn off fuel supplies. An ignition officer is designated to start the fire using an approved ignition device, sometimes with an electrically energized source. Medium to large props include the following:

- Vehicle fire props
- Dumpster fire props **(Figure 8.15)**
- Shipboard props
- Aircraft fire simulators **(Figure 8.16)**
- Railcars
- Propane tanks
- Mobile props specifically designed to simulate flammable liquid and gas fires using propane

Flammable/combustible liquid fires. Simulating fires in flammable/combustible liquid, LPG, and natural gas storage, production, and pipeline facilities is usually restricted to purpose-built props in permanent training facilities **(Figure 8.17)**. They require piped fuel supplies, control valves, product- and water-containment diking, high-capacity water supplies, and water-decontamination capabilities. Evolutions generally involve multiple-company training.

Wildland fires. Training fire personnel to control wildland and/or **wildland/urban interface** (areas where structures abut wildland fuels) fires can take the following two approaches:

1. Specialized training for organizations that are responsible only for protecting forests and wildland areas.
2. General training for structural firefighters who may be responsible for limited areas of wildland, or areas that include the wildland/urban interface.

In some areas of North America and Canada, the training for both groups focuses on the extreme dangers that wildland fires can pose. In other areas, structural firefighters may only receive limited training, involving off-road driving and pumping skills, fire attack, and exposure protection.

Training evolutions vary, depending on the skills required to meet the local dangers that wildland fires create. The nature of wildland fires makes them unpredictable and highly hazardous. A controlled burn can be affected by weather, wind direction, and other factors beyond the instructor's control, so training students on safety is even more of a concern during this type of training. The National Wildland Coordinating Group (NWCG) provides national standards for wildland training.

Figure 8.15 Dumpster fire props are typically fueled by Class B materials.

Figure 8.16 Aircraft fire props are contained in burn pits.

Figure 8.17 Foam application training is practiced with flammable/combustible liquid fires.

Increased Hazard Exposure Training

As overall fire suppression incidents decline, the fire and emergency services have increased technical training evolutions in other fields. While live-fire training remains the most hazardous, there are many other types of hazardous training that require a high level of attention to safety. Examples of training evolutions with increased hazard exposures include:

- Hazardous materials technician skill exercises
- Emergency vehicle operations
- Vehicle and machinery extrications
- Power tool and equipment operations
- High-angle rescue operations
- Confined-space entry rescue operations
- Trench shoring and rescue operations
- Building collapse search and rescue operations
- Surface water, swift water, and dive rescues **(Figure 8.18)**
- Ice rescues

Figure 8.18 Swift water rescue training can present unique challenges due to increased hazard exposure.

High-hazard training can be accomplished safely when responders are trained and alert. The following IFSTA manuals provide guidance in the referenced high-hazard environments: **Hazardous Materials Technician**, **Pumping and Aerial Apparatus Driver/Operator Handbook**, **Principles of Vehicle Extrication**, and **Fire Service Technical Search and Rescue**.

Legal Liability

Liability is a broad term that encompasses all aspects of legal responsibility. A person or organization can be held liable for acts they take and actions they fail to take. A wrongful act resulting in harm is known as an *act of commission*, and the person or organization who commits such an act is responsible for correcting it. Neglecting to take an action that could have prevented harm is known as an *act of omission*, and the person or organization who failed to act is responsible for the consequences. Instructors and/or organizations may be considered liable for any of the following acts:

- Providing incorrect information or instruction
- Failing to instruct in a topic they are responsible for teaching
- Teaching a topic they are unqualified to teach

Instructors can reduce the potential for liability and legal action against themselves and their organizations in the following ways:

- Being aware of expectations
- Teaching to the established standards
- Teaching only topics for which they meet all qualifications
- Providing a safe learning environment

Instructors are expected to foresee (predict) potential injuries that could happen during training events and prevent injuries while training personnel for appropriate performance on the job. Students are trained in a non-emergency environment for skills that will take place in an emergency environment. Students are learning how to perform high-hazard activities that they are initially not fully competent to perform. Instructors must foresee instructional problems and ensure that they can appropriately perform and demonstrate the skills in a thorough step-by-step introduction and then properly supervise the practice of those skills.

Training should be realistic but controlled. Instructors who fail to properly control hazards during training leave themselves and the training organization liable for any injuries that result **(Figure 8.19)**. Such failure could be interpreted as negligence in a court.

NOTE: Injuries that result from emergency scene situations have special legal defenses that do not apply to training environment situations.

Figure 8.19 Training situations should be safe and controlled while teaching students skills they will need in similar situations.

Training in the fire and emergency services will always carry an element of risk, so even if the instructor takes every possible precaution, students may still be injured. However, if the instructor has done a good job of controlling risk, these injuries will likely be the result of unavoidable accidents or student misconduct, not of negligence.

When planning training evolutions or scenarios, instructors should consider the answers to the following questions:

- Can instructors be held liable for actions of students who irresponsibly act on their own?
- Can individuals be held personally liable for contributing to their own injuries and the injuries of others?
- To what extent are employers liable for injuries caused by their employees?

The sections that follow explain the concepts of vicarious liability and foreseeability, and list precautions to help prevent instructor liability. Along with the precautions for preventing instructor liability, steps are given that can reduce the possibility of personal liability.

Vicarious Liability

Vicarious liability means that the blame for the actions of one person can be placed on another. It is the liability that is placed on the employer for the acts and omissions of employees during the normal course of their employment. For instructors, the liability is placed on them for the acts and omissions of students. In training situations, individuals are responsible for their own negligent actions that cause injury to themselves or others if and only if it can be shown that the instructor made every effort to prevent the actions through proper training before the actions were taken.

Foreseeability

Foreseeability is the legal concept that reasonable people should be able to foresee the consequences of their actions and take reasonable precautions. Failure to take reasonable precautions could be considered negligence. For instructors, this means that based on their knowledge of the hazard/risk analysis, they should be able to predict hazardous training conditions and take steps to reduce the risk of injury. If they do not, they may be liable for any resulting injuries. Foreseeability also applies to the risks that students face once they have left the training environment and perform their daily jobs. Instructors must foresee the risk students will face there and prepare them to properly handle that risk.

What This Means to You

Consider a scenario in which you are scheduled to conduct a training session on ladder raises. On the day before training, it rains, and the standing water freezes on the ground overnight at the training structure. The training had been scheduled for months, and it would be difficult for the students to reschedule, so you decide to hold the training anyway. While raising a ladder, one of the students slips on the ice, falls, and dislocates an elbow. Because a reasonable person would have foreseen that icy conditions were a hazard during this kind of training, you as the instructor could be held liable for the student's injury. The foreseeable danger posed by the ice means that you should have cancelled training or moved the training to another location without the ice hazard.

Liability Reduction

Generally speaking, the best way to reduce liability for any individual training scenario is to develop a training and safety plan, then follow it during training. If necessary, a safety or planning committee can help the instructor to develop the plan. More specific precautions an instructor can take to avoid liability include the following:

- Maintain written objectives and document each training session.
- Provide students with a written course syllabus so they can understand all requirements.
- Train all students how to safely operate equipment.
- Do not leave students unattended while they are practicing potentially dangerous skills.
- Do not exceed personal skill level when training students or working with other instructors.
- Do not ignore, shortcut, or exceed protocols or policies.
- Use as many ways as possible to ensure that students understand the intent and outcome of all directives or instructions, as well as the consequences for not following procedures. Never assume that they will understand without explanation. Tell them, show them, and give it to them in writing.
- Do not joke about serious situations or belittle the actions of others in any learning or service situation.
- Never disclose personal information (except to appropriate authorities) about students, other personnel, or any victim or patient who required emergency services.
- Follow your organization's policy on disclosing information to insurance companies, hospital personnel, legal representatives, news reporters, or other persons who want information about ongoing litigation. These policies also apply to issues that have the potential to be litigated.
- Maintain current certifications through credible refresher courses. *I didn't know* is never an acceptable defense.
- When in doubt, seek advice from a higher authority. Never attempt to make decisions that are beyond your personal knowledge or authority.
- Accurately document all issues of discrepancy, complaint, and injury, including details on dates, times, conversations, suggested resolutions, and follow-up plans.

Safety in Training

The instructor must always remember that training scenarios can present a high level of risk. Between 2001 and 2015, eighty-one training related fatalities were investigated by NIOSH as part of the Fire Fighter Fatality Investigation and Prevention Program. Of these fatalities, sixty-six were cardiac related and eleven were trauma related.

On March 7, 2014, a 51-year old male career fire department captain died while participating in his fire department's "rule of air management" training, wearing full PPE and carrying a 50-foot (15-m) section of 2-1/2-inch (65-mm) hose. After climbing the stairs of the five-story tower two successive times, the captain collapsed and later succumbed to cardiac related illness. NIOSH offered the following recommendations: exercise stress tests as part of the fire department medical evaluation program; medical monitoring in rehabilitation programs; and phasing in mandatory comprehensive wellness and fitness programs for firefighters. (*NIOSH report number: F2014-10*).

On October 23, 2005, a 47-year old male career fire department captain and instructor was severely burned during a live-fire training evolution at a training academy and two days later succumbed to his injuries. The instructor was in the basement of the live-fire burn building adding pallets to the fire prior to the students entering the building. The students found the instructor on the floor of the burn room as they advanced the hose line as part of the evolution. A mayday was immediately called and the instructor was taken outside and emergency medical care was provided. NIOSH offered the following recommendations: Use the minimum fuel load necessary to conduct live fire training; ensure that two training officers are present with a charged hose line during ignition and refueling of a training fire; and avoid having burn rooms below grade. (*NIOSH report number: F2005-31*).

On February 9, 2007, a 29-year old female probationary firefighter was fatally injured while participating in an instructor led live-fire training evolution at a three-story townhouse acquired structure as part of their NFPA 1001 Firefighter I training. The scenario had the four-person engine company pass a second floor fire and proceed to the third floor to extinguish a fire there. When heavy fire conditions were encountered on the second floor, they still proceeded to the third floor to extinguish the fire. The crew was met with heavy fire on the third floor with conditions that were untenable and firefighting efforts were not able to be accomplished. The instructor and a backup person were able to exit through a window with a 41-inch high sill; however, the victim was not able to get out of the window, and later being freed, succumbed to her injuries. NIOSH offered the following recommendations: conduct live-fire training in accordance with NFPA 1403; ensure that live-fire training is conducted under the supervision of qualified fire service instructors meeting the requirements of NFPA 1041; ensure that all students meet the physical performance requirements as established by the fire department. (*NIOSH report number: F2007-09*).

For an additional example of a training-related incident, see:
State of New York v. Baird – September 25, 2001, *NIOSH investigative report number: F2001-38*

Chapter Review

Answer the following questions to review the information provided in this chapter.

1. Where can instructors receive the most current information about safety guidelines and regulations?
2. How can instructors increase safety awareness and help prevent accidents?
3. What actions are required for planning for safe training?
4. What information is needed to present a skills demonstration?
5. What are the elements of evolution control?
6. What are some examples of a simple training evolution?
7. What are some examples of increased hazard exposure training?
8. When can an instructor be considered legally liable?

Discussion Questions

The following questions are intended to generate discussion, expand your understanding of the chapter text, and allow you to think critically about what you have learned. Answers to these questions may vary.

1. Why must instructors be aware of resources regarding safety guidelines, regulations, and information?
2. What specific experience have you had with increasing awareness and helping prevent accidents before or during training?
3. What responsibilities does the instructor hold when planning for safe training?
4. What challenges have you experienced as an instructor when demonstrating psychomotor skills?
5. How do simple training evolutions differ from psychomotor skills training?
6. What purpose does live-fire training serve?
7. What types of training evolutions with increased hazard exposures have you experienced?
8. How can instructors reduce the potential for liability and legal action against themselves and their organizations?

Key Terms

Accreditation — The process for an organization or entity to be recognized for upholding standards necessary to certify levels of competency in a given field or topic.

Acquired Structure — Structure acquired by the authority having jurisdiction from a property owner for the purpose of conducting live-fire training or rescue training evolutions. *Also known as* Acquired Building.

Foreseeability — Legal concept that states that reasonable people should be able to foresee the consequences of their actions and take reasonable precautions.

Liability — To be legally obligated or responsible for an act or physical condition; opposite of *immunity*.

Purpose-Built Structure — Building specially designed for live-fire training; fires can be ignited inside the building multiple times without major structural damage.

Vicarious Liability — Liability imposed on one person for the conduct of another, based solely on the relationship between the two persons; indirect legal responsibility for the acts of another, such as the liability of an employer for acts of an employee.

Wildland Fire — Unplanned, unwanted, and uncontrolled fire in vegetative fuels such as grass, brush, or timberland involving uncultivated lands; requires suppression action and may threaten structures or other improvements.

Wildland/Urban interface — Line, area, or zone where an undeveloped wildland area meets a human development area. *Also known as* Urban/Wildland Interface.

Chapter 9

Testing and Evaluation

SECTION A INSTRUCTOR I

Chapter Contents

- **Approaches to Student Assessment......185**
  - Norm-Referenced Assessments 186
  - Criterion-Referenced Assessments 186
- **Test Classifications..........................187**
  - Purpose Classification 187
  - Administration Classification 188
- **Test Bias.......................................189**
- **Test Administration 190**
  - Administering Written Tests...........190
  - Administering Performance Tests 191
- **Test Scoring 192**
  - Scoring Written Tests.................192
- Scoring Oral Tests 194
- Scoring Performance Tests 194
- Grading Fire and Emergency Services Tests 194
- Grading Bias.........................194
- **Grade Reporting 195**
- **Test Security..................................195**
- **Evaluation Feedback 196**
- **Chapter Review 196**
- **Discussion Questions 197**
- **Key Terms 197**
- **Skill Sheets 198**

JPRs addressed in this chapter

This chapter provides information that addresses the following job performance requirements of NFPA 1041, *Standard for Fire Service Instructor Professional Qualifications*, 2019 Edition.

- 4.2.5
- 4.5.2
- 4.5.3
- 4.5.4
- 4.5.5

Learning Objectives

1. Discuss two ways to assess a student's success. [4.5.2, 4.5.3, 4.5.4]
2. Identify the six classifications of tests. [4.5.2, 4.5.3]
3. Discuss test bias. [4.5.2]
4. Describe the process of test administration. [4.5.2]
5. Explain the different processes of scoring and grading tests. [4.5.3]
6. Explain the importance of accurate grade reporting. [4.5.3, 4.5.4]
7. Identify guidelines to reduce academic misconduct. [4.5.2, 4.5.3]
8. Discuss the benefits of instructor feedback on student outcomes. [4.5.5]
9. Skill Sheet 9-1: Give a test (oral, written, or performance). [4.5.2]
10. Skill Sheet 9-2: Grade a test (oral, written, or performance). [4.5.2, 4.5.3]
11. Skill Sheet 9-3: Report test results and training records. [4.2.5, 4.5.4]
12. Skill Sheet 9-4: Give feedback to Level I Instructor candidates on their test results. [4.5.4, 4.5.5]

Chapter 9
Testing and Evaluation

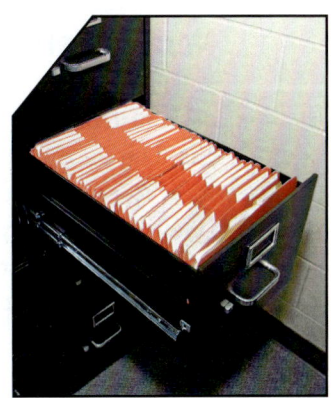

Evaluation is the last step in the four-step method of instruction and is accomplished through observation and testing. The purpose of student evaluations and testing is to determine how well students have learned and retained the course material. While evaluation is often used to assign grades or scores, it should also result in feedback for students that gives students a clear idea of what they have learned and what they still need to learn.

Evaluating student progress or performance requires a stated performance criterion. To determine whether this criterion has been met, instructors must either observe students directly or use testing instruments.

This chapter focuses on the formal evaluation and testing duties of Level I Instructors as defined in NFPA 1041, *Standard for Fire Service Instructor Professional Qualifications*. Level I Instructors should understand the following categories:

- Approaches to student assessment
- Test classifications
- Test bias
- Test administration
- Test scoring
- Grade reporting
- Test security
- Evaluation feedback

Approaches to Student Assessment

Certification examinations are intended to determine mastery of a subject or topic. To pass these exams, students must have a complete and thorough knowledge of the subject and the ability to recall and apply the knowledge. Assessing or evaluating student success may be accomplished either through norm-referenced or criterion-referenced assessment. These types of assessments are described in the following sections.

> **What This Means to You**
>
> Norm-referenced tests are common tools in academic settings and require the establishment of a ranking system. This approach is well suited to making those determinations because each student is compared to all the other students, based on preset criteria. The student with the most accurate answers on a written exam is judged to be the most knowledgeable.
>
> Criterion-referenced tests are used in routine training capacities where ranking is unimportant, but it is essential that all students be able to competently perform a task. All students who are able to perform the task within set parameters are judged to be competent and allowed to proceed, even if they do not all perform it the same way.

Norm-Referenced Assessments

Norm-referenced assessment measures the accomplishment of one student against that of another. At the end of an instructional unit, an exam is given, and the results are scored. The scores are translated into grades, often based on the distribution of scores. In most instances, the distribution is arranged in a bell curve, with the majority of scores in the middle, and a much smaller number of exceptional scores at the high and low ends of the range (**Figure 9.1**). This type of assessment is used for some **evaluations** in the fire and emergency services, particularly in cases where clear differentiations are made between top performers and others. For example, norm-referenced assessments are commonly used during promotional testing and determining ranking in recruit academies.

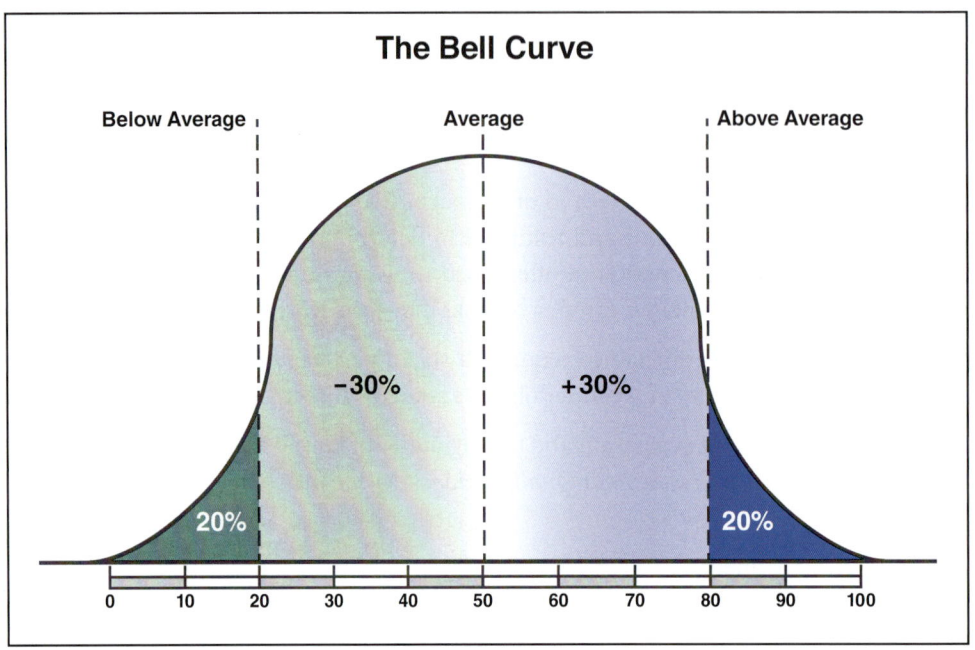

Figure 9.1 In norm-referenced assessments, exam grades are typically arranged in a bell curve to show the distribution of low- to high-range scores.

Criterion-Referenced Assessments

Criterion-referenced assessment compares student performance with the **criteria** stated in the learning objectives. In this type of assessment, any performance is considered acceptable as long as it meets the conditions stated in the criteria. Criterion-referenced assessment is primarily used where many students can succeed at the same level, such as in a classroom or skills training setting.

Using the criterion-referenced approach, test scores translate to either a passing or failing grade, depending on whether the student has met or failed to meet the criteria. For example, if the criterion for performance on an end-of-instruction exam is 70 percent, any grade of 70 or higher is a Pass, while anything below 70 is a No Pass/Fail.

A similar approach may be used to evaluate psychomotor skills, in which the student either passes or fails a manipulative skills test. In an SCBA test, for example, students are required to accurately don an SCBA in a fixed amount of time. Using a skills checklist and a timer, instructors can assess whether students have completed the skill quickly and accurately enough (**Figure 9.2**).

Figure 9.2 Using a skills checklist will help the instructor evaluate skills completely and consistently.

> **Training and Education Terms Used in the Fire Service**
>
> The terms *objectives* and *outcomes* are similar and should be carefully used in context. An objective is what you seek to achieve, while the outcome is what you actually achieved.
>
> Another example of terms that are very close in meaning are *psychomotor*, *performance*, and *skills*. Psychomotor refers to movement as a result of conscious mental activity; performance relates to the process of carrying out or accomplishing a task; and skill has come to mean a physical activity. They are all related back to actively accomplishing a physical task. The instructor is encouraged to be careful and specific in his or her use of language to avoid misunderstanding.

If they do, they pass, but if they do not, they fail. In some skills, a single step may be so critical that performing it incorrectly may result in a failing grade, even though the rest of the skill was performed perfectly.

Test Classifications

Both criterion-referenced and norm-referenced tests can be divided into more specific classifications, based on the reason for testing and the way the test is administered. No single test type is suitable for every situation. Instructors should be familiar with the following two test classifications and be able to select a test that will best measure the intended learning result:

- Purpose classification (the reason for testing)
- Administration classification (the way the test is administered)

Purpose Classification

The **purpose classification** clarifies the reasons a test is given and the point at which it is given during instruction. Test types include prescriptive, formative, and summative tests.

Prescriptive (Pretest)

Prescriptive tests are given at the beginning of instruction to establish a student's current level of knowledge, in order to compare it to a subsequent summative evaluation. When the tests' contents are the same or similar in both iterations, comparing the two scores measures the amount of learning that has occurred.

Formative (Progress)

Formative tests may be quizzes, pop tests, or question/answer periods that are given throughout the course or session. These tests typically measure improvement within a small scope of the class progression and give the instructor and students feedback on learning progress. When measuring improvement, the test answers the question: *Is the student achieving the objectives?* Formative tests can include the most important learning objectives, or all of them if possible. Each test item should be designed to a level of difficulty that matches the learning objective it is meant to measure. Doing so helps to ensure that the test is criterion-referenced.

Summative (Comprehensive)

Summative tests measure comprehensive knowledge and skills at the end of a course, or of a major segment of the course. These tests answer the question: *Has the student achieved the course objectives?*

Examples of comprehensive tests would be the written and/or practical exams given at the midpoint or end of emergency medical technician (EMT), basic fire fighting, or driver/operator courses. Students who are tested must demonstrate comprehensive knowledge and skills relating to all material from the beginning of the course to the testing point.

Administration Classification

The **administration classification** is based on how the test is administered, and includes oral, written, and performance tests. The Level II Instructor or AHJ SOPs may assign this type of test based on the type of learning that is being evaluated. The Level I Instructor may be assigned to administer any of these types of tests.

Oral Tests

During oral tests, the student gives verbal answers to spoken questions during a one-on-one interaction with the instructor (**Figure 9.3**). These tests are not commonly used in the fire and emergency services, but they may be useful under certain circumstances, such as determining a student's understanding at the end of a lesson.

Administering and scoring oral tests should follow an established **scoring rubric**. Instructors must listen carefully to student responses in order to prevent misunderstanding, because students will phrase the same answer in different ways. Instructors should also be careful not to make facial expressions that might confuse or mislead students. Although oral tests can be useful evaluative tools, they should rarely be used as the sole means of determining students' terminal performance for a course or course segment.

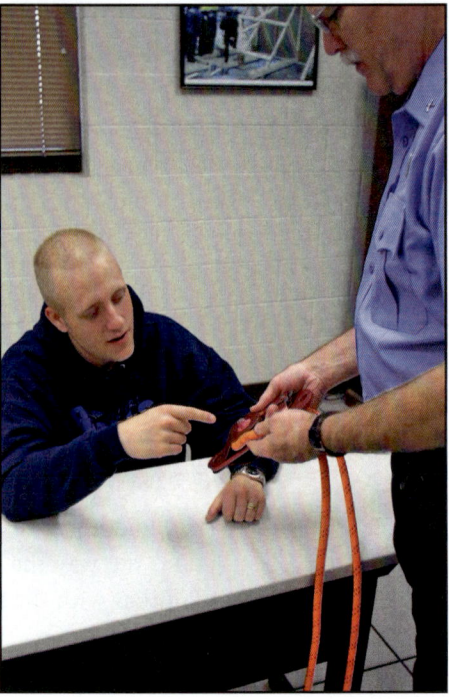

Figure 9.3 Occasionally, the instructor may employ an oral exam to evaluate a student's knowledge.

Written Tests

Written tests evaluate if students have met cognitive learning objectives from the lesson plan. They are useful for measuring retention and understanding of technical subjects, such as fire chemistry, laws and ordinances, hydraulic principles, and medical protocols. Written tests may have numerous question types, including the following:

- **Multiple-choice** — Single question followed by multiple possible answers, of which only one is the best answer.
- **True/false** — Students determine whether a statement is true or false.
- **Matching** — Students match dates, events, or items from one column with appropriate definitions from a second column.
- **Fill in the blank/completion** — Statement that is missing a word or several words that students must provide.
- **Short-answer** — Question that requires a brief, factual answer.
- **Essay** — Question that requires a lengthy, primarily subjective answer.

Written tests can be administered either by reading them aloud, providing students with a sheet containing the questions, or through electronic media (**Figure 9.4**). Students write their answers on a blank page, on a formal answer sheet, or in the electronic medium through which the test is administered. The legibility of students' handwriting can be a factor during scoring. Tests that require students to circle, check, or fill in a block reduce the potential for misinterpretation based on handwriting. Whenever possible, use answer sheets that can be read, scored, and recorded electronically.

Figure 9.4 Written test booklets are a common evaluation tool.

Computers are an increasingly popular method of administering written tests. Students take the test in a testing center or computer lab, and the testing program scores the results and records them in a database. This method can be used for any type of training course, and is the only way to administer tests for Web-based and Internet training programs. One advantage of this method is that it allows the instructor to create realistic scenarios that require the student to evaluate a situation and determine the correct response.

Performance Tests

Performance tests measure students' ability to perform skills and tasks as they would on the job, based on standardized criteria and performance objectives. They are tested on their current ability to complete the skill. When preparing to administer a performance test, instructors should perform the following tasks:

- Determine the materials, tools, or equipment that students must have available to perform the skills or activities. Make sure that they are available and in working order (**Figure 9.5**).

- Prepare a skill checklist, along with appropriate time limits for each of the steps necessary to perform the skill. The checklist provides/establishes/defines the criteria that must be met for the student to pass. Review the checklist with all instructors and test evaluators to determine whether any points may be misunderstood.

- Determine the number of test evaluators needed to observe and measure (by checklist) the performances of students.

- Review the operation of unfamiliar equipment with all instructors and test evaluators.

Figure 9.5 Instructors should ensure that all equipment needed for a performance evaluation is organized and available before a performance test.

Test Bias

A test is considered biased when members of different groups (age, cultural and ethnic background, gender) with the same ability level routinely score differently on a test. Bias can be difficult to prove because tests are designed to discriminate between those that know the information in the test and those that do not. In this application, discrimination is based on ability. Bias, in contrast, is an indication that the test puts one particular group at a disadvantage to another.

Level I Instructors have an obligation to eliminate testing bias wherever they can. The best way to eliminate bias is to recognize it and to request that potentially biased questions be rewritten or more closely examined. Level I Instructors may find biases in tests that have any of the following qualities:

- Gender references (personal pronouns and proper names) are all from the same gender.
- Ethnic references are stereotypical and/or irrelevant to the test.
- Cultural references do not reflect the cultural knowledge of the likely testing group or are irrelevant to the test.
- Regional jargon or dialects used in the questions could put students from a different region at a disadvantage.
- Terminology and vocabulary in the test are unfamiliar to the students taking the test.

In legal terms, certification skills tests and promotional tests are grouped together under the larger title *employment tests* or *selection tests*. Typically, employment tests are either entry tests for getting a job or tests for reaching a promotion after training. In the United States, the Equal Employment Opportunity Commission (EEOC) has the responsibility of investigating any bias on employment tests. In order to avoid any legal liability, Level I Instructors should follow any scoring rubrics and answer keys that have been provided to them. They should also teach to the learning objectives in provided lesson plans. See **Appendix D** for a sample scoring rubric.

Test Administration

Regardless of the platform or medium used to present a test, test administration begins before the test is given. Instructors should inform students what type of test they will take, what content from the lesson plan it will cover, the location of the test environment, and which materials they will need to bring (paper, pencil, pen, notes, books, PPE, etc.). The course syllabus should also contain the dates and times of major tests or periodic quizzes. On the day of the test, the instructor should consider the physical environment of the test location, making sure that students have the following (**Figure 9.6**):

- Appropriate lighting
- Comfortable seating
- Quiet surroundings
- Regulated temperature
- Proactive responses to performance barriers

Administering Written Tests

Instructors should adhere to the following guidelines when administering written tests:

- Before the test:
  — Ask the testing organization or AHJ to determine whether there are any specific instructions or protocols for administering the test.
  — Arrive at the testing location an appropriate amount of time before testing begins.
  — Maintain test security at all times.
  — Rearrange classroom seating when necessary so that it is conducive for taking written tests.
  — Eliminate loud talking or noises outside the testing room.
  — Eliminate all potential distractions within the room.
  — Have students place all backpacks, purses, books, and other unnecessary items out of the way or in an assigned area.
  — Ensure that all electronic devices are turned off and secured.
  — Inform students of expectations, including misconduct rules.
  — Make sure that students do not bring written or electronic notes into the testing room, unless these are specifically approved by the examination guidelines.
  — Remind students to follow instructions on the answer sheet correctly.

Figure 9.6 The physical classroom environment may affect students' learning ability.

- During the test:
  - Watch for signs of academic misconduct, such as:
    - Students whose eyes "wander" around the room.
    - Answers written on clothing, skin, or shoes, or on papers lying on the floor near students' desks.
    - Students talking to each other during the test.
  - Ensure that the environment remains quiet and safe for students.
  - If students are allowed to leave the testing area to use the restroom, refreshment area, or other rehabilitation facilities, make sure they do not take the test instrument or paper with them (verify with AHJ prior to test for policy).
  - Supply extra paper, writing implements, and answer sheets.
  - If possible, have testing aides on hand to assist with difficult students or situations, or to perform tasks such as handing out or collecting tests and evaluation forms.
- After the test:
  - Do not allow students to change their answers once answer sheets have been submitted.
  - Review the test (at an appropriate time) with students to clarify any objectives they may not have understood. This may occur at the following class in the case of formative tests but may not be possible with summative tests.
  - Maintain the security of tests and answer sheets, especially when scoring tests and recording grades in accordance with AHJ guidelines (**Figure 9.7**).
  - Return test materials to the proper authority if another instructor will be scoring and grading them.
  - Identify answer sheets with identical incorrect answers or sequences of answers on multiple test sheets. These could be indications of academic misconduct.

Figure 9.7 Follow AHJ guidelines to ensure that all tests, answer sheets, and recorded grades are secure.

Administering Performance Tests

Specific guidelines for administering performance tests are as follows:

- Before the test:
  - Ensure that the subject matter of the test matches the subject matter that is being tested.
  - Provide students with adequate practice time during class time leading up to the test date.
  - Provide rehabilitation facilities, such as restrooms or refreshment areas.
  - Ensure that tests are not biased through wording, timing, or unattainable criteria.
  - Include all test administration rules in the test instructions.
  - Read the instructions aloud to the students exactly as they are written. Do not paraphrase because this may alter the test results (**Figure 9.8**).

Figure 9.8 Read the test instructions to the students as written to ensure they understand what is required.

- Provide students with the time limits for each test and clearly state the times for emphasis.
- Explain the purpose of each test.
- Encourage students to ask questions if they do not understand something.
- During the test:
  - Give the test to each student in exactly the same manner.
  - Record students' scores on performance checklists as each one takes the test. Do not wait until the end of the testing period to record scores.
- After the test — Keep all test scores confidential.

NOTE: Skills associated with giving a test (oral, written, or performance) are shown in **Skill Sheet 9-1**.

Test Scoring

As indicated earlier in this chapter, after administering a test, instructors collect answer sheets or performance checklists, then score and grade the tests. *Scoring* is the act of identifying which answers are right and which are wrong. Scoring should always be objective and based on established criteria. *Grading* is the act of assigning a value to the score. For example, the AHJ may determine that a score of 75 points is a C grade. The sections that follow describe these two steps as they relate to evaluation of student performance on oral, written, and performance tests.

Scoring Written Tests

Scoring methods for written tests can be very simple, such as assigning 1 point per test item on a test containing 100 test items. Tests may also give more weight to some test items than others. For example, a test with 52 test items (50 multiple-choice, 2 essay items) might have the points divided with 1 point for each multiple-choice item and 25 points for each essay. Tests may also assign a higher point value to questions that address more critical learning objectives. Regardless of how tests are scored, the scoring method should always be indicated on the exam.

Composite scoring used for criterion-referenced tests is based on a point system that encompasses an entire course; each scored item in the course is added to a grand total to determine a final score for the entire course. Lesson or course outlines establish the point value of all activities, assignments, or tests. Instructors then add the points to determine the total points earned by a student during the course.

Written tests can be scored either manually or using an electronic scoring device. Tests administered on computers are scored automatically and the results are posted to student files in the course database. While the electronic scorer is easy and fast to use, it can be costly to purchase and require the use of specific types of answer forms. Electronic scorers can also make errors that require the instructor to rescore the tests manually.

Some guidelines for scoring written tests manually include the following:

- Score the same question on all of the tests before proceeding to the next question on short-answer or essay exams. This guideline improves scoring consistency and permits instructors to identify any questions that have resulted in a large number of incorrect answers.
- Read through several responses before scoring essay or short-answer tests. This technique is especially helpful when variations on answers are possible.
- Shuffle the papers before scoring the next question after scoring one set of questions on all the tests. This guideline helps offset the potential fatigue that instructors can experience and reduce the potential for scoring errors.
- Do NOT attempt to score large numbers of tests at one time. Take breaks or work on other projects between sets of papers.
- Use a checklist to identify the key points that should be addressed in each essay or short-answer response **(Figure 9.9)**.

Fire Instructor I   Student Presentation Evaluation Form

Name:_____   Date:_____

This is the evaluation form for the Fire Instructor I student presentation. A 'Satisfactory' must be obtained on all nine (9) tasks for a successful presentation. Any 'Unsatisfactory' rating will require the student to repeat the presentation. Any notes or comments about the presentation are attached.

Satisfactory: The task is clearly performed without error or hesitation and the task enhanced the presentation and the communication process.

Unsatisfactory: The task is not achieved, or performed with error; the task hampered the communication process and the student's understanding of the concepts.

1. Clearly states purpose of presentation and establishes relevance and benefit to the learner.
   ☐ Satisfactory   ☐ Unsatisfactory

2. Implements motivation strategy according to the lesson plan.
   ☐ Satisfactory   ☐ Unsatisfactory

3. Incorporates appropriate verbal and non-verbal communication techniques in presentation.
   ☐ Satisfactory   ☐ Unsatisfactory

4. Asks appropriate questions to facilitate 2-way communication during presentation.
   ☐ Satisfactory   ☐ Unsatisfactory

5. Engages students in active learning process.
   ☐ Satisfactory   ☐ Unsatisfactory

6. Presents and applies all content points identified in the lesson plan.
   ☐ Satisfactory   ☐ Unsatisfactory

7. Properly uses audiovisual materials and equipment.
   ☐ Satisfactory   ☐ Unsatisfactory

8. Facilitates application of manipulative skills as identified in lesson plan in a safe manner.
   ☐ Satisfactory   ☐ Unsatisfactory

9. Summarizes presentation.
   ☐ Satisfactory   ☐ Unsatisfactory

Overall Presentation was: ☐ **Satisfactory**   ☐ **Unsatisfactory**

OSU Fire Service Training
09/2018

Figure 9.9 An evaluation based on a scoring rubric will identify key points that should be addressed in a student response. *Courtesy of Oklahoma State Fire Service Training.*

- Add comments to essay or short-answer questions to indicate what is missing or congratulate the student on exceptionally good work. Be specific, and avoid sarcastic or overly negative comments. Scoring is an extension of teaching, and this practice acts as positive reinforcement and motivation for students.

- Using lines or arrows, indicate which comments refer to which words or portions of the answer.

- Inform students about the meaning of scoring marks that are used in essay or short-answer question tests.

Instructors who teach distance-learning courses need to take into consideration the limited contact they have with their students. Some factors to consider are:

- Classes in a distance-learning format take much more preparation time for educators than the same classes in a face-to-face setting.

- Over the length of a course, student improvement takes longer because feedback is not as immediate. Instructors have to work diligently to provide effective and timely feedback to the students.

- The lack of direct contact between students and instructors makes scoring written work an even more critical task. Scoring criteria, marking notations, and instructor comments must be clear, concise, and constructive. Students should have no doubt about what is expected or how to correct weaknesses in a test answer or assignment.

Scoring Oral Tests

Scoring oral tests depends on the purpose of the test, such as promotional exams, and the type of questions asked. The instructor should present the questions as they have been provided on the test and then follow the scoring guidelines provided. The scoring rubric for an oral test is often very similar to performance tests.

Scoring Performance Tests

Scoring performance (psychomotor) tests can be very subjective, so instructors should closely follow the scoring sheet guidelines. These list the tasks students must perform to complete the skill, and assign a point value to each task. When the student completes the task, the evaluator gives the student credit for that task. In some cases, the instructor may give partial credit. For example, a test for an equipment inspection might give some but not all points to a student who found most, but not all, of the flaws in the equipment.

Tasks that directly relate to the life or safety of a patient, student, or other personnel cannot be skipped or performed improperly. These tasks are pass/fail items on the scoring sheets; students must be able to complete all the steps correctly in order to receive credit. In some cases, scoring for these items should also be weighted so that students who fail to perform this task adequately will fail the entire exam, even if they performed perfectly in all other aspects of the test. There are some skills for which none of the tasks can be failed. In this case, the test has a passing grade of 100 percent and should be scored as such.

When possible, use multiple instructors or test evaluators to observe each student during the test. Having multiple instructors or test evaluators observe each student will result in more consistent and accurate scoring. When evaluating in this fashion, the instructors or evaluators should come to a consensus before assigning the student a score. If they disagree, the student may be asked to perform the skill again or to explain how or why the activity should be performed.

Students must be given a clearly stated set of objectives and the scoring criteria. Instructors should provide immediate feedback while observing the project, especially when safety is a concern.

NOTE: Skills associated with grading a test (oral, written, or performance) are shown in **Skill Sheet 9-2**.

Grading Fire and Emergency Services Tests

Assigning a grade is very simple after scoring a test. The total number of points the student earned is divided by the total number of points possible to arrive at a percentage grade. This method usually applies to written tests but may also be applicable to performance tests.

As indicated earlier in this section, some situations may require students to get every test item correct in order to pass. In these situations, tests should include questions that attentive students will answer easily. Students are disadvantaged when every question on such a test has a high level of difficulty.

Grading Bias

Grading bias is the practice of assigning grades based upon a factor other than the student's scores. On objective tests such as multiple choice tests, grading bias is very easy to prove and is considered unethical conduct. Instructors must be careful to follow scoring sheets on objective tests.

Subjective tests are more challenging for evaluators to grade. Because the evaluation is open to some interpretation, it is also an area where evaluators may inadvertently influence grades for students that they favor. Level I

Instructors who evaluate subjective tests should follow provided scoring rubrics and grading checklists. Should instructors feel that they are pressured to show bias toward some students, or perceive that another instructor is showing bias, they should report that opinion to the AHJ for further evaluation.

Grade Reporting

Once instructors have scored oral, written, or performance tests, the scores must be recorded and reported in accordance with local procedures or AHJ policies. Because fire and emergency services students are graded against a set of specific criteria and not against each other, grades are recorded in individual student records and used as feedback for students. Care must be taken to accurately record the grades in student records.

An instructor should follow the provided scoring guidelines to determine a final grade for the course. Students who fail to achieve the minimum required grade should not be recognized as having completed the training.

Testing records are private and confidential, so only instructors, training division administrators, and the relevant student should have access to them. Testing results should be retained in individual student files for the period of time that the AHJ requires and kept private in accordance with specific department/AHJ policies and applicable laws.

Reporting test results to the training division or organization's administration is necessary for two reasons:

1. The organization must be able to prove whether a student has met the minimum requirements to effectively perform a duty or task, especially where certification is the final intended outcome of training.
2. The test results for all participants in a course provide the training division or administration with an idea of the effectiveness of the course or curriculum. When test results indicate an abnormal or unusual number of students did or did not pass the course, the teaching style, course curriculum, or testing system should then be reevaluated and altered as appropriate.

NOTE: Skills associated with reporting test results and training records are shown in **Skill Sheet 9-3**.

Test Security

Security of all types of tests is essential to an effective training program. While the security of the test results has been emphasized in terms of a student's privacy requirements, security of testing instruments is equally important to prevent **academic misconduct**. Academic misconduct has a broad definition in a training environment, including cheating during exams, acquiring test answers before the exam, allowing a student to claim credit for work completed by another, or copying assignments from other students. On tests, academic misconduct may be reduced by adhering to the following guidelines:

- Follow security measures when writing, duplicating, and storing test materials.
- Never rely solely on questions published in the textbook or study guide.
- Revise test questions regularly.
- Use secure data storage systems to prevent unauthorized access to tests or grades.
- Require students to use assigned usernames in order to limit access to computer-administered tests **(Figure 9.10)**.

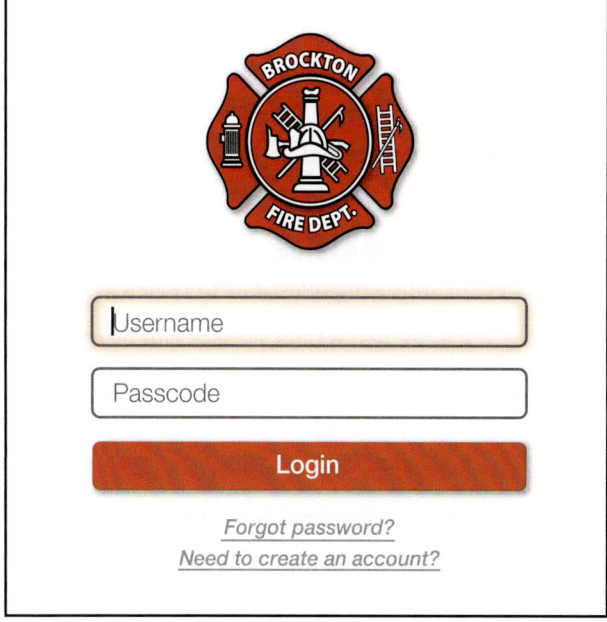

Figure 9.10 To reduce academic misconduct, instructors can require students to log into computer-administered tests with usernames and passwords.

- Number all test sheets, booklets, and answer sheets, and take inventory after each use.
- Use two or more versions of a single exam to prevent students from copying from each other during the test. The simplest way of doing this is to change the order or the wording of individual questions. When using alternate versions, instructors must take extra care when distributing, collecting, and scoring the tests.
- Store all old test sheets in a secure location.
- Destroy outdated testing materials.

Evaluation Feedback

Level I Instructors must not only possess the ability to accurately record test results, they must also have the ability to interpret the subsequent data in order to gauge the students' levels of knowledge and understanding of the course topics. Instructors are encouraged to formally evaluate the test results to determine which pieces of information should be reinforced. These evaluations can be conducted in two ways:

1. **Meet individually with students to discuss the test** — Individual coaching enables instructors to help a student understand the correct answers to missed questions and to determine whether a student's learning or studying styles were barriers to success **(Figure 9.11)**. Remedial instruction or practice can be recommended for the student or alterations in the presentation or testing methods can be made to assist the student in reaching the desired goal.

2. **Review incorrect answers with the entire class** — Class review of the questions that were answered incorrectly is an opportunity to review and reinforce the correct answers. Examples:

    — Questions that a majority of students answered incorrectly can be used for group discussions.

Figure 9.11 Reviewing the test with the student allows the instructor to provide additional assistance.

    — Class review may help an instructor to rephrase a question so that it is easier for students to understand.

    — Review of performance tests may indicate areas where additional practice can be assigned to ensure that skills are learned properly.

NOTE: Skills associated with giving feedback to students on their test results are shown in **Skill Sheet 9-4**.

Chapter Review

Answer the following questions to review the information provided in this chapter.

1. What are two ways a student can be assessed?
2. How are tests classified?
3. What is test bias?
4. What steps should an instructor take to properly administer a test?
5. What are some methods are used to score and grade tests?
6. Why is accurate grade reporting important?
7. Why is test security an important instructor priority?
8. Why should an instructor give purposeful feedback on test results?

Discussion Questions

The following questions are intended to generate discussion, expand your understanding of the chapter text, and allow you to think critically about what you have learned. Answers to these questions may vary.

1. What is the intent of a certification examination?
2. How are norm-referenced and criterion-referenced assessments different/alike?
3. How can test biases be eliminated?
4. What challenges have you experienced when administering a written or performance test?
5. What methods of scoring and grading do you find most challenging as an instructor?
6. Why is reporting test results to the training division or organization's administration necessary?
7. How can academic misconduct be reduced?
8. What methods have you used for evaluating test results?

Key Terms

Academic Misconduct — Any unethical behavior in which students present another student's work as their own, or gain an unfair advantage on a test by bringing answers into the testing area, copying another student's answers, or acquiring test questions in advance.

Administration Classification — Test classification based on how a test is administered.

Criteria — Plural of *criterion*. See Criterion

Criterion — The standard against which learning is compared after instruction.

Criterion-Referenced Assessment — Measurement of individual performance against a set standard or criteria, not against other students. Mastery learning is the key element to criterion-referenced testing.

Evaluation — Systematic and thoughtful collection of information for decision-making; consists of criteria, evidence, and judgment.

Formative Test — Ongoing, repeated assessment during a course to evaluate student progress; may also help determine any needed changes in instructional content, methods, training aids, and testing techniques.

Norm-Referenced Assessment — Form of assessment in which a student's performance is compared to that of other students. Grades are determined by comparing scores to the class average, and assigning grades based on where students scored compared to that average.

Prescriptive Test — Test given at the beginning of instruction to determine what students already know; alternatively, a test that is given remedially.

Purpose Classification — Means of classifying tests based on when the test occurs during a course.

Scoring Rubric — Scoring tool that outlines criteria that must be present on exams that are more subjective such as short-answer tests, essay tests, or oral tests; the criteria should be tied to learning objectives.

Summative Test — Evaluation that measures students' learning at the conclusion of a training session or course; the test results can also be used to measure the effects and effectiveness of a course or program.

9-1
Give a test (oral, written, or performance). [NFPA 1041 4.5.2]

Task Steps

Oral Assessment

Step 1: Give oral assessment to individual students one-on-one **(Figure 9.12)**.

Step 2: Speak in clear, articulated voice.

Step 3: Maintain neutral facial expression; limit gestures.

Step 4: Listen carefully to student's answers, asking for clarification as necessary.

Step 5: Record student's answers accurately.

Written Assessment

Step 1: Arrange classroom/facility to be suitable for written assessment.

Step 2: Explain test procedures to students.
   a. Time permitted for assessment
   b. Filling out answer sheets correctly
   c. Standards for passing mark
   d. Cheating policy
   e. What to do when assessment is complete

Step 3: Monitor assessment.

Step 4: Ensure all testing materials are collected at end of assessment **(Figure 9.13)**.

Step 5: Record student's answers accurately.

Performance Assessment

Step 1: Arrange classroom/facility to be suitable for performance assessment.

Step 2: Gather all necessary training aids/equipment appropriate for assessment **(Figure 9.14)**.

Step 3: Explain procedures to students.
   a. Task required to be completed
   b. Conditions of the assessment
   c. Time permitted for the skill
   d. Standards for passing mark
   e. Cheating policy
   f. What to do when assessment is complete

Step 4: Observe skill being performed.

Step 5: Ensure AHJ safety policies are followed at all times.

Step 6: Use checklist to accurately record skill being performed.

Figure 9.12

Figure 9.13

Figure 9.14

9-2
Grade a test (oral, written, or performance). [NFPA 1041 4.5.2, 4.5.3]

Task Steps
Oral Assessment
Step 1: Check student answer sheet against answer key.
Step 2: Count number of correct answers.
Step 3: Check number of correct answers against criteria for passing.
Step 4: Assign passing or failing grade based on criteria for passing.
Step 5: Secure results in envelope; seal envelope **(Figure 9.15)**.
Step 6: Provide results to appropriate testing authority according to AHJ policies and procedures.

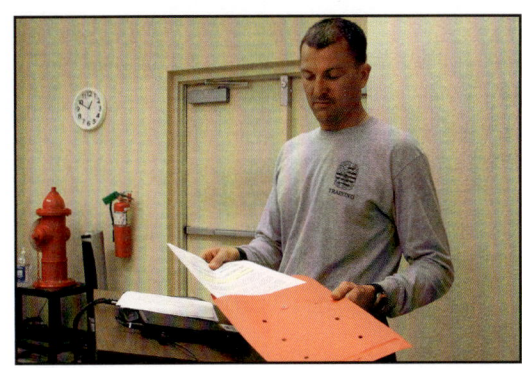

Figure 9.15

Written Assessment
Step 1: Check student answer sheet against answer key.
Step 2: Count number of correct answers.
Step 3: Check number of correct answers against criteria for passing.
Step 4: Assign passing or failing grade based on criteria for passing.
Step 5: Secure results in envelope; seal envelope.
Step 6: Provide results to appropriate testing authority according to AHJ policies and procedures.

Performance Assessment
Step 1: Review skills checklist.
Step 2: Count number of items performed correctly.
Step 3: Check number of correct answers against criteria for passing.
Step 4: Assign passing or failing grade based on criteria for passing.
Step 5: Secure results in envelope; seal envelope.
Step 6: Provide results to appropriate testing authority according to AHJ policies and procedures.

9-3
Report test results and training records. [NFPA 1041 4.2.5, 4.5.4]

Task Steps

Step 1: Using attendance sheets, skills checklists, and test results, complete a training report.

Step 2: Submit/forward reports to proper administrative organization(s).

Step 3: Confirm receipt of reports.

Step 4: Ensure all assessment materials/information remain confidential.

9-4
Give feedback to Level I Instructor candidates on their test results.
[NFPA 1041 4.5.4, 4.5.5]

Task Steps
Individual Feedback

Step 1: After testing, meet with each student on a one-on-one basis as soon as possible **(Figure 9.16)**.

Step 2: Speak in clear, articulated voice.

Step 3: Inform student of their test result.

Step 4 Discuss possible source of student errors.

Step 5: Allow for student feedback.

Step 6: Coach student how to improve to meet training/assessment objectives (additional reading, additional practice, tutoring, etc.).

Class Feedback

Step 1: After testing, meet with entire class as soon as possible.

Step 2: Review answers with entire class.

Step 3: Discuss questions the majority of students answered incorrectly **(Figure 9.17)**.

Step 4: Discuss possible source of errors.

Step 5: Rephrase questions to help students understand.

Step 6: Allow for student feedback.

Step 7: Coach students how to better meet training/assessment objectives (additional reading, additional practice, tutoring, etc.).

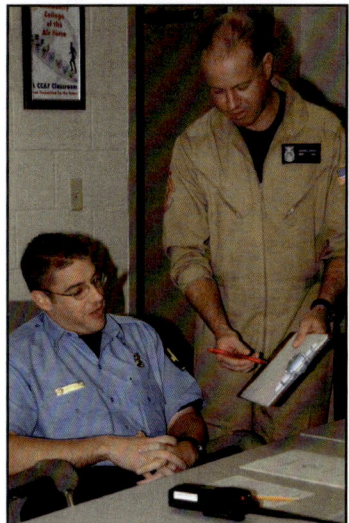

Figure 9.16

Figure 9.17

Records, Reports, and Scheduling

Chapter 10

SECTION A INSTRUCTOR I

Chapter Contents

Difference Between Records and Reports 205	Report Writing 208
Types of Training Records and Reports 205	Scheduling Training Sessions 209
Training Records 206	Chapter Review 211
Training Reports 207	Discussion Questions 211
	Key Terms 211
	Skill Sheet 212

JPRs addressed in this chapter

This chapter provides information that addresses the following job performance requirements of NFPA 1041, *Standard for Fire Service Instructor Professional Qualifications*, 2019 Edition.

4.2.3
4.2.4
4.2.5

Learning Objectives

1. Explain the difference between records and reports. [4.2.3, 4.2.5]
2. Identify the types of training records required by training agencies. [4.2.5]
3. Describe the parts of a written report. [4.2.5]
4. Identify Instructor I responsibilities for scheduling training. [4.2.4]
5. Skill Sheet 10-1: Schedule a training session using resources provided by the authority having jurisdiction (AHJ). [4.2.4]

Chapter 10
Records, Reports, and Scheduling

Instructors spend a significant part of their on-duty time writing reports and maintaining records, which provide a history of the organization's training practices. Instructors must be able to write concise, accurate reports about training sessions and incidents. They must also keep complete, accurate records given the proper record forms and the policies and procedures set forth by the AHJ, and they must schedule individual training sessions.

Difference Between Records and Reports

Training **records** are permanent accounts of events or actions taken by an individual, unit, or organization. From these records, raw data can be used to help develop reports or demonstrate the effectiveness of the AHJ's training program. Training records include:

- Types and hours of training provided
- Names of personnel in attendance
- Learning outcomes achieved
- Training resources expended

Training **reports** are official accounts of a training events, presented verbally or in writing. Following a training event, a written report details pertinent activities required to control it. Training reports inform administrators of the accomplishments, problems, and daily training activities of an AHJ's members and divisions. Reports also provide data that help an organization to make informed decisions about operations and strategic planning.

Records can be standardized forms, narratives, or a simple list of names. They may be handwritten, typed, or computer-generated. Reports are generally in essay format and typed or computer-generated (**Figure 10.1, p. 206**).

Reports and records must be stored in a secure location. Appropriate personnel must be able to access necessary reports and records.

Types of Training Records and Reports

Training documentation is a critical part of an instructor's day-to-day activities. This documentation benefits students, instructors, and training organizations in many ways. This information can be divided into two main categories: training records and training reports.

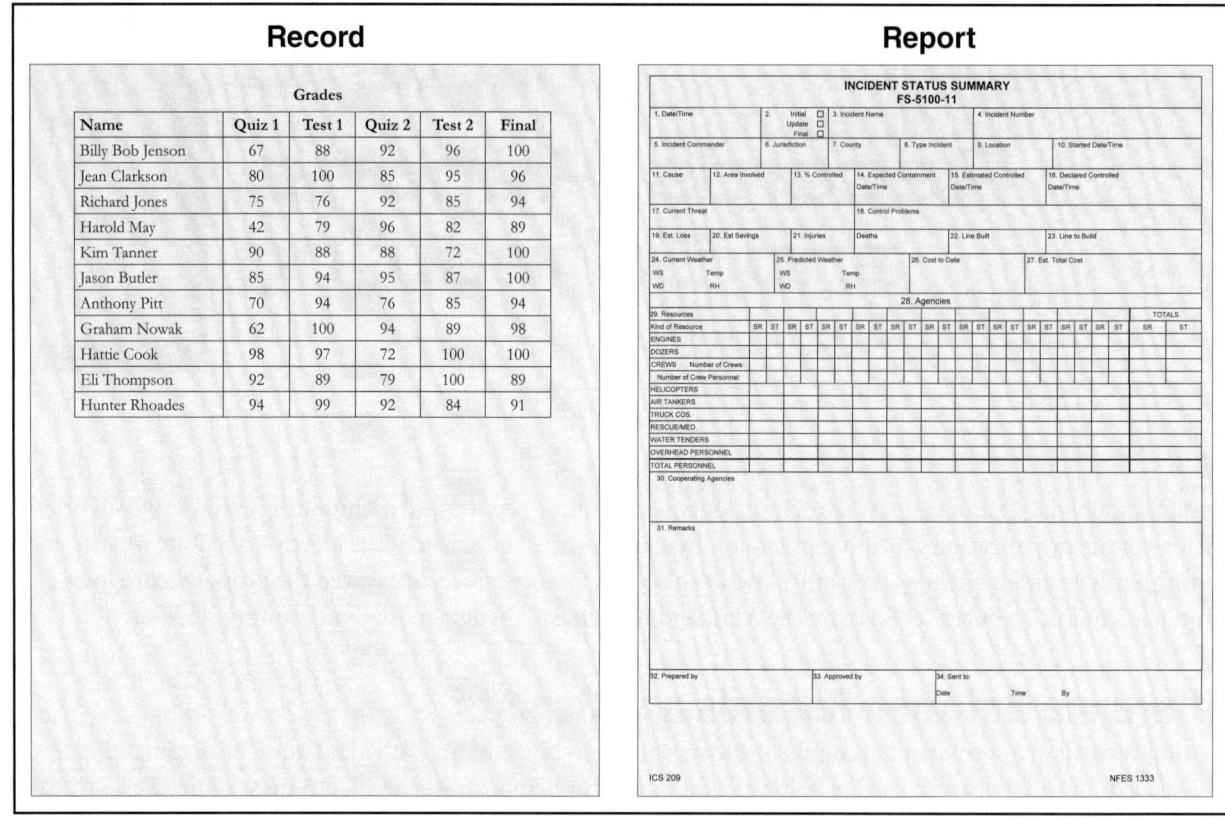

Figure 10.1 Both records and reports are used to store information in an established format, but they differ in how the information is presented and organized.

Training Records

Level I Instructors must be able to accurately complete many types of record forms. Sometimes, this may be as simple as keeping a daily attendance sheet during a course. Instructors may need to complete the following training records:

- **Attendance Records** — Evidence that an individual or unit completed a specified number of training hours on a topic, such as respiratory protection or hazardous materials incident response.

- **Certification Applications** — Forms that students submit after a certification course; students may require an instructor's assistance and/or signature to complete these forms.

- **Incident/Injury Records** — Documentation of student injuries during training. These records may be completed by the instructor or by an investigative team. These records may be requested for public view after they have been completed but should not be shared without proper authorization or cause.

- **Test Records** — Documentation of scores on individual tests given during coursework; test results must remain secure after becoming part of a student's personal training records and/or personnel files.

- **Scheduling Records** — Documentation about what training has been approved by the organization. Usually completed as part of the training scheduling process and may be kept as a history of training offered by the organization.

- **Resource Request** — Documentation of what instructional resources and equipment an instructor needs, as well as documentation of the sources for acquiring these materials, in order to teach a certain lesson. The AHJ may have a specific form that the instructor needs to complete to obtain the needed resources. Instructors should follow procedures in their organizations for requesting the resources they need for training.

In addition, instructors may be required to collect basic medical information from students as part of a course. For example, instructors may document students' medical vital signs before live-fire training. This information should be confidential. If this falls under the instructor's responsibility, he or she should follow all AHJ policies and procedures for completing and maintaining these records.

All training records should be considered private. Only the instructor and student should have access to test scores. Test scores should be communicated directly to the student in writing. The same rules apply to certification records.

Training Reports

Level I Instructors must be able to complete many types of reports, and do so according to organizational policies and procedures. Instructors may need to complete the following training reports:

- **Attendance Reports** — Can be used as a transcript of students who attended a class or an individual's transcript for the training they attended within the organization.
- **Certification/Qualification/License** — Track members' certifications, qualifications, or licenses, and may include issue dates and dates for recertification.
- **Test Report** — Generated from test scores for a class.
- **Instructional/Contact Hours** — Generated for all classes taught or supported (external training) by the organization. Contact hours are the exact number of hours the student had in the class. Instructional hours are generated by taking the contact hours multiplied by the total number of students in the class.
- **Competency Reports** — Can be provided on individuals from the setup of the initial course. These reports could be used to for specific tasks or JPR's addressed in a particular year for hazmat or technical rescue; could be applicable to any requirement.

All training records may be subpoenaed under open records laws. Simply put, records and reports can be considered legal documents that track a student's training career. Although Level I Instructors are not responsible for managing their organization's record system, they should become familiar with the AHJ's filing system and its managers. Any records created during a course should be submitted to managers.

> ### Level II Instructor Duties
> Level I Instructors may be the only personnel available to maintain training records for the entire organization. If so, they must assume duties normally assigned to Level II Instructors. Instructors in this situation may wish to consult Chapter 14, Supervisory and Administrative Duties, for information on maintaining records, legal requirements, privacy information, and records management systems.

> ### Assessment Materials Security
> Regardless of their source (developed in-house or purchased commercially), assessment materials must be kept secure to ensure validity. If students access test materials prior to assessment, the results may reflect the students' memorization abilities rather than their actual understanding and knowledge. Additionally, test materials may be distributed widely via the Internet, potentially destroying the assessment instrument's validity for many organizations.
>
> Once the student has taken the assessment, its security and confidentiality becomes more critical, in many cases being protected by law under the requirements of the Family Educational Rights and Privacy Act (FERPA) and regulations adopted at the state or local level. Beyond the regulatory implications, test result confidentiality is critical in maintaining student dignity and confidence in the organization regarding their best interest and well-being.

Report Writing

Instructors must be able to write clear, concise, and accurate reports based upon witnessed events and the records available to them. They must also ensure that all written reports are accurately completed, properly filed, delivered in a timely fashion, and securely stored. NFPA 1041, *Standard for Fire Service Instructor Professional Qualifications*, assigns report-writing responsibilities to instructors at each classification level. Because these responsibilities may vary between AHJs, all instructors should be familiar with the report types for which each level is responsible. Level I Instructors are generally responsible only for training reports. Examples of types of reports that instructors at each level may be required to complete are shown in **Table 10.1**.

Table 10.1
Report Writing Responsibilities

Instructor Level I	Instructor Level II	Instructor Level III
• Injury reports • Training activity • Lesson outcome (grade)	• Budget requests • Budget administration • Purchase requests • Specifications • Training activities • Facility and equipment usage • Facility and equipment repair requests • Course outcome (grade) • Staff attendance • Postincident critiques • Personnel evaluations	• Division annual budget • Training activity • Accident investigation • Executive summaries • Budget justifications • Disciplinary documentation

The narrative section of a report is often the most difficult part to complete. A report narrative should answer the five questions important to the report: who, what, when, where, and why **(Figure 10.2)**. To keep report narratives simple and concise, organize them based upon the following parts **(Figure 10.3, p. 210)**:

- **Heading** — Contains basic information similar to a memo heading or e-mail. It includes the date, name of the recipient(s), name of the sender or author, and subject of the report. Some organizations have a formal template or format that includes these items.

- **Introduction** — Provides a single-paragraph overview of the report. Includes the report's purpose, time period covered, and name or names of the people involved in writing the report.

- **Body** — Contains all information relating to the report, including the following:
  — Reason for the report
  — Specific and concrete facts based on accurate figures and data; may include visually effective graphs and tables
  — Problems that were discovered
  — Proposed solutions

- **Conclusion/summary** — A final paragraph (or two) that summarizes main points and recommends changes or other actions.

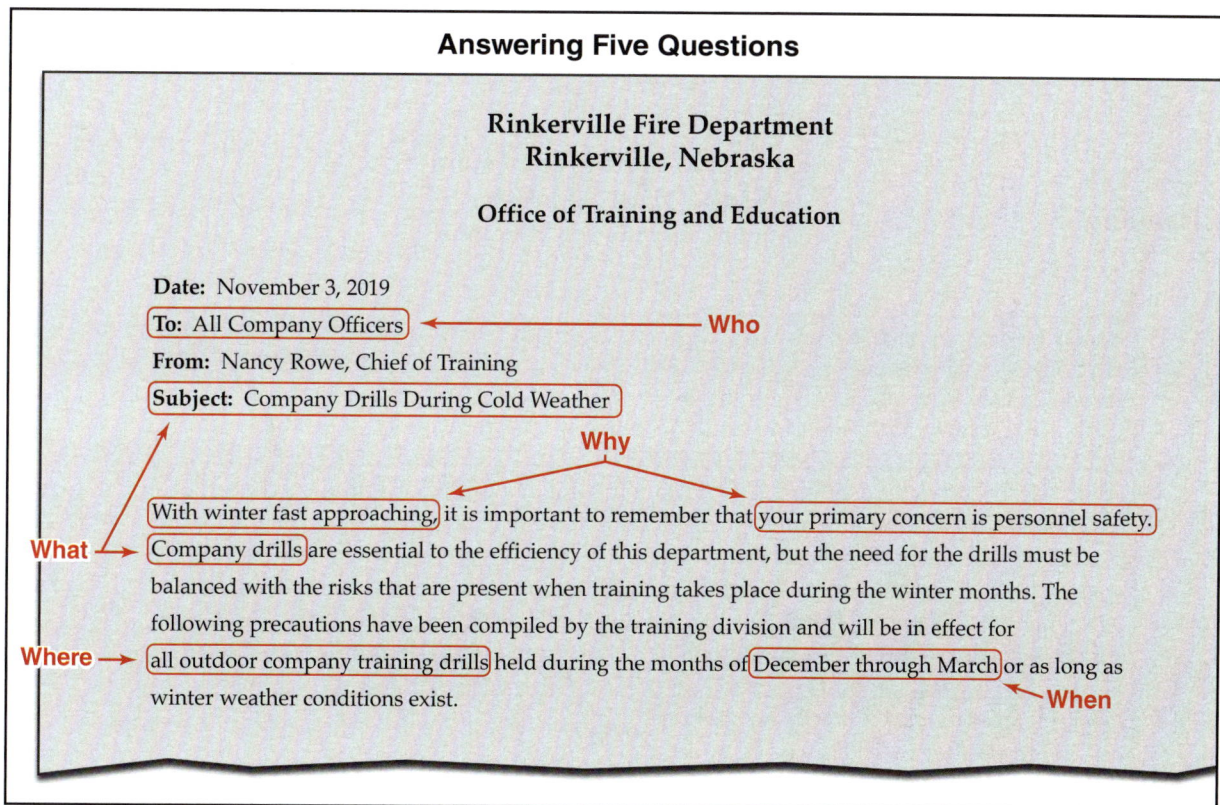

Figure 10.2 A report should answer the five journalistic questions: who, what, when, where, and why.

After Action Reports

The following four questions taken from after action reports can be used to help organize the body of a narrative:

1. What was planned?
2. What actually happened?
3. Why did it happen?
4. What can we do differently next time?

Scheduling Training Sessions

Level I Instructors should be able to schedule individual training sessions for their organizations. Typically, the instructor will have an idea of the resources that can be used to facilitate training. Whatever the needs entail, instructors should become familiar with the AHJ's policies and procedures for scheduling and delivering training. Considerations for scheduling training sessions include the following:

- **Sufficient planning period** — Instructors should schedule long enough in advance to ensure that the session is well planned with a timeline for delivery.
- **Student availability** — Attempt to schedule training sessions at a time when as many students as possible are available.
- **Mandated staffing levels** — Work within AHJ requirements for keeping minimum personnel levels available for emergency response. It may be necessary to schedule the training session more than once in order to accommodate all students and still meet staffing levels.
- **Facilities availability** — Ensure the availability of proper learning and training environments. This could include arranging for use of the apparatus bay or scheduling the use of a local training facility. Scheduling a facility could require planning months in advance.

Parts of a Report

Rinkerville Fire Department
Rinkerville, Nebraska
Office of Training and Education

Heading →

Date: November 3, 2019
To: All Company Officers
From: Nancy Rowe, Chief of Training
Subject: Company Drills During Cold Weather

← *Introduction*

With winter fast approaching, it is important to remember that your primary concern is personnel safety. Company drills are essential to the efficiency of this department, but the need for the drills must be balanced with the risks that are present when training takes place during the winter months. The following precautions have been compiled by the training division and will be in effect for all outdoor company training drills held during the months of December through March or as long as winter weather conditions exist.

Training evolutions will not be held when:
- Wind velocity exceeds 20 mph.
- Snow has accumulated to a depth of 3 inches.
- Horizontal surfaces are ice-coated.
- Temperatures are at or below freezing.
- Lightening, hail, or tornadoes have been predicted within the next 4 hours.

Training may be held:
- During light to medium rain conditions.
- When temperatures are above freezing.
- When snow is less than 3 inches in depth.
- During sunny conditions with temperature above freezing.

Precautions that must be taken for all winter weather company drills:
- Full personal protective clothing must be worn.
- Rehabilitation must be provided at 30-minute intervals.
- Traction devices (chains) must be applied to all apparatus in snow conditions.
- Ground ladders and aerial devices must be clean and free of ice or mud.

← *Body*

The training division believes that these guidelines will provide company officers with a practical approach to winter weather training. Training during winter weather conditions will provide personnel with near-realistic conditions and still reduce the risk inherent during these conditions.

↑ *Conclusion/Summary*

Figure 10.3 The four component parts of a report are: heading, introduction, body, and conclusion/summary.

- **Facility policies and procedures** — If another facility is selected as a location, instructors should learn the policies and procedures for the facility and follow them during training. Part of this discovery may be determining if there are adjunct instructors at the facility who can assist with the training.
- **Equipment availability** — Ensure that any equipment needed is available. This may include arranging for the use of an apparatus or could include the requisition of equipment or instructional technology tools from the department.

Scheduling policies may include the following:

- Which personnel are authorized to schedule training
- Which areas of training are authorized and/or needed
- Which instructors are responsible for which training (Example: policies that may require an EMS instructor rather than a fire instructor to teach CPR classes)
- Who approves any expenses for training

Procedures may include timelines for completion of training or mandated training at certain times of the year. Each AHJ has a way of announcing training to personnel, and instructors should become familiar with procedures. Finally, Level I Instructors should know to whom they make requests within their AHJs for any needed training resources. Skills associated with scheduling a training session using resources provided by the authority having jurisdiction (AHJ) are shown in **Skill Sheet 10-1**.

Chapter Review

Answer the following questions to review the information provided in this chapter.

1. What are the differences between records and reports?
2. What training records are the responsibility of the instructor?
3. What are the four major parts of a written report?
4. What factors must an Instructor I consider when scheduling training?

Discussion Questions

The following questions are intended to generate discussion, expand your understanding of the chapter text, and allow you to think critically about what you have learned. Answers to these questions may vary.

1. What is the purpose of having records and reports in fire and emergency services?
2. What types of challenges have you experienced when completing training records?
3. What types of challenges have you experienced when completing training reports?
4. What types of challenges have you experienced when scheduling training?

Key Terms

Records — Permanent accounts of past events or of actions taken by an individual, unit, or organization.

Reports — Official accounts of an incident, response, or training event, either verbally or in writing.

SKILL SHEET

10-1
Schedule a training session using resources provided by the authority having jurisdiction (AHJ). [NFPA 1041, 4.2.4]

Task Steps

Step 1: Schedule training session with supervisor at AHJ site.

Step 2: Complete necessary scheduling and resource request forms as per AHJ's policies and procedures **(Figure 10.4)**.

Step 3: Ensure training session is well organized and as required by the AHJ's policies and procedures by planning far enough in advance.

Step 4: Confirm minimum enrollment for training session is met.

Step 5: Confirm minimum staffing needs for training session are met.

Step 6: Follow up with applicable facility to confirm booking (training classroom, burn building, etc.).

Step 7: Follow up to confirm all needed equipment will be ready and available for the training session.

Figure 10.4

Chapter 11

Lesson Plan Development

SECTION B INSTRUCTOR II

Chapter Contents

Laws of Learning Applicable to Lesson Plan Development 215
Lesson Plan Creation 216
    Eliminating Bias in Instructional Materials 218
    Learning Objective Development 219
    Conducting Basic Research 223
    Lesson Outline Development 227
    Instructional Method Selection 227
    Lesson Activity Development......... 227

    Ancillary Components 229
    Technology Tools 231
Lesson Plan First Use 232
Lesson Plan Evaluation and Revision 233
    Lesson Plan Evaluation 233
    Lesson Plan Revision 234
Chapter Review 234
Discussion Questions 234
Key Terms 234
Skill Sheets 235

JPRs addressed in this chapter

This chapter provides information that addresses the following job performance requirements of NFPA 1041, *Standard for Fire Service Instructor Professional Qualifications*, 2019 Edition.

5.3.2
5.4.2

Learning Objectives

1. Discuss effects the laws of learning have on developing a lesson plan. [5.3.2]
2. Identify components and steps used to create a lesson plan. [5.3.2]
3. Describe considerations when teaching from a newly developed lesson plan. [5.4.2]
4. Discuss lesson plan evaluation and revision. [5.3.2, 5.4.2]
5. Skill Sheet 11-1: Create a lesson plan. [5.3.2]
6. Skill Sheet 11-2: Write a learning objective. [5.3.2]
7. Skill Sheet 11-3: Create a lesson outline. [5.3.2]
8. Skill Sheet 11-4: Create an information sheet (handout). [5.3.2]
9. Skill Sheet 11-5: Create a skill sheet. [5.3.2]
10. Skill Sheet 11-6: Create a worksheet. [5.3.2]
11. Skill Sheet 11-7: Create a study sheet. [5.3.2]
12. Skill Sheet 11-8: Create an assignment sheet. [5.3.2]
13. Skill Sheet 11-9: Create a lesson plan evaluation plan. [5.3.2]
14. Skill Sheet 11-10: Evaluate and revise a prepared lesson plan. [5.3.2, 5.4.2]
15. Skill Sheet 11-11: Conduct a class using a lesson plan created by the Level II Instructor candidate. [5.4.2]

Chapter 11
Lesson Plan Development

Section B. Instructor II

Section A (Chapters 1 through 10) of this manual emphasized Level I Instructor job performance requirements (JPRs). Section B (Chapters 11 through 15) will address Level II Instructor while Section C (Chapters 16 through 18) will address Level III Instructor. The previous chapters contain prerequisite information for understanding the remaining chapters.

Several short review sections will be included in the upcoming chapters, but with a shift in emphasis toward JPRs for Instructor Levels II or III. Students or instructors who do not feel confident in their Level I Instructor information knowledge should review any portions of Chapters 1-10 that may help them to learn the new material.

According to NFPA 1041, lesson plan development is a basic duty of Level II Instructors. EMS instructors share these responsibilities. This chapter provides instructor candidates with the information needed to develop lesson plans, including the following elements:

- Laws of learning applicable to lesson plan development
- Lesson plan creation
- Lesson plan first use
- Lesson plan evaluation and revision

Laws of Learning Applicable to Lesson Plan Development

Chapter 2, Principles of Learning, introduced Thorndike's Laws of Learning. When developing lesson plans, instructors should reacquaint themselves with the Laws of Learning and how they relate to developing effective lesson plans as follows:

- **Readiness** — Readiness means a person is emotionally, mentally, and physically prepared to learn new knowledge or skills. If students are ready to learn, instructors may choose not to include unrelated "ice-breaker" or "warm-up" exercises in their lesson plans that are not related to the lesson.
- **Exercise** — Adults learn best when they are allowed to exercise skills; the more an act is practiced, the faster and surer the learning becomes. Instructors should make sure that their lesson plans include enough practice time.
- **Effect** — Adult learners need to see the positive effect of what they are learning. When developing a lesson plan, reinforce how the information is useful to the student.
- **Disuse** — Among adult learners it can be assumed that habits and memories used repeatedly are strengthened, and habits not reinforced are weakened through disuse. Training programs should force students to repeat skills at regular intervals.
- **Association** — Instructors can assume that adult learners tend to try to associate new information with information they have already learned. Lesson planners should consider what students know and connect it to the material they need to learn.

- **Recency** — Skills and information practiced or learned most recently are also the best remembered. Whatever students have learned most recently will be most prominent in their minds. If a lesson requires that students recall information from an earlier stage of the course, the lesson plan should include time for review.
- **Primacy** — Primacy is similar to recency. Primacy assumes that the first of a series of learned acts will be remembered better than others. In lesson plan development, especially skills training, the first portion of a process will be best remembered. As a result, there may need to be additional emphasis on the middle and ending of a series of skills steps.
- **Intensity** — The principle of intensity states that if a stimulus (experience) is vivid and real, it will more likely change or have an effect on the behavior (learning). The necessary equipment recommended in a lesson should be the same as equipment used on the job. Similarly, the lesson should require training that is as similar to a real world experience as is safely possible.

Lesson Plan Creation

A lesson may vary in length from a few minutes to several hours, depending on the required learning objectives, number of students, and complexity of topic to be covered, to name a few. When instructors create a lesson plan, determining the learning outcomes is the first step, which leads to the writing of the learning objectives. Using the learning objectives as a starting point, instructors then create the rest of the lesson plan. All of the following will need to be completed as part of lesson plan creation (**Figure 11.1**):

- Analyze the job to be performed and identify all of the expected job skills and tasks that a person would perform when doing the job (see What This Means to You: Identifying Learning Outcomes box).
- Consult the appropriate standards or subject matter experts to assist you in organizing and sequencing tasks.
- Divide the tasks into the basic knowledge and skills steps required to perform them. Use this list to identify the essential knowledge and skills required to perform the task.
- Generate learning objectives from the knowledge and skills needed to complete each task. JPRs in standards can be a source for developing learning objectives.

> ### What This Means to You: Identifying Learning Outcomes
>
> Let's assume that you are responsible for writing a lesson plan about inspecting self-contained breathing apparatus (SCBA) equipment as part of a refresher course on basic skills. Your first step is to decide precisely what your students need to know how to do when they complete the class. This skill is the learning outcome for the course, and can be described using a simple statement about student behavior. In this example, it might be something like: *Students will be able to inspect SCBA and identify any flaws or deficiencies that could lead to the device malfunctioning.*
>
> This learning outcome must now be divided into separate tasks. Think very methodically about the tasks involved in completing an SCBA inspection, and check that these tasks match the appropriate standards. Next, organize the tasks in an appropriate sequence, and divide each task into a series of steps. For each step, determine what information students will need to successfully complete it. For example, if students have to examine the SCBA's faceplate, they should be shown what damage to look for (information) and also what skills steps should be followed to find the damage (demonstration). Next, create a learning objective for each step, such as: *Students will identify the damage on the SCBA's faceplate*.
>
> These early steps of the lesson plan creation process require the most thought. Matching outcomes, objectives, tasks, and steps to the appropriate standard involves abstract thinking, which can be difficult and confusing. However, doing this hard work at the process's beginning makes the entire lesson more effective.

- Develop test items or practical assessments for lesson objectives. Make sure that these match the requirements from the relevant standard and reflect the skills that the student will have to perform on the job.
- Determine the order in which knowledge and skills will be taught. Identify the prerequisite knowledge and skills students need to learn the new material.
- Familiarize yourself with the topic by conducting research. Use sources such as the following:
  — Standard operating guidelines and procedures
  — Current literature
  — Manufacturer instructions
  — Current accepted practices
  — National consensus standards
- Develop the lesson plan using any one of the following formats (see What This Means to You: Creating Lesson Outlines box):
  — Outline with only major points
  — Detailed outline
  — Outline featuring major points supported with explanatory information (considered the best type of format for the fire and emergency services Level I Instructor to use; see Chapter 4, Instructional Materials and Equipment)

Figure 11.1 Following a step-by-step process is an excellent way to create effective lesson plans.

What This Means to You: Creating Lesson Outlines

Instructors may incorrectly confuse creating computer-generated slides with creating a lesson outline. An outline provides the lesson's structure, while computer-generated slides are developed from the outline and are technology tools that help teach the learning objectives — they serve the lesson plan, but are not the plan itself.

If all you create for a lesson is a series of slides, you will probably not give enough consideration to other important issues, such as the learning objectives, additional activities that could aid learning, or safety. Though it seems like the simplest way to create a training lesson, simply creating slides for viewing and then reading the slides does not provide students with the best training opportunity possible.

- Develop lesson activities that reinforce objectives and provide students the opportunity to apply what they have learned. Indicate where during the lesson the activity should occur.
- Identify and develop teaching aids and technology tools to support instruction. List all the appropriate media, props, equipment, materials, facilities, costs, and time that will be required to present the lesson.
- Develop the planning components of the lesson plan. Include title, level of instruction, list of instructional references, and a list of required resources including human, physical, and instructional elements.
- Write ancillary and reference materials:
  — Identify and develop assignments (when required).

- Develop course and instructor evaluation instruments that will be used to determine the effectiveness of the course.
- Write a bibliography of references that you referred to when creating the lesson plan. Use an accepted method of citing sources, such as American Psychological Association (APA) formatting.

• Write the lesson summary. Emphasize important, critical, or key information, especially by reviewing or previewing it. Provide a logical, effective conclusion to the lesson.

Some of these steps are self-evident, such as listing the needed materials, but others require more detail to understand. Skills associated with creating a lesson plan are shown in **Skill Sheet 11-1**. The sections that follow discuss the steps that require more explanation.

> **Review: Lesson Plan Components**
>
> The components of a lesson plan were introduced in Chapter 4, Instructional Materials and Equipment. It is important for instructors to review these components before developing lesson plans. In the most basic format, lesson plans consist of the following components:
>
> - **Job or topic** — Short descriptive title of the information covered
> - **Time frame** — Estimated time it takes to teach the lesson
> - **Level of instruction** — Desired learning level that students will reach by the end of the lesson
> - **Learning objectives** — Descriptions of the minimum acceptable knowledge and behaviors that students must display by the end of the lesson.
> - **Resources/materials needed** — List of all items (including quantity) needed to teach the number of students in the class.
> - **Prerequisites** — List of information, skills, or previous requirements that students must have completed or mastered before starting this lesson.
> - **References** — List of specific references and resources (textbooks and other instructional materials) that will be required reading during the course of the lesson.
> - **Lesson summary** — Restatement or reemphasis of the key points (sometimes referred to as the conclusion) of the lesson.
> - **Assignments** — Readings, practice, research, or other outside-of-class requirements for students.
> - **Lesson outline** — Summary of the information to be taught.
> - **Evaluations** — Type of evaluation instrument the instructor will use to determine whether students have met lesson objectives.
>
> In addition to this information, Chapter 4 also has information on sources for additional resources to be used in the classroom. This information can also be useful in the lesson plan development process.

Eliminating Bias in Instructional Materials

Lesson plans should be designed so that a wide variety of students can learn from the materials. If the materials – lesson plans, learning activities, ancillary components — show bias to one group or another, then only some students will learn well from those materials. The materials may offend other students or otherwise hinder their learning experience. Bias in instructional materials may be based on gender, racial, or cultural stereotypes. It may also be a result of using terminology or regional jargon that students do not understand.

The best way to avoid bias when creating instructional materials is to adhere very closely to the source material that is being taught. If the source material is an NFPA standard, for example, whatever terminology is used in the standard is what should appear in the materials. This same guideline applies if the materials are created from an approved textbook or training manual.

In addition to terminology, instructors creating instructional materials should be very careful in the wording that they use. For example, gender specific names and pronouns should be used in equal amounts throughout the materials. When in doubt, words like *firefighter, fire officer,* or *paramedic* can be used instead of *he* or *she*. Any references to specific cultural backgrounds should be avoided unless they are relevant to the materials. If they are included, stereotypes are strictly prohibited.

Learning Objective Development

As previously discussed in Chapter 4, Instructional Materials and Equipment, learning objectives are specific statements (also referred to as performance objectives, behavioral objectives, or competencies) that describe desired learning results. They describe the knowledge or skills that students should have acquired by the conclusion of a lesson. Written learning objectives represent the learning outcomes and, therefore, learning objectives and learning outcomes are directly related. Learning objectives also provide students with a self-assessment tool. By having a list of the learning objectives with the lesson plan or lesson outline, students are able to determine whether they are accomplishing the requirements of each learning objective as they complete the course. They are also better prepared for success in comprehensive final or summative tests.

Learning objectives focus on the specific, measurable results of instruction. Learning objectives are basic instructional development components and have the following key purposes:

- Provide a foundation for instructional design, and aid in overall course development.
- Help instructors select content and instructional materials and develop an appropriate instructional strategy.
- Provide a basis for measuring and evaluating student learning through appropriate assessment and testing.
- Inform students of expected performance standards and criteria.
- Allow instructors flexibility in teaching and make teaching more efficient.
- Help the instructor facilitate instruction.

Learning objectives can be developed or written in various ways. Although each individual instructor may approach developing learning objectives with a different perspective, all learning objectives should communicate the intended learning outcome and be:

- Clearly stated
- Measurable
- Specific
- Detailed

The learning objective statement may also be student-centered so that the learning objective focuses on the student as the person displaying the observable behavior. The learning objective may be written to include the phrase, *The student will ...*, although the active participation of the student is understood when the phrase is not used.

Effective learning objectives should adhere to the following guidelines:

- Avoid terminology such as understand, know, comprehend, or learn. For example, *The student will understand the principles of fire behavior* does not indicate a measurable result. Learning objectives must contain an action verb and a specific description of the lesson content. A measurable outcome example: *The student will state the principles of fire behavior.*
- Make learning objectives short and focused on a single result. Example: *Match U.S. Department of Transportation (DOT) symbols to their meanings.*
- Make learning objectives specific and objective. Example: *Apply an occlusive dressing to a sucking chest wound.*

In addition to the guidelines already described, instructors should understand how the levels of learning apply to developing learning objectives in the cognitive, psychomotor, and affective areas. Instructors should also understand how internationally-standardized job performance requirements (JPRs) can be used as a model for creating learning objectives. **Skill Sheet 11-2** provides steps for writing a learning objective.

Review: Mager Model of Learning Objective Development

The Mager Model for writing learning objectives was introduced in Chapter 4, Instructional Materials and Equipment. Whenever an instructor is writing learning objectives it is important to review this method. According to Mager, learning objective statements should contain the following three components:

- **Performance (behavior) statement** — Identify what the student is expected to do in clearly observable terms using clear action verbs.
- **Conditions description** — Describe the situation, tools, or materials required for a student to perform a single specific action or behavior.
- **Standards criteria** — State the acceptable level of student performance which may be based upon measurable criteria from an existing standard.

SMART Model for Writing Learning Objectives

One model for writing learning objectives that has become popular is the SMART model. With this model the word SMART is an acronym to remind the user of the key characteristics that an objective should possess.

S - Specific — The objective needs to be specific and observable in nature. An example would be, "The student shall be able to…".

M - Measurable — The objective needs to be measurable and observable. Remember to use action verbs to focus the objective on what is expected and how it will be measured. Clear, action-oriented language will reduce confusion and increase understanding.

A - Attainable — Remember the student's skill level, what you are teaching, and what is a reasonable level of attainment that the student can be expected to achieve during the course.

R - Relevant — The objectives should be relative to the topic being discussed and information being taught.

T - Time Focused — The objective should be finite in time frame. An example would be, "By the end of the class the student shall be able to…".

Cognitive Levels of Learning

The levels of learning in the cognitive domain follow an ordered progression or hierarchy of instructional outcomes. Each level builds upon the previous level and is progressive in its format. Emphasizing one level over another in a learning objective indicates the specific outcome desired from that objective. The levels of learning in the cognitive domain are as follows **(Figure 11.2)**:

- **Remember (Knowledge)** — Students remember, recall, and recognize previously learned facts and theories. They can describe, define, label, list, and match terms and items.
- **Understand (Comprehension)** — Students understand, compare, and contrast information, and estimate future trends. They give examples and explanations, make predictions, and summarize information and ideas.

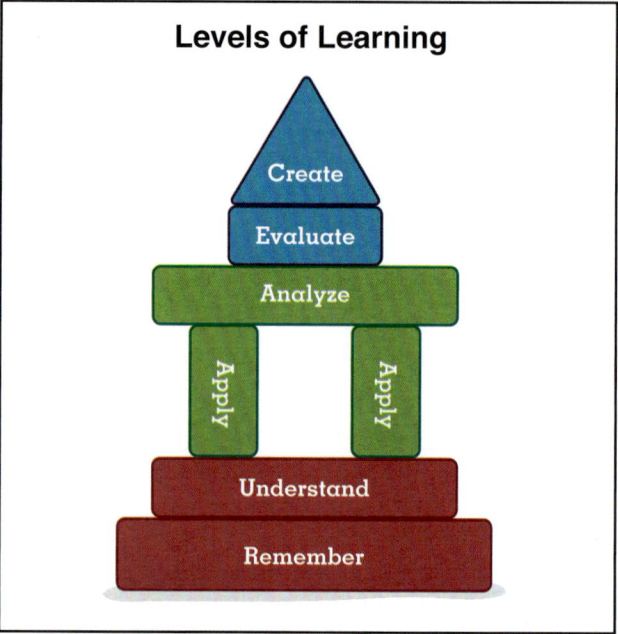

Figure 11.2 The levels of learning begin with the foundational (Remember/Understand) and build toward the more complex (Evaluate/Create).

- **Apply (Application)** — Students apply information, rules, and concepts that they have learned to new situations. They compute, demonstrate abilities, solve problems, modify ideas and actions, and operate equipment.
- **Analyze (Analysis)** — Students divide information into its component parts to understand how the parts relate to one another and to the whole.
- **Evaluate (Evaluation)** — Students judge the value of materials or actions based on defined criteria using elements from all other levels. They compare, conclude, contrast, discriminate, and justify decisions based on standards and criteria.
- **Create (Synthesis)** — Students put parts together to form a new whole. They categorize, create, design, organize, revise, and integrate parts to invent something new.

Because effective learning objectives depend on the use of action verbs, instructors should find or create a list of appropriate verbs to help them develop learning objectives in a variety of learning levels. Some appropriate action verbs for use in the cognitive domain include those in **Table 11.1, p. 222**. Note that some words are applicable to a variety of learning domains and emphasis.

Phrasing of Cognitive Domain Objectives

The following are examples of the way the fire service and the EMS environment might respectively approach the phrasing of learning objectives:

- **Remember (Knowledge)**

  FIRE SERVICE — *Define the combustion elements of the fire tetrahedron.*

  EMS — *List the possible locations to obtain a pulse on a patient.*

- **Understand (Comprehension)**

  FIRE SERVICE — *Explain how the fire tetrahedron combustion elements interact to create or sustain combustion.*

  EMS — *Explain how a pulse is generated.*

- **Apply (Application)**

  FIRE SERVICE — *Demonstrate how fire extinguishment can occur when one of the combustion elements of the fire tetrahedron is eliminated.*

  EMS — *Demonstrate how to obtain a patient's pulse.*

- **Analyze (Analysis)**

  FIRE SERVICE — *Analyze the relationships among the combustion elements of the fire tetrahedron.*

  EMS — *Analyze the pulse rate and strength of the patient and compare findings with patient history and overall presentation.*

- **Evaluate (Evaluation)**

  FIRE SERVICE — *Justify the use of various extinguishing agents to disrupt the combustion elements of the fire tetrahedron.*

  EMS — *Determine if an intervention is necessary based on patient's history, presentation, and assessment.*

- **Create (Synthesis)**

  FIRE SERVICE — *Show how combining various chemical compounds in the absence of oxygen can result in combustion.*

  EMS — *Create a treatment plan based on findings, and periodically reassess to determine if intervention is achieving the desired effect.*

Table 11.1
Useful Words for Expressing Objectives

Remember	cite cite rule count spell define state trace find identify	list name quote read repeat recite recognize review point	indicate write tell recall describe select gather data show
Understand	associate conclude convey meaning of compute deal with describe in own words distinguish	extrapolate give reasons interpret predict differentiate discriminate translate	reformulate restate rewrite summarize tell why discuss
Apply	employ illustrate compare classify administer adopt a plan utilize demonstrate	contrast solve construct put in action apply carry out plot complete	use result perform make use of calculate practice detect use
Analyze	analyze check detect	explain deduct separate	infer designate criticize
Evaluate	assess appraise ascertain assay assize	determine diagnose recommend test grade	judge fix value of evaluate
Create	combine compose design develop devise	fabricate form specify propose prescribe	prepare plan integrate formulate

Source: Permission given to reprint from *Instructional Techniques For Company Officers,* 1983, National Fire Academy, Emmitsburg, Maryland: Government Printing Office, pp. 1–12.

Psychomotor Levels of Learning

Like the cognitive domain, the levels of learning in the psychomotor domain progress through a series of steps, with each one more complex than the previous. Action verbs, such as those listed in **Table 11.2,** are

**Table 11.2
Possible Action Verbs for Psychomotor Objectives**

stand	grasp	watch
sit	crawl	catch
raise	write	open
turn	balance	close
select	stop	run

usually included to define the activity in a psychomotor learning objective. Learning objectives may be written based upon any of the following psychomotor levels:

- **Observation** — Witness the motor activity as the instructor demonstrates it.
- **Imitation** — Imitate this activity in a step-by-step process.
- **Adaptation** — Modify and personalize the motor activity.
- **Performance** — Practice the activity until the steps become habit.
- **Perfection** — Improve the performance until it can be completed without error.

Instructors first demonstrate the skill correctly as the students watch. Students develop a sense for performing the motions and mentally prepare themselves to take action. As they begin to practice, they imitate the instructor's motions. Instructors guide the students, correct their mistakes, and reinforce desired performance.

Affective Domain

Desired changes in attitude (affective domain) must be determined in advance and planned into the lesson. Instructors then emphasize the correct behavior that the student must learn and exhibit. When phrasing learning objectives, use of words in the affective domain such as *confident* or *satisfied* can help to emphasize correct attitude or emotional state: *The student will be confident navigating the driving course in the time required.* Affective changes in attitude likely develop alongside cognitive and psychomotor learning. Learning objectives in this domain are difficult to measure objectively.

Conducting Basic Research

Level II Instructors use research to develop lesson plans and training courses, perform administrative duties such as purchasing equipment, developing budgets and programs, evaluating personnel, and supporting training content **(Figure 11.3)**. The five basic steps of research are:

Step 1: Identify the topic that is to be researched.
Step 2: List all possible topics that are similar to the main topic.
Step 3: List the various types of data that may support the topic, such as internal reports, regional or national reports, legislation, or NFPA standards.
Step 4: List possible sources for each type of information.
Step 5: Read the gathered research and note any relevant information on the topic.

Instructional development policies and procedures should be based on sound research that has been gathered for one of two purposes:

Figure 11.3 Research benefits training programs in many ways.

1. To obtain knowledge about an open-ended question, such as: *How effective is the current training curriculum?*
2. To obtain data that supports or contradicts a proposal, such as: *Increasing the training budget will reduce on-the-job injuries.*

The sections that follow describe information and skills that instructors can use to perform basic research.

Data Collection

Useful data may include the following:

- **Expert opinions** — Statements by credible experts or someone who has analyzed or experienced a similar situation. The Level II Instructor's personal knowledge is an acceptable starting point when looking for this type of information.
- **Trends** — Patterns that can be traced over time and used to forecast the future; trends may be developed from raw data such as hazardous materials incident rates. Raw data must be of the same type for each entry on the timeline.
- **Models** — Frameworks of accepted practices that an organization can adapt to its own needs, such as a model building or fire code.
- **Similar situations** — Interviews with other professionals on how they handled particular situations or courses; sources for this type of research include industry journals, newspaper articles, or peers in professional organizations.
- **Statistics** — Raw data, such as a unit's response times or average staffing for units.
- **Examples** — Representations of processes, situations, items, or models that a researcher can use to illustrate concepts. A good example can be used to create an informal standard or goal that an organization is attempting to attain. An organization can determine how well it has done in reaching or surpassing the standard by using the example situation as a benchmark.
- **Analyses** — Third-party testing organizations evaluate equipment, procedures, or trends, then publish an analytical report of their findings. By reading these reports, researchers have access to objective reviews without having to perform the tests themselves.
- **Recommended/suggested practices** — Suggestions on how to do something, usually based on scientific research or trial-and-error experiences. The Hazen-Williams Water Flow Tests are an example of this type of information because they provide a basis for determining nozzle pressures on hoselines.
- **Industry standards** — Methods for accomplishing some task or function that has formalized standardization by a recognized testing or professional organization. For example, the NFPA provides the industry standards for fire and life safety. Other types of standards are developed by engineering and scientific testing organizations such as the American Society of Mechanical Engineers (ASME) or the member organizations accredited by American National Standards Institute (ANSI).
- **Legal requirements** — Laws, codes, ordinances, and decrees that are legally binding requirements created at most levels of government to statutes that ensure the safety and welfare of a society. In the U.S., the *Code of Federal Regulations (CFR)* applies to many areas of the fire and emergency services while the *Americans with Disabilities Act (ADA)* mandates that public facilities be accessible to people with physical impairments.
- **Sources of supplemental funds** — Most fire and emergency service organizations need to have information on types of grants and loans, where to find them, how to apply for them, and how to administer them.

Information Sources

In general, data may be located from the following sources:

- Internet
- Government agencies
- Libraries **(Figure 11.4)**
- Educational institutions

- Professional organizations
- Testing and standards organizations
- Vendors/manufacturers
- Nonprofit organizations

It is often a challenge to locate reliable and credible data. While credible information will be available to the Level II Instructor, instructors are encouraged to scrutinize all sources closely.

Figure 11.4 Libraries are a print and digital data source.

Sources for Research Material

It is a mistake to believe everything one reads because most information (regardless of the source) is biased in some way. For example:

- Information provided by a government agency may be written in such a way as to justify that agency's actions, such as increasing taxes or proposing a noise abatement ordinance.
- Information obtained from vendors tends to show the positive results of using their products rather than providing an unbiased view, such as one provided by third-party testing agencies.
- While more and more organizations are putting credible information online, instructors are encouraged to scrutinize Internet sources more closely than others. Information that is posted online may not have been reviewed before publication.

It is important to follow accepted guidelines for determining the credibility of research material. Instructors should always attempt to find the primary source for information, for example the first time a study was reported rather than a later article that reproduces only parts of the study. The following terms are used to describe levels of research source material:

- **Primary literature** — Source material written by someone with direct knowledge of the event or topic. The material is original and considered to be the most credible type of literature to use.
- **Secondary literature** — Source material that is based on primary literature but was written or compiled by someone without direct knowledge of the event or topic. This type of literature is also acceptable as long as the primary source material was accurately recorded and evaluated.
- **Tertiary literature** — Literature that has been twice removed from the original source. This literature is usually not a good reference source and may contain errors of translation, interpretation, or context. Tertiary literature is often found on the Internet, such as an article that has been linked to multiple sites.

Researchers should always assess the quality of any source, especially one from the Internet. To do so, they must be sure the source possesses the following characteristics:

- **Credibility/Authority** — Author's credentials, quality controls used in collecting the data, and reputation of the source for providing reliable information
- **Accuracy** — The factuality, recency, and comprehensiveness of the information
- **Reasonableness** — The information's objectivity and freedom from bias
- **Support** — Quality of the supporting evidence and research methodology used to develop the data
- **Review** — Independent verification of the information before it is published

The Internet is a common place to find primary source material, because it contains so much readily available information. However, much of this information may be inaccurate or out of date. Instructors should consider the following aspects of data found on the Internet before assuming it is credible:

- **Authority** — Verify the authority of information found on Internet sites by considering the following:
  - Determine who is producing the document or site by examining the headers, footers, and the site address.
  - Recognize that the domains *.edu*, *.org*, *.gov*, and *.mil* are generally more reliable than *.com* domains.
  - Read the *About This Page* or *About This Company/Organization* link for more information on the sponsoring organization.
  - Look for clues to the author of a document such as links to that person's home page.
  - Look for a date stamp to see when the information was created or last updated.
  - E-mail the creator of a page (when an address is provided), and ask about the author's experience, education, or credentials.
  - Look for evidence that the source for the website would be an authority on the topic.
  - Consider the credibility of an organization that refers or links to another source.
  - Consider an author's biases, especially when consulting *.com* sites.

- **Functionality and User Interface** — Professionally created websites are typically more reliable than those that appear to be poorly constructed or difficult to use. Dead links, pages that will not load, and home pages that are difficult to navigate may also indicate that the site is updated infrequently, which should be a warning sign that the information may be out-of-date or inaccurate.

- **Relinked Information** — Many Internet sites exist to gather information from other sources and report that information to their users. Such sites are not original sources of information. Information may have been changed, quoted inaccurately, taken out of context, or presented only in part. Researchers should follow links from these sites to the original information.

There are many specific agencies or organizations that can provide research information, and many of them have an Internet presence that may be the first place to look.

Level II Instructors who use researched material in a report or proposal must make reference to the original authors in order to give them proper credit. Where applicable, they should also be sure to use the appropriate citation format if one is required.

Reference Material Citations

A Level II Instructor may be required to adhere to a certain citation style, depending on how the material is used. Citations are used in research papers, books, articles, and electronic media to indicate the source of quotes, statistics, and other information that is not original to the writer. Citations can also be used in lesson plans to help instructors answer student questions or refer students to the source material. There are many accepted citation styles. The Level II Instructor should be familiar with the style used by his or her organization and use it consistently.

Two popular citation styles frequently used in the fire and emergency services include the American Psychological Association (APA) and Chicago styles. Instructors should consult their organization's style guide or select the style that seems most appropriate. Whichever style instructors choose, they should use it consistently. For a book with an editor but no author, a reference page citation is as follows:

1. **APA Style** — Goodson, C. and Murnane, L. (Eds.). (2008) *Essentials of fire fighting and fire department operations 5th ed.* Stillwater, OK: Fire Protection Publications.

2. **Chicago Style** — Carl Goodson and Lynne Murnane, eds. *Essentials of Fire Fighting and Fire Department Operations 5th Ed.* Stillwater, OK: Fire Protection Publications, 2008.

For a book with a single author, a reference page citation is as follows:

1. **APA style** — Covey, S. R. (1990). *Principle-centered leadership.* New York: Fire-side.

2. **Chicago Manual of Style** — Covey, Stephen R. *Principle-Centered Leadership.* New York: Fireside, 1990.

Lesson Outline Development

Writing a basic lesson outline is a skill that all Level I Instructors should be able to perform (see Chapter 4, Instructional Materials and Equipment). Level II Instructors are required to go beyond writing outlines to writing lesson plans. For them, the outline is only one step in the process.

After establishing learning objectives, the development of an outline is the next step in the lesson planning process. A good way to create an outline is to organize the learning objectives into the order that the instructor should teach them. For each objective, the instructor determines an instructional method and time frame, and possibly a learning activity or discussion questions.

When developing a course for other instructors, remember that a lesson outline is not enough. A complete lesson plan based on the learning objectives is required, as the other instructors may not be qualified to develop an effective plan. **Skill Sheet 11-3** provides steps for creating a lesson outline.

Instructional Method Selection

When writing lesson plans, instructors should consider which instructional methods best match the learning objectives in the lesson. An illustrated lecture, for example, might be the best method for conveying information while a demonstration is necessary when teaching skills. If the instructor attempts to teach the skill with a lecture format without the use of demonstration, the instruction may not be as effective.

Lessons should also employ a number of different instructional methods in order to reach a variety of learning styles. **Table 11.3** provides a cross-reference depicting how different instructional methods emphasize interactions, skills, and learning styles.

Table 11.3
Instructional Method Characteristics

	Interaction	Visual	Auditory	Kinesthetic	Skills
Illustrated Lecture	✔	✔	✔		
Discussion	✔		✔		
Demonstration		✔	✔	✔	✔
1-on-1	✔		✔	✔	✔
Case Study	✔		✔	✔	✔
Simulation	✔	✔	✔	✔	✔
Mentoring	✔		✔		
Company Drill		✔		✔	✔

Lesson Activity Development

Activities that address specific learning objectives should always be included in lesson plans. The sections that follow describe some learning activities that instructors may want to include in their lesson plans.

NOTE: Skills practice is one variety of lesson activity. When a lesson requires students to successfully complete a skill, the practice is considered a mandatory lesson activity that the instructor must include.

Whole Group Discussion Development

Preparing for classroom discussions and including them in lesson plans requires careful planning. When including a whole group discussion in a lesson plan, the instructor should take the following actions:

- **Define the purpose** — Know in what direction the discussion should go and what students should understand after completing the discussion.

- **Set goals** — Establish the discussion's goals or outcomes. If possible, ensure that goals are measurable so that the discussion can be evaluated for its effectiveness.
- **Establish ground rules** — Establish rules that govern interpersonal relations, because discussions are social as well as educational. Explain in the lesson plan any rules that instructors need to communicate to students. Include how students who wish to speak are recognized, how to be respectful of others, and how long each speaker may control the discussion.
- **Recommend questions** — In the lesson plan, provide opening questions for the instructor to ask.

Small Group Discussion Development

Small group discussions are different from whole group discussions because the instructor is not part of the group. A student is selected to facilitate or lead the discussion in each of the small groups **(Figure 11.5)**. Small group discussions work best under the following conditions:

- The task is structured.
- Students are experienced in working with others.
- The outcome is clearly defined.
- Students have time to prepare for the discussion.

The following actions are important for planning small group discussions:

- **Select a topic** — Select a topic for the discussion. As students become more familiar with small group discussions, more controversial topics can be chosen.
- **Define group goals** — Define goals for the group such as:
  — Generate a new process or policy, or create a task completion plan.
  — Determine an appropriate course of action or a solution to a problem.
  — Negotiate a dispute and come to a consensus.
  — Compete with other groups in a planned activity.

Figure 11.5 Small group discussions allow students to practice leadership skills.

- **Establish time frames** — Set time limits to help students stay on task. Divide class time between research (if necessary), discussion, and summary. Remind the groups a few minutes before they should move into the next work phase.
- **Gather closing summaries** — Reconvene into the larger group when each of the small groups has completed its work. Have a member from each small group present their conclusions and summarize their discussion. The instructor should post the small groups' conclusions and summarize the results, making connections between each group's findings.

Case Study Development

When selecting or developing a case study, the instructor should ensure that the problem is similar to one that students will face in performing their duties. The case study should also be relevant to the lesson plan in which it appears and support the lesson's learning objectives. Whether selecting an existing case study or developing a new case study, instructors should take the following steps:

- **Identify a story or event** — Locate an event that relates to the lesson's learning objectives.
- **Research the story or event** — Locate as much information about the event as possible and ensure that all information is readily available to students. For example:
  — If the event occurred locally, the instructor may choose to interview participants to gain an insight

into their decision-making process. This information can be used following student presentations to compare the actual approach taken to the students' suggestions.

— If the story is hypothetical, the instructor should research similar actual events and use pertinent elements to create the fictional event.

- **Develop an outline** — An outline provides students with all the facts in the story and ties the details and visual aids (if any are used) to the timeline of the story. Key elements in the outline should include those that had/will have a direct effect on the event's outcome.
- **Determine presentation factors** — Determine how to present the case study to students, how to conduct the question-and-answer process, and whether or not any visual aids will be necessary. It is also important to set a time frame; otherwise presentation and discussion can take up much more time than is necessary.
- **Write the case study** — The final step includes the following actions:

  — Write the story.

  — Write questions that students must answer to guide their research and analysis.

  — Prepare any audiovisual aids.

  — Write the instructions for presenting the case study during the lesson (time frame, rules for discussion, explanations of audiovisual aids, etc.).

  Case studies can come from any of the following settings:

- Private business community
- Local, regional, state/provincial, or national training agencies
- Professional training associations
- Other government agencies, such as National Institute for Occupational Safety and Health (NIOSH) investigations and warnings
- Fire and EMS trade journals
- National Fire Fighter Near-Miss Reporting System

Role Play Development

Role-playing is particularly effective for teaching or reinforcing concepts in the affective domain (values, beliefs, and emotions) or basic interpersonal communication. Instructors should take the following steps when creating a role-play activity:

- Ensure that scenarios apply to the course materials, and clearly explain learning objectives.
- Explain the purpose of the activity at the beginning of the role-play, and ensure that students fully understand the scenario, character roles, time frame, and expected results.
- Limit the number of character roles, but involve all students in some part of the activity. Students who do not play a character should be assigned to critique, take notes, or observe, and prepare to discuss their observations or conclusions with the rest of the class.
- Have several role-plays prepared if there will be a large number of students to provide many (if not all) individuals the opportunity to act a role. These role plays may be performed in small groups to save time.

Ancillary Components

Ancillary components are any supplementary materials, such as informational handouts, study guides, skill sheets, work or activity sheets, and assignment sheets. Ancillary components are aids that an instructor can use as desired, and often serve as important and useful instructional adjuncts as well as helpful guides that reinforce learning.

NOTE: Ancillary components may be presented under a variety of names in different lesson plans. They may be referred to under the broader term learning activities, for example.

In addition, some ancillary components list performance steps that students can follow while practicing skills or evolutions. The overall purpose of ancillary materials is to enable students to apply, study, and practice the lesson content. The following sections describe several types of ancillary components, including information sheets (handouts), skills sheets, worksheets, study sheets, and assignment sheets.

Handout

A handout may be something that addresses a topic not covered in the course textbook or other course materials. It might be an outline summarizing key ideas, or a reproduction of computer-generated slides used during a lecture. It might also provide a reference list or include suggestions for further research. It also may be something that would emphasize what is being covered in a different way.

Handouts are usually created for one of the following reasons:

- The information is unavailable to some students because texts or other learning resources are limited.
- To get the information, students would have to find and consult a number of texts, which may be time-consuming.
- The information is not available in any text.

Handouts should be designed to encourage students to learn. Skills associated with developing handouts for use in a class are shown in **Skill Sheet 11-4**.

Skill Sheet

Skill sheets are appropriate for tasks that require both psychomotor skills and cognitive knowledge. They divide a task into **operational steps**, **critical criteria**, and the **key points** or steps for completing each operation. Skill sheets give instructors the information they need to teach the task successfully. Depending upon the skill, students may be able to use an approved skill sheet as a guide for unsupervised practice when an instructor allows it. Unsupervised practice time is important because even during supervised practice time, instructors may not effectively be able to oversee every student's efforts.

Other skills, especially those in which there are hazards that could lead to injury or fatality, should never be practiced unsupervised. Instructors should follow AHJ regulations, training safety standards, and/or departmental procedures to determine which skills may be practiced without supervision.

Skill sheets provide the steps students need to know and practice. Students can use them to practice in groups on their own, coach each other, discuss and think about the activities, and develop higher level (analytical and synthesis) cognitive skills. These self-practice exercises allow students to prepare for performance evaluations where they perform without instructor guidance, exercise thinking skills, and perform at the mastery level for an evaluator. Students can use a skill sheet to prepare for a performance evaluation. Skills associated with developing skill sheets are shown in **Skill Sheet 11-5**.

Worksheet

Worksheets are assignments that students will complete during class. The assignments will vary widely but could include answering open-ended questions, writing a short narrative, or researching a topic.

A worksheet or activity sheet provides students opportunities to apply rules, analyze and evaluate objects and situations, or use multiple skills while completing activities. Instructors should create student worksheets from the content of the lesson plan. Any worksheets that the instructor develops must support the learning objectives and provide activities that enable students to meet those objectives.

The completion of a worksheet may also be a learning objective. In this case, the knowledge and skills learned in class are applied to the particular task of completing the worksheet. Worksheets can be used to generate discussions on a topic and generally contain optional activities. Worksheets typically do not need to be scored or graded.

Worksheets that require students to exercise abilities in the affective domain may support more than one learning objective. Recall that the affective domain has students change or adjust, develop, practice, and adapt attitudes, values, and beliefs. Skills associated with developing worksheets for use in a class are shown in **Skill Sheet 11-6**.

Study Sheet

A study sheet explains the specific areas students will need to study before an exam or certification test. Instructors may want to distribute study sheets for students to use during instruction or for them to use as self-study aides. It is also helpful to include a practice test with the study sheet, which enables the instructor to measure and provide feedback on how well students understood the material. Skills associated with developing study sheets for use in a class are shown in **Skill Sheet 11-7.**

Assignment Sheet

An assignment sheet contains information about a specific activity or project that the student is expected to complete without supervision. The activity may occur within the class period or outside of class. Generally, an assignment sheet contains the three components of the Mager Model (performance, conditions, and criteria) and some of the same material listed for the lesson plan. Assignment sheets differ from worksheets in that the assignment is required and will be graded. Skills associated with developing assignment sheets are shown in **Skill Sheet 11-8**.

Technology Tools

In addition to creating lesson plans, the Level II Instructor may also be required to create or select technology tools to use with lesson plans during training evolutions. The instructor may have to do research to find illustrations, photos, video recordings, or audio recordings that enhance a lesson. Instructors rarely create these items by themselves, but sometimes they are able to use photographs that they took during training evolutions or at the scene of an incident. Usually, instructors will be responsible for creating computer-generated slides to accompany lectures.

When selecting technology tools, instructors should remember that good audiovisual teaching aids have the following purposes:

- Show abstract concepts through the use of charts or diagrams.
- Aid or help clarify memory through the use of eye-catching or colorful images.
- Illustrate real environments using plans, maps, photographs, or videos.
- Reinforce key points through the use of quotes, tables, or figures.
- Tie complex ideas together through diagrams, outlines, and headings.
- Compare information through the use of charts and graphs.
- Introduce the lesson through the use of a title slide or image **(Figure 11.6)**.
- Illustrate a process through the use of artwork, photographs, or cutaway models.

New instructors often have difficulty making effective computer-generated slides. For example, they have a tendency to create slides that contain too many words or overuse animation effects. To create effective computer-generated slides, instructors should apply some generally accepted presentation guidelines, such as the following **(Figure 11.7)**:

Figure 11.6 A title slide can reinforce important facts.

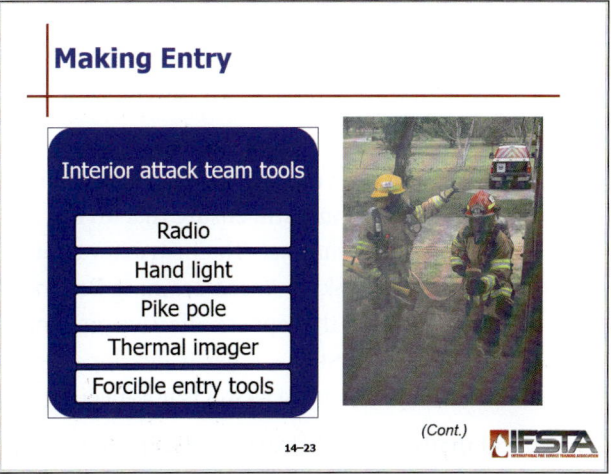
Figure 11.7 Computer-generated slides can be very effective teaching aids.

- Keep visual aids simple and easy to understand.
- Take the time to ensure that all slides look professional and convey their messages effectively.
- Use typeface or fonts that are consistent, easy to read, and large enough to read at a distance. Font size depends on the size of the presentation room and projection screen. One guideline is to never use less than 24-point type.
- Make text concise, emphasizing phrases and lists when possible. The text helps the learner focus on the presentation's key points. The oral presentation expands on these phrases.
- Create one heading for each slide or image. Use subheadings or illustrations when appropriate.
- Keep the backgrounds simple so they do not conflict with the text or graphics.
- Use a background color that contrasts with the color of the text or graphic, but does not clash with it. Never use colors that will distract the learner's attention.
- Use transition effects sparingly because they can overpower the message that a slide or image is attempting to convey.
- Use graphs, charts, photographs, and clipart to create interest.
- Use parallel structure on each slide or image; starting phrases with nouns and bullets with verbs make points easier to link together.
- Use one style of transition effect for the major topics and a different one for the subtopics.
- Provide handouts of some slides or images. Handouts can be particularly helpful when presenting complex or detailed concepts. The learner can make appropriate notes on the handouts; however, handouts can also distract learners by taking their attention away from the presentation.

> **What This Means to You**
>
> Simple computer-generated slides are the most effective. You may find it tempting to create fancy, involved, and visually stimulating transitions and animations. But this can be distracting, causing students to focus on the visual effects and not the information. The sound effects that are included in most presentation software are also inappropriate for training sessions and should not be used unless they are necessary to convey the message.
>
> You might feel as though you aren't working hard enough if your slides are straightforward and avoid sound and visual effects. But simpler slides lead to better instruction. Make sure your slides meet the guidelines included in this section, but don't waste time building elaborate transitions, sound effects, and animations.

Lesson Plan First Use

Once a lesson plan has been created, it can be put into use in the learning environment. Level II Instructors should be able to teach from a newly developed lesson plan while at the same time evaluating their work in a real world situation.

Teaching from a newly created lesson plan is no different than teaching from a previously prepared lesson plan. The instructor must still ensure that all learning objectives have been achieved and that the lesson is delivered in a safe, effective manner. The instructor should still use the skills included in Chapter 6, Classroom Instruction, and Chapter 8, Skills-Based Training Beyond the Classroom. The only difference is that there may be items in a new lesson plan that appear effective on paper but do not meet expectations in practice. Experienced instructors should be the first to use newly developed lesson plans because they are more likely to make smooth transitions when one portion of a lesson is not meeting expectations.

Finally, Level II Instructors using a newly developed lesson plan for the first time should closely examine the instruction to ensure that the content and delivery was sound. Instructors should examine areas such as:

- Instructional content was as effective as possible.
- Teaching approaches and materials were unbiased and well-informed.
- Technology tools performed smoothly and safely and all audiovisual aids relayed content to students according to expectations.
- The instructor eliminated as many distractions and potential hazards as possible.
- The instructor ensured student engagement by choosing instructional methodologies that increased student discussion and buy-in.
- The instructor successfully facilitated all student or group discussions and kept the class focused on the subject matter.
- The instructor used proper transitions to help students connect ideas and content.

Such detailed examination provides feedback for lesson plan evaluations and can be early indicators of lesson plan areas that require revision. Noting problematic approaches or content can also help hone the Level II Instructor's teaching styles and techniques.

Lesson Plan Evaluation and Revision

Course evaluations evaluate the knowledge and skills of the students, the performance of the instructor, and the effectiveness of the lesson plan. Administrators, employers, course or curriculum developers, and instructors can judge whether the course or curriculum accomplished its objectives by assessing the results of instruction. After this evaluation, it may be necessary to revise the lesson plans to correct any deficiencies that are discovered. A thorough course evaluation is based on input from students, instructors, and administrators. Skills associated with creating a lesson plan evaluation plan are shown in **Skill Sheet 11-9**. Skills associated with evaluating and revising a prepared lesson plan are shown in **Skill Sheet 11-10**. Skills associated with conducting a class using a lesson plan created by the Level II Instructor are shown in **Skill Sheet 11-11**.

Lesson Plan Evaluation

Review of student and instructor evaluations can help course planners determine how effective a lesson plan has been. The following steps assist the instructor in evaluating a lesson plan:

Step 1: Review student test/course scores — Analyze test results. Situations:
- When the majority of students met the evaluation criteria, lesson plan alterations may not be required.
- When the majority of students did not perform satisfactorily, a review of the instructor and course evaluations may help determine the cause.
- To determine the cause of unsatisfactory performances, review the students' training records and interview those who did not meet the criteria.

Step 2: Review the instructor and course evaluations — Look for consistency in comments concerning the presentation style. Considerations:
- Determine whether environmental factors such as lighting, noise, or temperature could have created a barrier to learning.
- Determine whether any other factors could have affected the teaching or learning environment.
- Determine whether training aids and devices were appropriate to the topic.

Step 3: Review the lesson plan — Determine whether the learning objectives were clear, concise, and attainable. Factors:
- Time frame was sufficient to cover the required material in sufficient detail.
- Testing criteria were appropriate to the topic and teaching style.
- Testing criteria and learning objectives were properly explained to the students.
- Support materials and personnel were adequate to meet the lesson plan requirements.
- The instructor was familiar with the topic and lesson plan.
- Unforeseen elements (such as weather, equipment malfunction, or site conditions) caused a problem in the presentation.

If the lesson plan has significant flaws, the instructor or the training division has two possible options. The lesson plan may be revised or instructors may seek prepared lesson plans that better help students meet the desired learning objectives.

Lesson Plan Revision

Revising a lesson plan generally follows the steps shown in the Lesson Plan Creation section of this chapter. The evaluation should provide a definite list of revisions to be made. The instructor should make the revisions, and then compare the revised lesson plan to the revision list and course curriculum requirements. It may be necessary to further refine the changes to meet all requirements. The revision of fire or EMS lesson plans may require approval from the AHJ.

The revised lesson plan should be reviewed by other instructors, the training division or AHJ administration, or other experts. When possible, it should be presented to the original group of students to determine whether the revisions were effective. Ultimately, the revised lesson plan will be evaluated based upon student test scores and course evaluations after the revised plan has been taught for the first time.

Chapter Review

Answer the following questions to review the information provided in this chapter.

1. What effect do the laws of learning have on the development of a lesson plan?
2. When creating a lesson plan, what steps should a Level II Instructor take and what components should be included?
3. What might a Level II Instructor anticipate when teaching a newly developed lesson plan for the first time?
4. How does a Level II Instructor evaluate and revise a lesson plan?

Discussion Questions

The following questions are intended to generate discussion, expand your understanding of the chapter text, and allow you to think critically about what you have learned. Answers to these questions may vary.

1. What are Thorndike's Laws of Learning and how do they relate to developing effective lesson plans?
2. Why is it important for a Level II Instructor to develop a lesson plan?
3. What experiences have you had as an instructor with lesson plan first use?
4. What are some reasons you have had to evaluate and revise a lesson plan?

Key Terms

Ancillary Components — Supplemental written materials that help students meet the learning objectives; may include information sheets, study guides, skills sheets, work or activity sheets, and assignment sheets.

Critical Criteria — Step or steps on a practical skills test that must be completed accurately in order for the student to pass the test.

Key Point — Important cognitive information on a skill sheet that students need to know in order to perform a task or operational step; generally appears on right-hand side of a skills sheet.

Operational Step — The smallest aspect of performing a task; to complete the task, students perform a series of operational steps in sequential order.

11-1
Create a lesson plan. [NFPA 1041, 5.3.2]

Task Steps

Step 1: If your AHJ uses a template for lesson plans, follow the template provided.

Step 2: Determine the topic of the lesson.

Step 3: List prerequisites that students must have completed or mastered before starting this lesson.

Step 4: Estimate the time and order in which knowledge and skills will be taught **(Figure 11.8)**.

Step 5: List specific resources, textbooks, and other reference materials used to help enhance the lesson.

Step 6: Determine the level of instruction students will reach by the end of the lesson.

Step 7: Determine learning objectives.

Step 8: Create a lesson outline with the information to be taught. Determine the format you will use for the outline:
  a. Outline with only major points
  b. Detailed outline
  c. Outline featuring major points supported with explanatory information

Step 9: Identify lesson assignments/activities that reinforce objectives and provide students the opportunity to apply what they have learned.

Step 10: Write the lesson summary/conclusion to emphasize important, critical, or key information. Allow time for a review of today's lesson and a preview of the next lesson.

Step 11: Determine type of evaluation instruments to use to determine whether students have met the lesson objectives.

Figure 11.8

11-2
Write a learning objective. [NFPA 1041, 5.3.2]

Task Steps

Step 1: Review the Mager Model of learning objective development and the SMART Model for writing learning objectives **(Figure 11.9)**.

Step 2: Identify the specific performance (behavior) the student should be able to accomplish. Use a clear action verb that reflects the required level of learning.

Step 3: Identify the conditions under which the student will perform the specific action or behavior.

Step 4: Identify the standards criteria indicating acceptable level of student performance.

Step 5: Write the learning objective, ensuring that the behavior, condition, and measurable criteria are clearly identified, or implied.

Figure 11.9

11-3
Create a lesson outline. [NFPA 1041, 5.3.2]

Task Steps

NOTE: Figure 11.10 provides an example of a lesson outline.

Step 1: Identify lesson topic.
Step 2: Identify intended learning objective(s) of the lesson.
Step 3: Determine the format you will use for the outline:

a. Outline with only major points
b. Detailed outline
c. Outline featuring major points supported with explanatory information

Step 4: Review and adjust outline as needed.

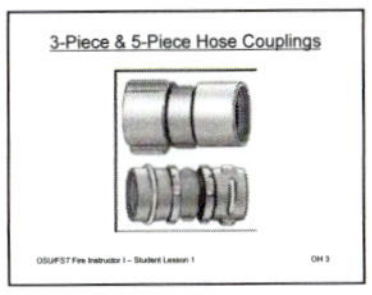

Figure 11.10

Chapter 11 • Lesson Plan Development 237

SKILL SHEETS

11-4 Create an information sheet (handout). [NFPA 1041, 5.3.2]

Task Steps

NOTE: Figure 11.11 provides an example of an information sheet.

Step 1: Create a title that indicates the subject area and relates the title to the lesson.

Step 2: Introduce the information with a brief description that explains its importance, relating it to the course textbook or part of the lesson.

Step 3: Present the information in a form that creates interest, motivating the student to read, study, and learn.

Step 4: Present the information so that it is easy to read and follow. Include appropriate charts, tables, or illustrations, and label them for easy referral.

Essentials of Community Risk Reduction
Preventing Drowning Tip Sheet

Protect Children Inside the Home
- Never leave a young child unattended near water.
- Empty all tubs, buckets, containers and kiddie pools immediately after use. Store them upside down so they don't collect water.
- Close toilet lids and use toilet seat locks to prevent drowning.
- Keep doors to bathrooms and laundry rooms closed.
- Once bath time is over, immediately drain the tub.

Protect Children in and Around Pools
- Practice active supervision when children are in or around the water, even if lifeguards are present. Don't be distracted by electronic devices, reading, or conversation.
- Keep young children within arm's reach of an adult.
- Don't trust your child's safety to another child.
- Know how and when to call 9-1-1 or the local emergency number.
- If you own a home pool or hot tub, have appropriate equipment, such as reaching or throwing equipment, a cell phone, life jackets and a first aid kit.
- Enroll in home pool safety, water safety, first aid and CPR/AED courses to learn how to prevent and respond to emergencies.
- Install four-sided isolation fencing around your home pool or hot tub. It should be at least five feet high, equipped with self-closing and self-latching gates that completely surround the pool and prevents direct access from the house and yard.
- If you have an above-ground or inflatable pool, remove access ladders and secure the safety cover whenever the pool is not in use.

Protect Teens and Young Adults
- Make sure everyone wears a U.S. Coast Guard-approved life jacket when boating.
- Never consume alcohol when swimming or operating a boat.
- Don't underestimate the power of water; even rivers and lakes can have undertows.
- Always have a first aid kit and emergency contacts handy.
- Don't dive in unfamiliar areas.

Figure 11.11

11-5
Create a skill sheet. [NFPA 1041, 5.3.2]

Task Steps

NOTE: Figure 11.12 provides an example of skill sheet steps.

Step 1: Identify the skill (title). (What the student will be assessed on)

Example: *Lesson Plan Development*

Step 2: Select the learning objective.

Example: *Create a lesson plan.*

Step 3: Briefly describe the information and resources needed to complete the task/skill. A resources and/or information list could include:

  a. AHJ SOPs
  b. Information sheets
  c. Equipment
  d. Conditions
  e. Warnings
  f. Safety factors
  g. Time frames

Step 4: Develop a list of the steps to complete the task/skill. This list is a sequential procedure for completing the task, ending with the final objective outcome.

Step 5: Develop a rating system (using a scale or rubric with points) that includes a description of what will be considered pass or fail.

  a. Provide space for student and evaluator(s) comments/scores/signatures/date.
  b. Include evaluation information for record-keeping requirements (student name, ID number, date, department, etc.)

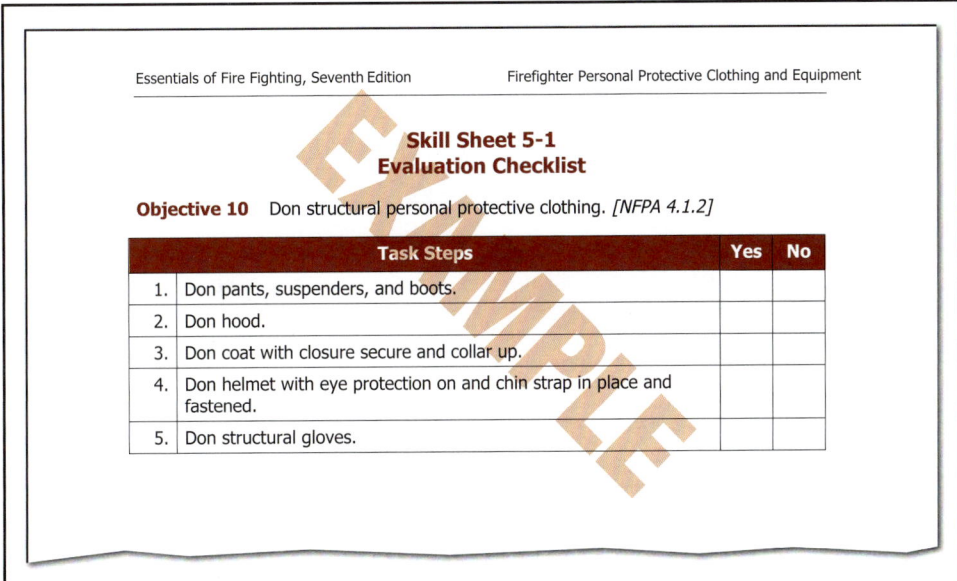

Figure 11.12

11-6
Create a worksheet. [NFPA 1041, 5.3.2]

Task Steps

NOTE: Figure 11.13 provides an example of a worksheet.

Step 1: Create a title that reflects the subject or topic.

Step 2: List all the materials and resources that students need in order to complete the activity.

  a. List titles and page numbers of books, journals, or other reference material.

  b. Provide enough information so that students can locate resources quickly and easily.

Step 3: Write a brief introduction that generates interest and motivates students to complete the activity.

  a. Explain how the skill or activity relates to the topic, the learning objectives, and students' job performance.

  b. Show how the activity will help students to master the relevant skill.

Step 4: Provide clear directions that explain how to complete the worksheet.

Step 5: Provide answers or solutions on a separate page.

Figure 11.13

240 Chapter 11 • Lesson Plan Development

11-7
Create a study sheet. [NFPA 1041, 5.3.2]

Task Steps

NOTE: **Figure 11.14 provides a sample study sheet.**

Step 1: Create a title that reflects the subject or topic and is related to the lesson.

Step 2: List all the materials and resources that students need to complete the study sheet.

    a. List titles and page numbers of books, journals, or other reference material.

    b. Provide enough information so that students can locate resources quickly and easily.

Step 3: Write a brief introduction that generates interest and motivates students to complete the study sheet.

Step 4: Present the information in a format that enables students to use and learn the material.

    a. Design study questions to make students think about all aspects of the topic.

    b. Include enough questions to thoroughly address the material.

Step 5: Create an answer sheet (if applicable) on a separate sheet of paper.

Figure 11.14

11-8
Create an assignment sheet. [NFPA 1041, 5.3.2]

Task Steps

Step 1: Identify/determine the topic and what students are to do.

Step 2: Create a title that reflects the subject or topic and is related to the lesson.

Step 3: List all the materials and resources that students need to complete the assignment.

    a. List titles and page numbers of books, journals, or other reference material.

    b. Provide enough information so that students can locate resources quickly and easily.

Step 4: Write a brief introduction that generates interest and motivates students to complete the assignment.

Step 5: Present the information in a format that enables students to use and learn the material.

Step 6: Include the scoring and grading criteria for the assignment.

Step 7: Describe the form that the final product (essay, table, graphic, etc.) should be in. Make sure that all students understand that the following must be addressed or included:

    a. Format (handwritten, typed, model, etc.)

    b. Due date

    c. Student information (name, ID number, department, etc.) for record keeping requirements

11-9
Create a lesson plan evaluation plan. [NFPA 1041, 5.3.2]

Task Steps

Step 1: Identify lesson plan to be evaluated.

Considerations:

a. Is this a topic that can/should be continued in future classes?

b. Is the topic meaningful to the students?

c. Does the topic match the course material?

d. Does the topic match the interest, age, language, learning styles of the students?

e. Are the learning objectives observable/measurable?

f. Does the introduction get the students' attention and prepare them for the lesson to follow?

g. How are the students being taught? Can the students play a larger role in their learning?

h. Are the tasks/skills relevant to the material?

i. Are students given time to practice the objective?

j. Are the evaluations/assessments meaningful?

k. Are there barriers to learning that can be modified or controlled? Analyze items such as:

1. Timeframe was sufficient.
2. Testing criteria was appropriate.
3. Test directions were properly explained.
4. Support materials and personnel were adequate to meet the lesson plan requirements.
5. Instructor was familiar with topic and lesson plan.
6. Planning for unforseen elements (weather, equipment malfunction, site conditions, etc.) was conducted.

l. What follow-up should be addressed?

Step 2: Review instructor and course evaluations.

Step 3: Develop evaluation plan form to include:

a. Name of lesson plan

b. Name of instructor

c. Date/time lesson taught

d. Name of course

e. Length of class/course

f. Name of department where course was taught

g. Content of the lesson presented

h. List of the lesson's strong points

i. List of areas of improvement for the lesson

j. List of classroom management changes/modifications that can be made

k. List of assessments/evaluations and any changes/modifications that can be made

l. List of resources/materials used

m. List of expectations to be met by the instructor and students

11-10
Evaluate and revise a prepared lesson plan.
[NFPA 1041, 5.3.2, 5.4.2]

Task Steps

Step 1: Revise lesson plan to ensure modified or adapted instructional materials (props, equipment) are appropriate to the learning levels and needs of given audience. The revision should account for:

    a. Attention span

    b. Learning level(s)

    c. Material appeal

    d. Visual aids

Step 2: Ensure modified or adapted instructional materials include the proper amount of information to meet the learning objectives for the given audience.

Step 3: Ensure modified or adapted lesson appropriately explains the actions learners are to take.

Step 4: Ensure modified or adapted materials are appropriate length or duration for lesson taught to given audience.

Step 5: Explain why modified or adapted materials will meet needs of audience.

Step 6: Ensure modified or adapted materials are properly completed and organized.

Step 7: Create an evaluation plan for the revised lesson plan.

11-11
Conduct a class using a lesson plan created by the Level II Instructor candidate.
[NFPA 1041, 5.4.2]

NOTE: Prior to conducting a class, the instructor should:
- **a.** Arrive at least thirty minutes prior to class start time.
- **b.** Ensure the classroom/training area is appropriate.
- **c.** Confirm fire and emergency procedures and other facility-related policies and procedures (parking, food and beverage use, etc.)
- **d.** Confirm audiovisual/technology tools, HVAC, lighting, and electrical outlets are in working order and ready to use.

Task Steps

Step 1: Select two or more teaching methods (lecture, demonstration, discussion, etc**.**) for subject being taught.

Step 2: Teach a class consistent with Level I Instructor criteria **(Figure 11.15)**:
- **a.** State learning objective(s) of the lesson.
- **b.** Use an introduction that will catch the students' interest.
- **c.** Provide a brief overview of the lesson's content.
- **d.** State student expectations.
- **e.** Define or explain unfamiliar terminology.
- **f.** Present the lesson.
- **g.** Provide examples to clarify and emphasize key points.
- **h.** Use appropriate transitions between key ideas.
- **i.** Incorporate technology tools or support materials.
- **j.** Check for student understanding throughout the lesson presentation.
- **k.** Answer questions throughout lesson presentation.
- **l.** Provide closure for the lesson. Restate what you expect the students to gain from the lesson/materials.
- **m.** Ask for and answer student questions.
- **n.** Provide assessments/tests, if applicable.
- **o.** Provide preview for next lesson topic.
- **p.** Complete any follow-up items necessary after lesson presentation is complete.
- **q.** Complete all course records/reports necessary per AHJ policies and procedures.

Figure 11.15

Chapter 12: Training Evolution Supervision

SECTION B — INSTRUCTOR II

Chapter Contents

The Safety Challenge 249	**Environmental Issues at**
Organizational and Administrative Support 249	**Training Evolutions** 257
Unsafe Behavior 251	Water 257
Hazard and Risk Analysis 253	Atmosphere 258
Using IMS to Supervise Training 254	Soil 258
Incident Management System Duties and Functions 254	**Accident Investigation** 259
Training Plan or Incident Action Plans (IAP) 254	**Chapter Review** 259
Training Evolution Evaluation 255	**Discussion Questions** 259
	Key Terms 259
	Skill Sheet 261

JPRs addressed in this chapter

This chapter provides information that addresses the following job performance requirements of NFPA 1041, *Standard for Fire Service Instructor Professional Qualifications*, 2019 Edition.

5.4.3

Learning Objectives

1. Describe the safety challenges a Level II Instructor faces during a training evolution. [5.4.3]
2. Summarize the use of the Incident Management System (IMS) model to supervise training. [5.4.3]
3. Discuss environmental regulations that affect training evolutions. [5.4.3]
4. Discuss the roles and responsibilities of the Level II Instructor during an accident investigation. [5.4.3]
5. Skill Sheet 12-1: Supervise multiple instructors and candidates during increased hazard exposure training. [5.4.3]

Chapter 12
Training Evolution Supervision

Chapter 8, Skills-Based Training Beyond the Classroom, introduced the basics of smaller-scale training situations. Level II Instructors should be able to supervise these as part of their job. This chapter expands on the information in Chapter 8 to include instructors' supervisory duties during large-scale training evolutions. Specifically, this chapter expounds on the following topics:

- The safety challenge
- Using the Incident Management System (IMS) to supervise training
- Environmental issues at training evolutions
- Accident investigation

The Safety Challenge

Level II Instructors are encouraged to provide realistic training situations that resemble actual emergencies while still maximizing safety. The sections that follow describe various aspects of the challenge of providing safe training evolutions for students and instructors.

> **Sources for Casualty and Injury Statistics**
> From time to time, Level II Instructors may need to obtain data on injuries and/or casualties incurred during training. Consult the following organizations or agencies for statistical information:
> - National Fire Protection Association (NFPA)
> - U.S. Fire Administration (USFA)
> - International Association of Fire Chiefs (IAFC)
> - International Association of Fire Fighters (IAFF)
> - National Volunteer Fire Council (NVFC)
> - Other U.S. and Canadian government agencies

Organizational and Administrative Support

In addition to instructors incorporating safety measures into the training curriculum, changes in organizational policy can reduce training accidents. Some USFA recommendations include:

- Follow established guidelines and accepted organizational procedures as well as training and safety standards.
- Conduct live-burn evolutions in a variety of structures to provide realistic fire fighting experiences.
- Train firefighters and emergency responders to recognize the visual and physical clues of impending danger (such as changes in smoke conditions) and anticipate fire behavior in a variety of building types (**Figure 12.1, p. 250**).

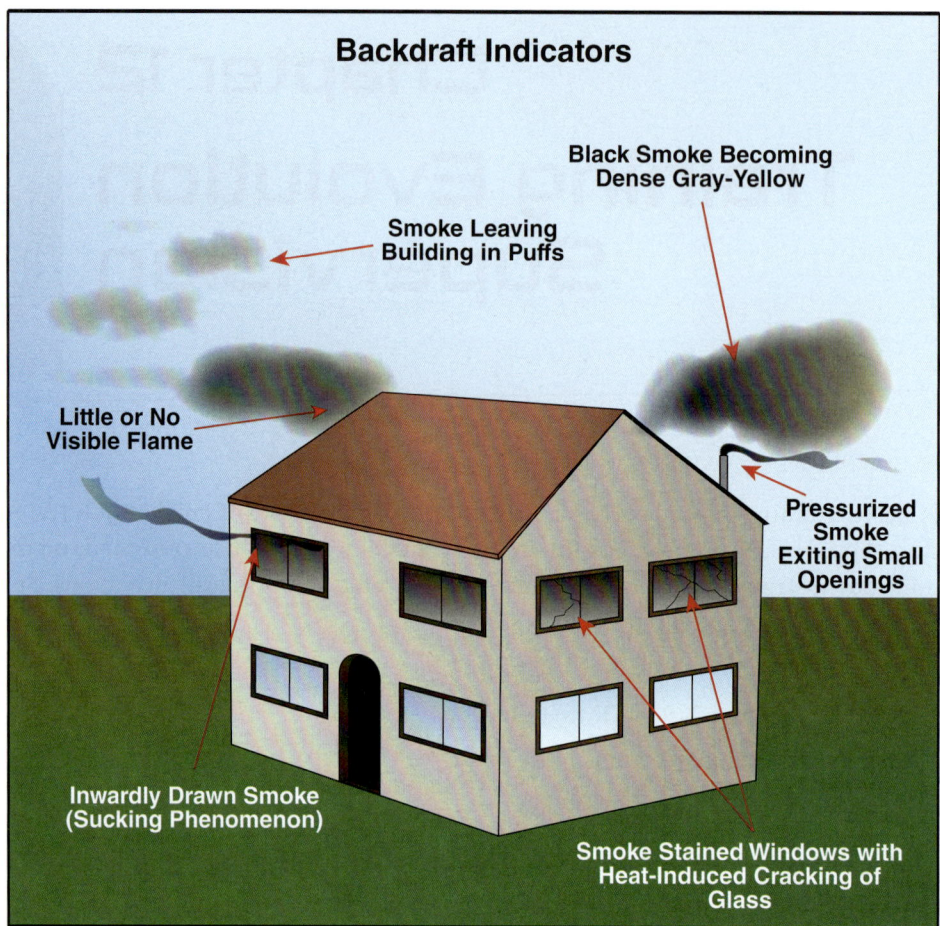

Figure 12.1 Firefighters must be able to recognize the visual indicators of backdraft conditions.

- To reduce the risks to personnel, all fire and emergency services organizations regulated by NFPA standards must have a risk-management plan. After implementing the plan, instructors must monitor its effectiveness. Risk-management plans accomplish the following objectives:
  — Identify risks.
  — Evaluate the potential for injury or damage, based on the frequency and severity of risk.
  — Establish appropriate controls to minimize or eliminate the risk.

The **risk-management plan** includes all job-related activities in which fire and emergency services personnel normally participate, including emergency, nonemergency, training, and support activities. NFPA 1500™, *Standard on Fire Department Occupational Safety, Health and Wellness Program*, provides detailed guidelines for developing a risk-management plan. IFSTA's **Occupational Safety, Health, and Wellness** manual can be an excellent resource for establishing safety and health programs.

An organization's administration plays a role in supporting and enforcing safety, fitness, health, and wellness programs in the organization's operations. As part of the leadership of the administration, the Level II Instructor's responsibility is to perform the following actions:

- Provide adequate personal protective equipment (PPE).
- Ensure all proper equipment is provided.
- Ensure that all apparatus and equipment are maintained.
- Ensure that all safety equipment is properly installed and operational.

Figures 12.2 a-c Fitness, health, and wellness programs are an important aspect of firefighter safety and should be supported by organizations.

- Provide policies and procedures for the safe use of the apparatus and equipment.
- Address the fitness, health, and wellness of personnel by providing (**Figures 12.2 a-c**):
  — Employee assistance programs
  — Job-related physical fitness testing
  — Annual medical evaluations and periodic examinations
  — Health- and wellness-related information and training

> **Review: NFPA 1403, *Standard on Live Fire Training Evolutions***
>
> An essential element for planning live-fire training is NFPA 1403. Instructors must be familiar with the requirements of this standard for all live-fire training in purpose-built burn buildings, acquired structures, and burn props. Safety requirements must be enforced by the instructor in charge of any training evolution, the designated safety officer, and the organization's administration. For more information on NFPA 1403, refer to Chapter 8, Skills-Based Training Beyond the Classroom.

Unsafe Behavior

Injuries and fatalities during training usually result from unsafe acts by persons who are unaware of potential hazards, who are ignorant of the safety policies, or who fail to follow safety procedures. Casualties may be caused by conditions in the physical environment that were not examined or considered as potential hazards. Almost all accidents are predictable and preventable.

Normalization of deviance can compromise safety over time **(Figure 12.3)**. When deviating from SOPs becomes the normal behavior, the value of the SOPs becomes marginalized.

Human factors generally lead to unsafe behavior in fire and emergency services training. Before allowing participation in a training scenario, an instructor should determine whether any of the following factors apply to a student and take the appropriate action:

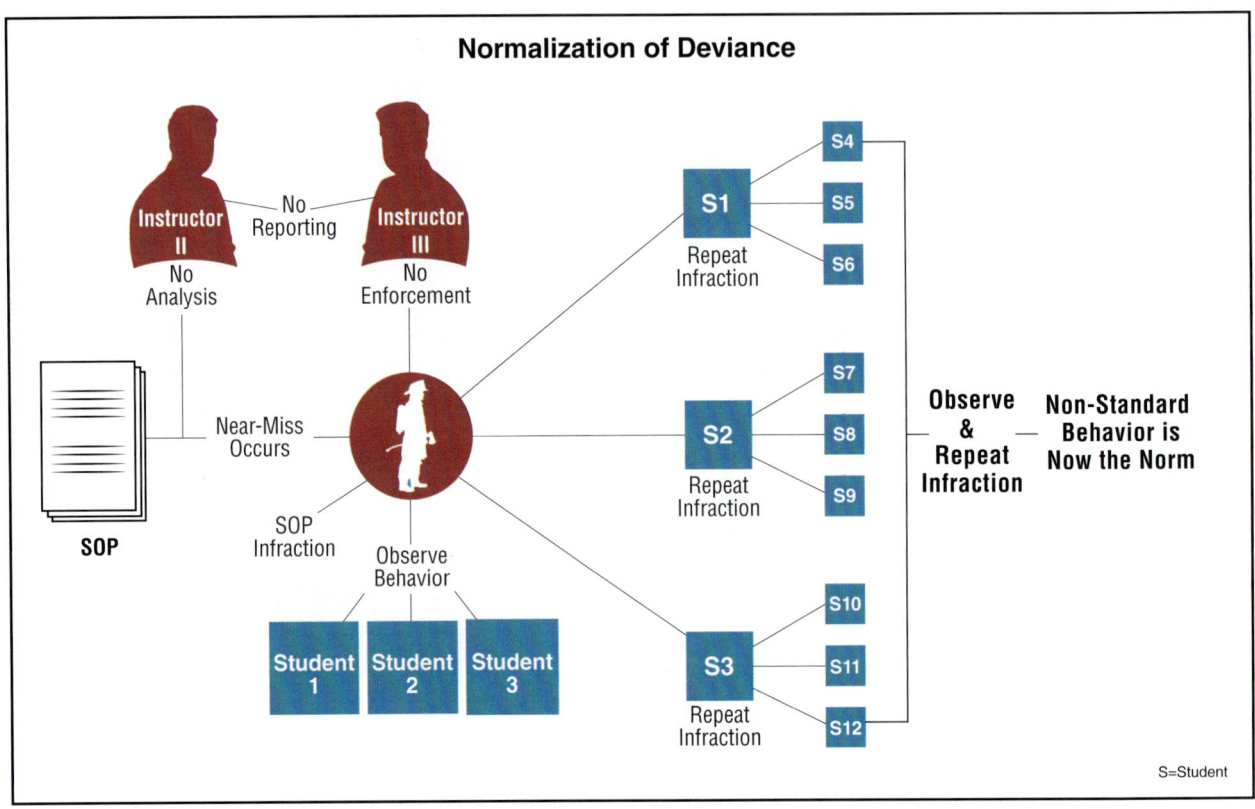

Figure 12.3 To prevent normalization of deviance, instructors should immediately address unsafe behaviors.

- **Improper attitude** — Address unsafe attitudes or behaviors so an individual does not create or become involved in an accident. Instructors should monitor their classes for signs of high-risk behavior in students, such as behavior that is:
    — Irresponsible or reckless
    — Inconsiderate or uncooperative
    — Fearful or phobic about a situation
    — Egotistical or jealous
    — Intolerant or impatient
    — Excitable or oversensitive
    — Obsessive or absentminded
- **Complacency** — Students may perceive some safety procedures as unnecessary, perhaps because the tactics have been so well-practiced. Instructors should ensure that students are not avoiding or overlooking safety steps during training.
- **Lack of knowledge or skill** — Instructors must address students who are unprepared for an evolution by providing additional training, including supervised practice time. Instructors must be sure students are:
    — Sufficiently informed about the training

— Capable of interpreting the training and convinced of its need

— Experienced in requisite knowledge and skills, and capable of decisive actions

— Properly trained and able to recognize potential hazards/risks

- **Physical limitations** — Instructors should be aware of any physical limitations that could lead to training injuries or fatalities. These limitations could include any of the following:

  — Inability to see or hear well enough for a situation

  — Physical characteristics that reduce a student's ability to perform safely

  — Limited strength or aerobic capacity

  — Effects of a medical condition, allergy, illness, or mental condition

  — Reduced reaction times due to substance abuse or legally prescribed medications

Hazard and Risk Analysis

A **hazard and risk analysis** identifies potential problem areas and is the foundation for any risk-management plan. For example, a planned driver/operator training evolution may require novice or inexperienced personnel to drive apparatus on public streets or highways. This situation creates a risk to personnel and the public. The instructor should search for a better location and use cones to simulate traffic. After personnel have gained experience, subsequent evolutions can be held on public streets.

When creating a lesson plan for skills training, instructors perform a task analysis to determine the necessary tasks and their order. An instructor should examine these tasks and ensure that they can be safely performed on the training ground. For example, if the roof at an acquired structure looks unlikely to support the weight of firefighters, then roof ventilation training cannot occur without reinforcing the roof or acquiring a new structure or prop.

Instructors should follow all aspects of risk-management plans established by their organizations. They should also heed any risks outlined in prepared lesson plans before teaching lessons and assess risks in the training environment. Instructors must eliminate hazardous environments near training areas. If hazards cannot be eliminated or minimized, instructors must move the training.

NOTE: Even with good policies in place and good instructions to work from, skills training can be unpredictable.

When reviewing injury records, investigators should prioritize potential hazards that have led to injuries while considering the following factors (**Figure 12.4**):

1. Frequency of the hazardous activity, and how this relates to the frequency of accidents or injuries
2. Relative severity of the potential loss

Ideally, all hazards are addressed as high priorities, but prioritizing hazards is often a necessity imposed by limited resources. See **Appendix E**, Risk Management Formulas, for the frequency/severity table and calculation formulas.

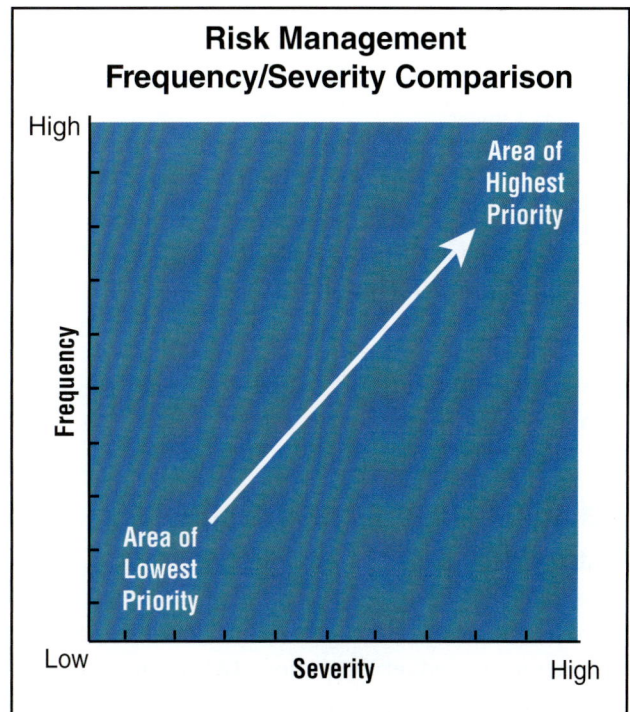

Figure 12.4 Give top priority to frequently-encountered hazards that may result in serious injuries.

> **Hazards and Risks**
>
> In everyday conversation, the terms *hazard* and *risk* are often used interchangeably; however, in the fire service, they describe two different things. *Hazard* usually refers to a condition, substance, or device that can cause an injury or loss. A *risk*, on the other hand, is the likelihood of suffering harm from a hazard. Risk can also be thought of as the potential for failure or loss. In other words, risk is the exposure to a hazard, and a hazard is the source of a risk.

Using IMS to Supervise Training

The IMS model adopted by many jurisdictions in North America is based on NFPA 1561, *Standard on Emergency Services Incident Management System and Command Safety*. It provides guidance and direction for the management and control of emergency incidents, ranging from single company responses to multiple agency and jurisdiction incidents. These guidelines can be applied to large-scale skills-training exercises and operational training evolutions. Using IMS also increases the safety and accountability of students.

The IMS model should be used at all training evolutions. Students and instructors will then become familiar with IMS roles and procedures. The sections that follow describe topics related to implementing an IMS during large-scale skills exercises and operational training evolutions. Skills associated with supervising multiple instructors and candidates during an increased hazard exposure training session are shown in **Skill Sheet 12-1**.

Incident Management System Duties and Functions

Level II Instructors should have a working knowledge of incident management systems, and they should implement the IMS their AHJ adopted. IMS adapts to a wide variety of situations and is a useful tool for organizing training evolutions.

The instructor in charge should staff the necessary IMS positions to effectively manage the training evolution. IMS positions that may need to be staffed include (**Figure 12.5**):

- Incident Commander (IC) or Lead Instructor
- Safety Officer
- Logistics Officer
- Staging Officer
- Division or Group Supervisor
- Communications Officer
- Ignition Officer
- Water Supply Officer
- Public Information Officer (PIO)

Figure 12.5 IMS positions should be chosen based on the scope of the training evolution.

Training Plan or Incident Action Plans (IAP)

Operational training evolutions can be described as large-scale lessons. The lesson plan for these evolutions must be based on plans used at actual incidents, an agency approved training plan, or an **incident action plan (IAP)**. Instructors can adapt the various IAP forms in the federal National Incident Management System (NIMS) to create the training evolution IAP. Like operational IAPs, training plans usually contain the following elements:

- **Objectives** — Clearly stated and measurable objectives designed to be achieved safely in a specific time interval
- **Organization** — IMS-defined units and agencies that are involved, and the roles they will play in a command structure

- **Assignments** — Specific unit tactical assignments for students and instructors; usually divided by branch, division, and group
- **Support materials** — Includes site plans, access or traffic plans, and locations of support activities such as staging, rehabilitation, and logistics
- **Safety message** — Information concerning personnel safety at the training incident, including a site-safety plan

At the end of the training evolution, the training plan is used as part of the postincident analysis and critique. It is an instrument for evaluating the students' learning achievements and overall effectiveness of the training.

What This Means to You

Skills Training with a Large Number of Students

You have been assigned to be the lead instructor for a weekend training exercise on basic skills for fifty students. To make the best use of the time needed to complete training, you divide the skills that will be taught into ten exercise sections, so that students can be divided into groups of five for each section.

When planning the exercise, you locate enough qualified instructors to supervise each of the ten sections. You also enlist three additional instructors for other command structure positions. The first is a safety officer whose task is to travel among the training sections, ensuring that safety procedures are being followed. This officer will report to you in the event of safety violations. The second is a logistics officer whose task is to ensure that the students and instructors have the equipment and supplies that they need to train safely. The third functions as a staging officer who maintains a rehab area and the common area where students go between training assignments. The instructors at each training section report problems up the chain of command to these command positions, who then report to you. The clear-cut roles and responsibilities dictated by NIMS and the IAP ensure that the training runs smoothly and that students focus solely on learning basic skills throughout the weekend **(Figure 12.6, p. 256)**.

Operational Training Evolutions

You are planning an operational training evolution that simulates a response to a three-alarm fire while applying IMS management methods. One of the learning objectives for the evolution is to teach IMS to students who meet the prerequisites for command training. All the students will assume IMS-dictated roles for extinguishing a fire, issuing commands, and conducting all communication within the IMS framework.

You first develop a training plan based upon the IAP that would normally be written for an operational incident, one that is similar to the one in the training evolution **(Figure 12.7, p. 256)**. You assign instructors to the various command positions needed for the evolution. You then identify students to act as command officers and pair them with instructors who will monitor their decisions during the training. Some will lead units who will enter the live-fire area, while others will function in support operations as safety officers or the communications officer.

At the time of the training, you review the training plan (IAP) with all the students and instructors who will be involved. Non-fire agencies get briefed on the communication structure you will use for the evolution. You tell command students who they will be paired with during the training. At the end of the training session, you use the IAP to debrief all participants and focus on how the training did or did not meet their expectations.

Training Evolution Evaluation

At the termination of the practical training evolution, a **postincident critique** or analysis should be held. The postincident critique fulfills the following purposes:

- Evaluates student skills and learning
- Evaluates the practical training evolution
- Determines safety problems that need to be corrected

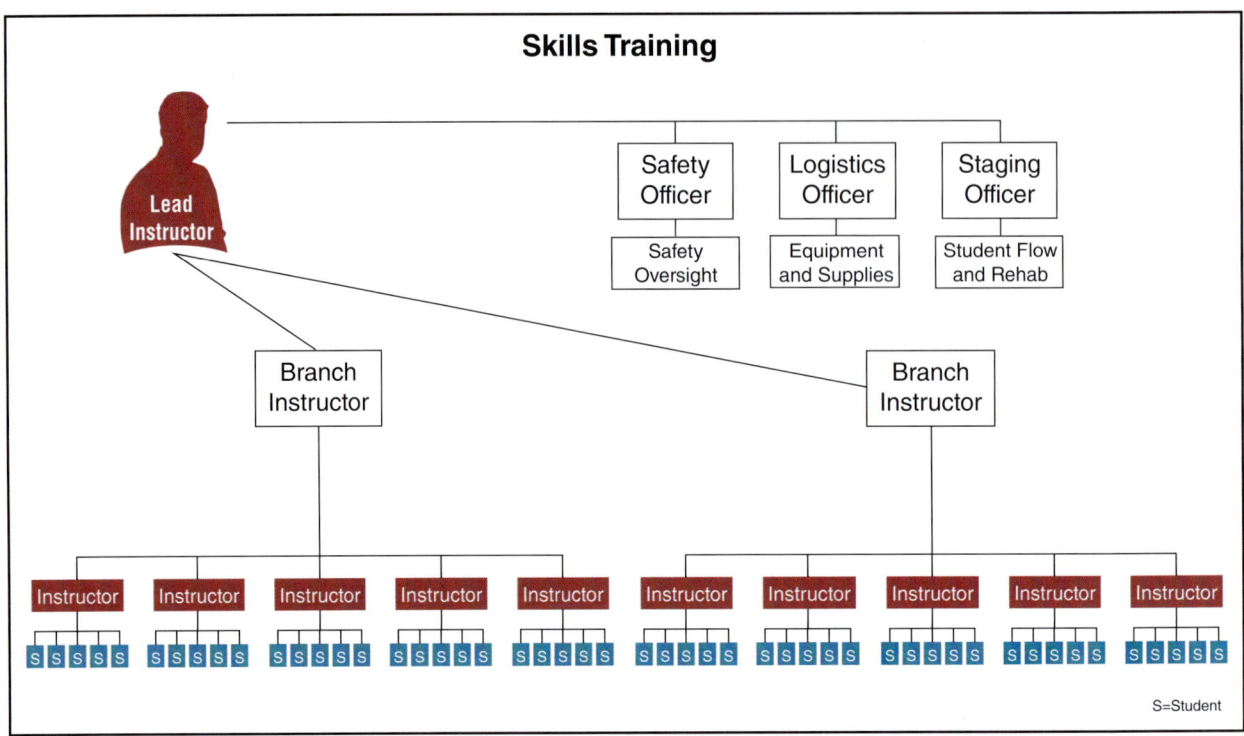

Figure 12.6 When large numbers of students are involved, skills training requires clear-cut roles.

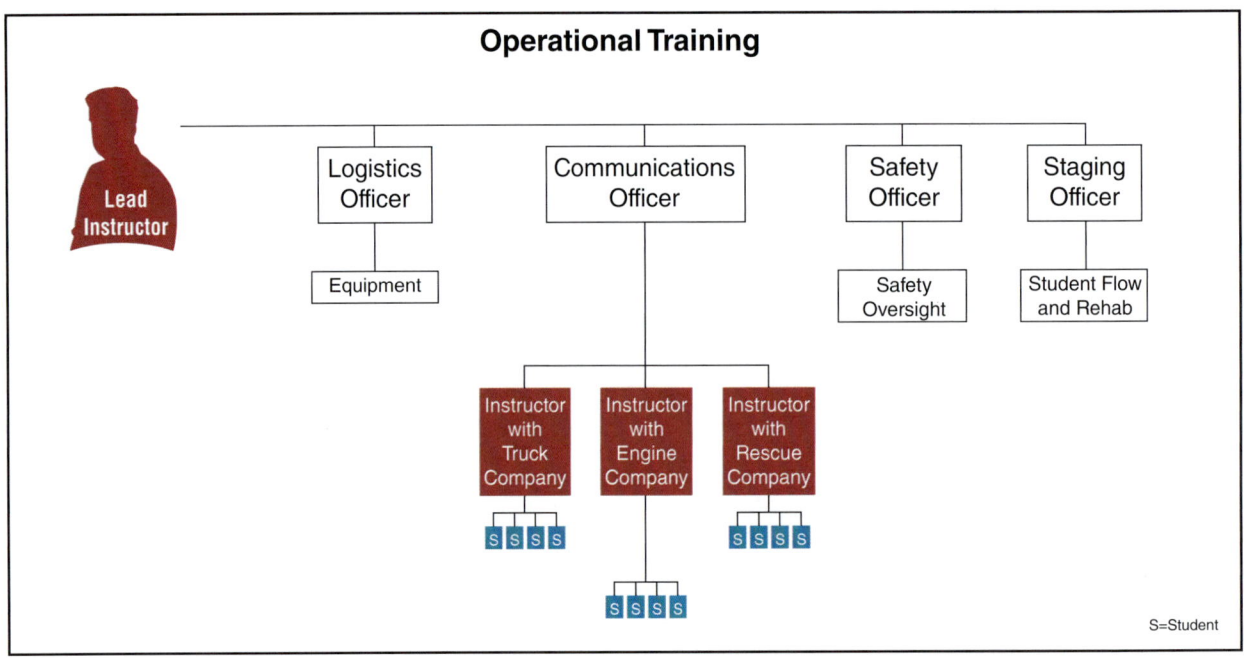

Figure 12.7 Roles and responsibilities assigned during operational training should reflect those required at the incident.

- Trains students in the postincident critique process
- Evaluates the instructor's supervisory and teaching skills

If the critique determines that students have not learned the objectives for the evolution, further training should be provided and the evolution restaged. If the evaluation determines that the evolution did not provide the level of training necessary to meet the learning objectives, then the evolution should be redesigned to provide that level.

If safety problems are discovered, they should be corrected before the evolution is reused. If an instructor did not provide adequate supervision or instruction, he or she should receive additional professional development.

Along with the IAP, the postincident critique should be used to generate a report on the training evolution. That report should contain the recommendations for changes to the evolution and be sent to the organization's leadership. All reports should be maintained in the organization's record system.

Environmental Issues at Training Evolutions

In the United States, the Environmental Protection Agency (EPA) regulates activities that affect the nation's water, atmosphere, and soil. In addition, in each of the states, similar agencies enforce their own rules and regulations that may affect training evolutions. Instructors and administrators must be familiar with federal, state, and local environmental regulations. That familiarity will allow them to follow laws and apply for necessary permits before conducting training.

NOTE: Canada also has laws that protect the environment. Instructors conducting training in Canada should become familiar with that nation's laws and take the appropriate measures to train safely and legally.

Water

Bodies of water can be contaminated by runoff water from training evolutions. Training facilities should have systems that trap, contain, and clean contaminated water and foams from training props, especially when using flammable liquids **(Figure 12.8)**. Some municipal fire and emergency services organizations may also have storm drains equipped with filtration systems. However, most training divisions do not have the ability to clean contaminated materials from runoff water before it reaches streams, rivers, or lakes.

Figure 12.8 Training facilities should include systems for trapping and cleaning contaminated water.

Instructors and administrators should contact local environmental AHJs or environmental protection agencies to determine the specific regulations they will have to follow. Sometimes an evolution will release such a low level of contaminant that it may not cause any environmental damage, but permits may still be required.

Atmosphere

Live-fire exercises will release hydrocarbons into the atmosphere. Some local open-burning ordinances may control or prohibit those releases **(Figures 12.9 a and b)**. These ordinances may require acquiring a permit and displaying it at the training site. To ensure that training does not threaten the environment, instructors should take the following steps:

Step 1: Adhere to all environmental rules and regulations.

Step 2: Meet environmental protection guidelines for the use of Class B (flammable/combustible) liquids.

Step 3: Ensure that the weather cannot spread contaminants into populated areas.

Step 4: Provide spark and cinder control for adjacent areas.

Figures 12.9 a and b Training fires that produce thick smoke (a) are generally considered atmospheric hazards, while fires that produce less smoke (b) are more likely to meet local open-burning ordinances. *Photo (a) courtesy of U.S. Air Force.*

Soil

Soil can be contaminated from water runoff that holds byproducts of combustion, fuels, and nonbiodegradable extinguishing agents. Large quantities of these materials are considered hazardous waste, but even small quantities must be removed after training, along with any contaminated soil. Because this can be costly and could result in litigation if done improperly, training agencies must take steps to avoid soil contamination. Instructors and administrators should consult the EPA or a local environmental agency about the relevant rules and regulations.

If it is impossible to meet the requirements for containing water runoff, training should be performed on a nonporous concrete surface that can be cleaned with inert materials. Training agencies should also consider using training-type foam extinguishing agents that are nontoxic and biodegradable.

NOTE: Specialized training foams can reduce the contamination runoff in water and soil. These foams would replace the standard foams that personnel would use at an incident.

NOTE: See Chapter 8, Skills-Based Training Beyond the Classroom, for more information about the environmental impact of fuels used in live-fire training. Additional information on Class B fuels can be found in IFSTA's **Aircraft Rescue and Fire Fighting** and **Industrial Exterior and Structural Fire Brigades** manuals.

Accident Investigation

Instructors and students who are involved in or witness an accident are often afraid to provide information, thinking that they may get someone in trouble. But an investigation cannot be resolved and accidents cannot be prevented when personnel withhold valuable information. After any injury or fatality, investigators must determine the sequence of events and their cause. Their job is about fact-finding, not fault-finding; they try to determine only what caused the accident. After any training accident, instructors should do the following:

- Report the accident according to their organizational policies.
- Answer any questions asked by investigators about the accident (**Figure 12.10**).
- Complete any appropriate forms.
- Decide whether or not the training evolution can continue after an accident.
- Obtain statements immediately after the accident from persons involved and potential witnesses.

Figure 12.10 Instructors must comply with investigators after any training accident.

Chapter Review

Answer the following questions to review the information provided in this chapter.

1. What safety challenges does a Level II Instructor face when supervising a training evolution?
2. How does a Level II Instructor use the Incident Management System (IMS) model to supervise training?
3. How do environmental regulations affect training evolutions?
4. What are the roles and responsibilities of the Level II Instructor during accident investigation?

Discussion Questions

The following questions are intended to generate discussion, expand your understanding of the chapter text, and allow you to think critically about what you have learned. Answers to these questions may vary.

1. What example of a safety challenge have you incurred as an instructor when supervising a training evolution?
2. Why should an Incident Management System (IMS) be used for all training evolutions?
3. Why should an instructor enforce environmental regulations during training evolutions?
4. Why is it important for instructors to report any training accident?

Key Terms

Hazard and Risk Analysis — Identification of hazards or risks and the determination of an appropriate response; combines the hazard assessment with risk management concepts.

Incident Action Plan (IAP) — Written or unwritten plan for the disposition of an incident; contains the overall strategic goals, tactical objectives, and support requirements for a given operational period during an incident.

Normalization of Deviance — State of a safety culture in which acting against SOPs becomes normal behavior rather than an exception.

Postincident Analysis — Overview and critique of an incident by members of all responding agencies, including dispatchers. Typically takes place within two weeks of the incident. In the training environment it may be used to evaluate student and instructor performance during a training evolution.

Risk-Management Plan — Written plan that identifies and analyzes the exposure to hazards, selects appropriate risk management techniques to handle exposures, implements those techniques, and monitors the results.

12-1
Supervise multiple instructors and candidates during increased hazard exposure training. [NFPA 1041, 5.4.3]

Task Steps

Step 1: Identify applicable department policies, regulations, standards, and recommended practices regarding the increased hazard exposure training session **(Figure 12.11)**.

Step 2: Identify resources (staff, equipment, facilities, apparatus, props, etc.) needed to address safety issues.

Step 3: Describe how the resources in Step 2 will be utilized.

Step 4: Identify the procedures that will be implemented to address identified safety issues.

Step 5: Describe procedures used to supervise multiple instructors and candidates to include:
   a. Observation techniques
   b. Evaluation form(s)
   c. Provision of feedback/coaching
   d. Assignment of Safety Officer position **(Figure 12.12)**

Step 6: Review the Incident Action Plan (IAP) that:
   a. Addresses safety issues
   b. Implements an Incident Management System (IMS)
   c. Ensures compliance with safety rules, policies, regulations, and appropriate practices that includes:
   - NIMS compliant
   - Incident command utilized
   - Water supply
   - Operations
   - Safety
   - Rehab
   - EMS
   - Logistics
   - Equipment positions

Step 7: Supervise instructors and candidates during the training, following all appropriate AHJ SOPs and the IAP **(Figure 12.13)**.

Step 8: Document that all training goals and learning objectives were accomplished **(Figure 12.14)**.

Figure 12.11

Figure 12.12

Figure 12.13

Figure 12.14

Test Item Construction

Chapter 13

SECTION B — INSTRUCTOR II

Chapter Contents

Common Considerations for all Tests 265	**Test Planning 281**
Test Formatting and Item Arrangement 267	Determining Test Purpose and Classification 281
Test Item Level of Cognition and Difficulty 268	Identifying Learning Objectives 282
Test Instructions and Time Requirements 268	Constructing Appropriate Test Items 283
Testing Bias 269	**Test Scoring Method Selection 285**
Student Evaluation Instruments 269	**Chapter Review 286**
Written Tests 270	**Discussion Questions 286**
Oral Tests 277	**Key Terms 287**
Performance (Skills) Tests 277	**Skill Sheets 288**

JPRs addressed in this chapter

This chapter provides information that addresses the following job performance requirements of NFPA 1041, *Standard for Fire Service Instructor Professional Qualifications*, 2019 Edition.

5.5.2

Learning Objectives

1. Describe common considerations for test instruments. [5.5.2]
2. Discuss various types of evaluation instruments used in fire and emergency service training. [5.5.2]
3. Explain the steps for test planning. [5.5.2]
4. Describe the process to select a test scoring method. [5.5.2]
5. Skill Sheet 13-1: Develop test items. [5.5.2]
6. Skill Sheet 13-2: Develop a performance skills evaluation. [5.5.2]

Chapter 13
Test Item Construction

This chapter expands on the information presented in Chapter 9, Testing and Evaluation. In general, the information discussed there concerning tests purchased or prepared by the AHJ applies to Level II Instructors as well. When prepared tests are not available or do not measure the learning objectives that the AHJ must evaluate, Level II Instructors may have the following two additional responsibilities as described in this chapter:

1. Creating testing instruments and test items for their organizations.
2. Modifying existing test items or creating replacement test items that more appropriately measure the learning taking place in their jurisdictions.

Level II Instructors should be able to select appropriate testing instruments that address all of the following:

- The three domains of learning — cognitive, psychomotor, and affective
- Levels of learning ranging from basic ("understand") to advanced ("create")
- The level of difficulty presented in a course — beginner, intermediate, or advanced

The sections that follow describe common considerations for all test types and provide specific information about student evaluation instruments, test planning, and test scoring method selection.

> **Evaluation Terminology**
>
> NFPA 1041 refers to testing instruments as "student evaluation instruments." In order to avoid confusion with other types of evaluation instruments referenced in NFPA 1041 (such as class evaluations, instructor evaluations, and course evaluations) this chapter uses the more common terms "testing instrument" and "test item" to describe student evaluation instruments that are designed to measure a student's knowledge or mastery of learning objectives.

Common Considerations for all Tests

Test items must always be based on specific learning objectives. Level II Instructors must consider the following additional criteria when designing any test:

- Test formatting and item arrangement
- Test item level of cognition and difficulty
- Test instructions and time requirements
- Testing bias

Donald Kirkpatrick's Four Levels of Evaluation

Donald Kirkpatrick, Ph.D., created The Four Levels of Evaluation, also referred to as the Kirkpatrick Model **(Figure 13.1)**. While developed in the 1950s, it is still recognized as the definitive training evaluation model for adult education. The four levels include:

Level 1 — Reaction. Commonly recognized by students and instructors as the end of course evaluation, this measures the satisfaction students had in the course. Correlation is shown between student satisfaction in a class and the level of learning the student felt they accomplished.

Level 2 — Learning. Commonly recognized by students as the "test." While this is normally the type of tool used in level 2 evaluation, note that anything that can measure learning occurred is considered level 2 evaluation.

Level 3 — Behavior. This is not commonly used by training agencies because of the process to meet this level of evaluation. In short, individuals who attend a class would then have a comprehensive appraisal of the learner completed to determine if the class helped them in doing their job better. Individuals surveyed could include supervisors, the students themselves, and peers.

Level 4 — Results. Final level again is difficult for training agencies to measure since it evaluates a particular training program may have been key in an individual growth or promotion in the organization.

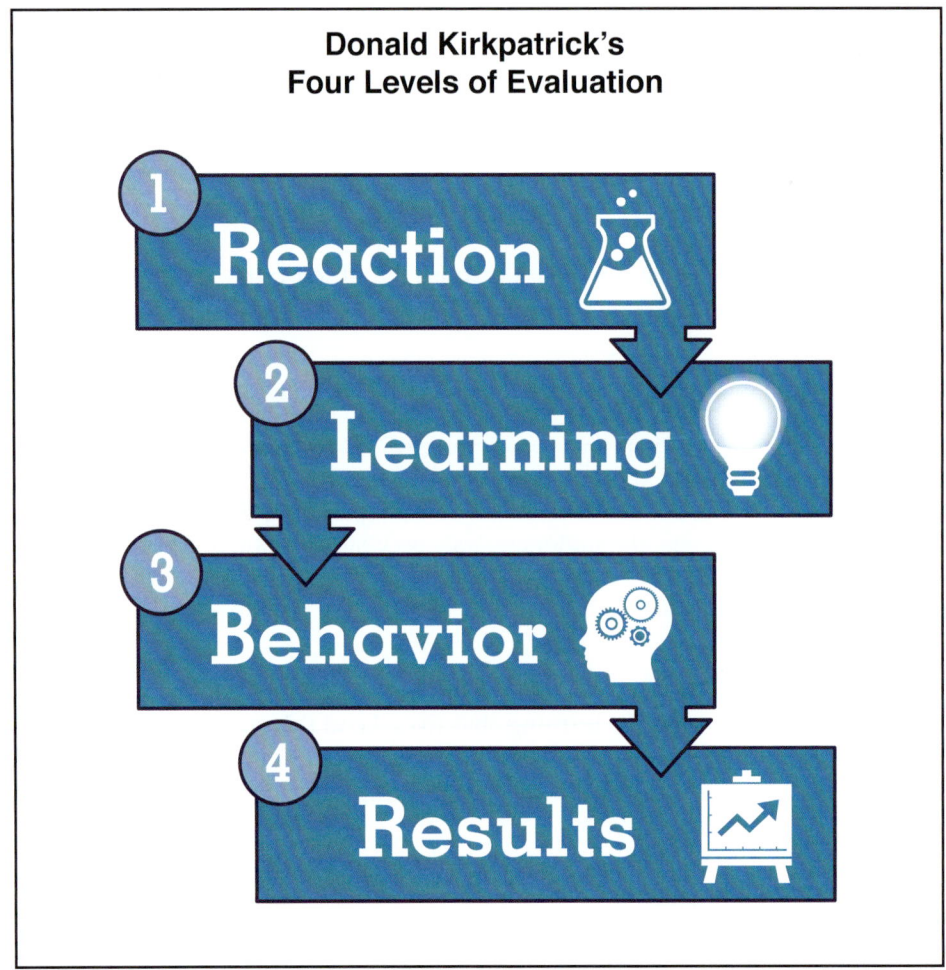

Figure 13.1 Kirkpatrick's Four Levels of Evaluation guide an instructor in determining whether a student has learned necessary material.

Test Formatting and Item Arrangement

Proper layout and formatting of the test sheet will make any test easier to take, administer, and score (**Figure 13.2**). Consider the following guidelines for test formatting:

- Provide space for students to write their name and the date on either the test sheet or a separate answer sheet.
- Provide a title or label at the top of the first page.
- Number all tests and label different versions of the test. This will help with score reporting and test security.

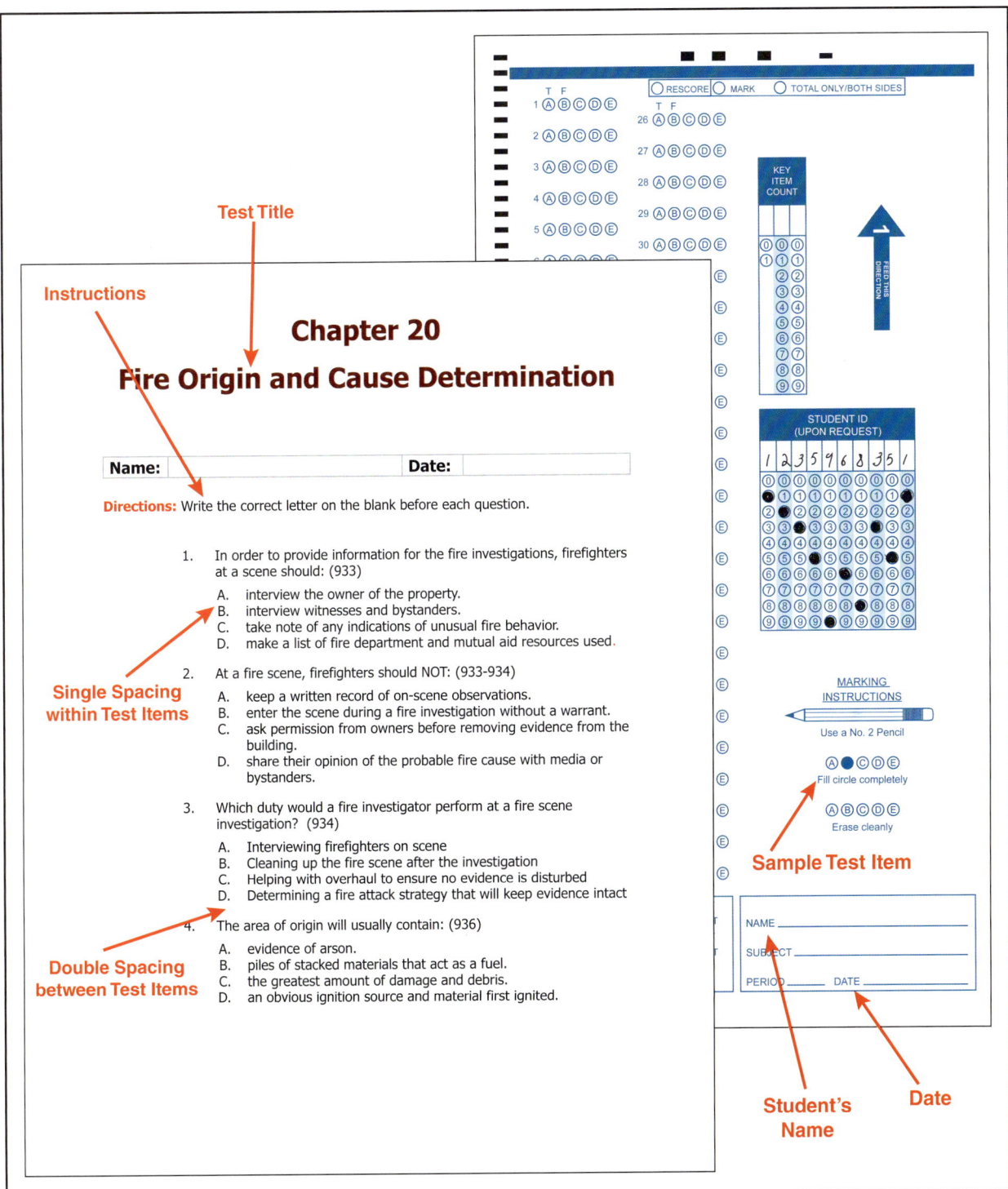

Figure 13.2 Both students' and instructors' work is easier when test sheets have proper formatting and layout.

- Number all pages of the test. Students will be able to budget their time more wisely if they can see the length of the test.
- Provide clear instructions at the beginning of the test and at the beginning of each section that uses a different type of test item (such as multiple-choice, matching, true-false, or fill-in-the-blank).
- Provide a sample test item, along with a sample answer, to show students how to respond to each item.
- Number all items consecutively.
- Single-space each test item, but double-space between items.
- State the point value of each test item — for example, *multiple-choice: 1 point each; short-answer: 2 points each*.
- Use commonly understood terms; for example, do not use abbreviations unless they are placed in parentheses following the common term.

Test items must be arranged in a logical sequence. They can be grouped into either of the two following categories:

- Learning domain outcome (knowledge, comprehension, or application)
- Type of test item (multiple-choice, matching, or short-answer).

Instructors must make sure that the wording of one test item does not reveal the answer to another test item. On computer-adaptive tests, such as those used in EMS, sequencing may be organized so that students can only progress to a more difficult question after correctly answering a simpler one.

Test Item Level of Cognition and Difficulty

The cognition levels found in Bloom's Taxonomy may be categorized from low to high as follows:

- Remember (knowledge)
- Understand (comprehension)
- Apply (application)
- Analyze (analysis)
- Evaluate (evaluation)
- Create (synthesis)

Test items should evaluate the student's ability at the level within the taxonomy that corresponds to the learning objective being evaluated. A variety of levels can exist within a course. For example, an advanced level class will still need test items written to lower levels of the taxonomy for new information that students are meant to remember and understand. Planning the purpose and scope of a test will be further reviewed later in this chapter.

The actual determination of test difficulty is the responsibility of a Level III Instructor. To complete that work, the instructor compiles scored tests, and evaluates students' performance on individual questions. The difficulty level of each question is then saved for future testing use and evaluation.

Test Instructions and Time Requirements

Test designers and presenters should include clear instructions at the beginning of the test that explain the following:

- Purpose of the test
- Method and means for recording answers
- Recommendation whether to guess when undecided on an answer (in some cases, incorrect answers are penalized more than not answering the question)
- Amount of time available to complete the test

In terms of time requirements, benchmarks for answering certain types of questions are as follows:

- **True-false (true answer where a student is only to label the statement)** — 15 seconds for true answers
- **True-false (false answer where a student is often asked to correct the statement)** — 30 to 45 seconds
- **Seven-item matching** — 60 to 90 seconds
- **Multiple-choice (with four possible responses)** — 30 to 60 seconds
- **Problem solving, analyze, create, or evaluate level questions** — 30 to 60 seconds
- **Short-answer** — 30 to 60 seconds
- **Essay** — 60 seconds for each major point that students must include

Instructors can use these estimates to calculate how much time students will need to complete the test. The instructor may take the test, or ask another professional member of the organization to take the test, to see how much time is needed. Tests should be an appropriate length to address the learning objectives that the test is intended to evaluate.

When time is a restrictive factor, tests can emphasize the most critical learning objectives and include a sampling of less important objectives. This method of test construction is called *sampling*. When the instructor uses sampling, the plan must be documented to prove that necessary components were addressed. The most critical objectives must be tested in each version of the test while the less critical objectives tested are included in a certain rotation.

Testing Bias

As described in Chapter 9, Testing and Evaluation, test items and **testing instruments** should not favor or penalize any particular group of students. Ensuring that test questions very closely reflect the materials being tested is the best way to avoid bias in testing materials. In addition, when students recognize that test items closely resemble the information that they have studied, they are more likely to perform confidently on tests regardless of their gender, cultural, ethnic, or regional backgrounds.

When instructors find evidence of biases, the questions associated with the bias should be revised or, if necessary, rewritten. In the fire and emergency services, bias is generally limited to use of regional jargon and differences in terminology. For example, local governments in the U.S. and Canada may be referred to as *counties* or *parishes*, or referred to in legal terms as *jurisdictions*. Similarly, some fire apparatus may be referred to as a *tanker* or a *tender* depending on geographical region or the differences between departments. The terminology on the test should reflect the terminology of the students and the materials from which they studied or received training.

Student Evaluation Instruments

Deciding how to evaluate students can be a confusing and difficult topic for the Level II Instructor. The instructor needs to establish the appropriate evaluation instrument to meet the parameters of the curriculum and course being presented. In addition, once the instructor determines the type of instrument to use, they must then design the instrument to be fair and unbiased. This process takes time, and new Level II Instructors may often underestimate the time commitment. The following sections present a number of different evaluation instruments (**Figures 13.3 a-c, p. 270**):

- Written tests, which measure cognitive ability
- Oral tests, which measure either cognitive or communicative ability
- Performance tests, which measure psychomotor skill ability

    NOTE: Skills associated with developing test items are shown in **Skill Sheet 13-1**.

Written Tests

Written tests measure students' understanding and retention of technical information and evaluate their accomplishment of the cognitive learning objectives. Written test items fall into one of two categories that are defined as follows:

- **Objective** — An objective test item is a question for which there is only one correct answer. The judgment of the instructor or evaluator is not relevant and has no effect on assessment. Objective items measure cognitive learning but typically only at the lower levels of remembering and understanding. However, properly constructed objective test items can also be used to measure higher levels of cognitive learning such as evaluation or creation. There are three main types of objective questions:

    — Multiple-choice

    — True or false

    — Matching

- **Subjective** — A subjective test item has no single correct answer. The evaluator's judgment may therefore affect assessment. Subjective items are an effective way of measuring higher cognitive levels because they allow students the freedom to organize, analyze, revise, redesign, or evaluate a problem. The strength of a student's response to these items depends on a variety of factors, such as how well they communicate their ideas, and the personal opinions of the evaluator. There are three main types of subjective test items:

    — Short-answer or completion

    — Essay

    — Interpretive exercise

The following sections describe both objective and subjective tests.

Figures 13.3 a-c The most common test types are written, oral, and performance.

Study Guides and Workbooks

Questions that are published in study guides, test preparation guides, textbooks, and manuals should not be used for certification or summative tests. These questions should only be used to help students determine how well they understand the material in a particular coursebook.

Multiple-Choice

A multiple-choice test item consists of either a question or an incomplete statement, called the **stem**, plus a list of several possible responses, which are referred to as *choices* or **alternatives** (**Figure 13.4**). Students are tasked to read the stem and select the correct response from the list of alternatives. The correct choice is known as the

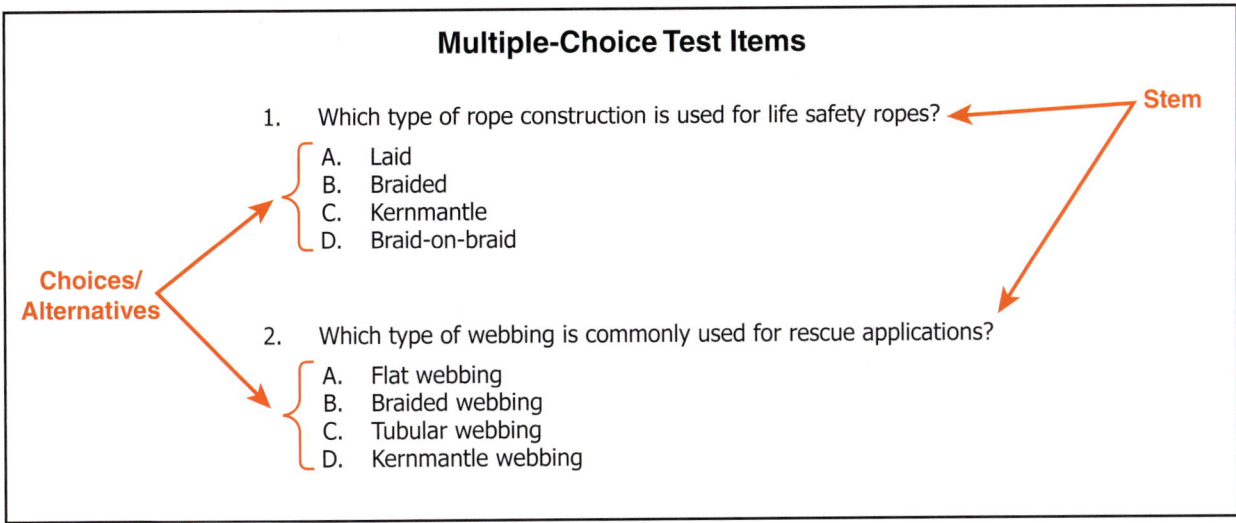

Figure 13.4 Multiple-choice test items consist of a stem and choices. Incorrect choices are known as distractors.

answer and the remaining choices are called **distractors**. Distractors are used to discriminate between students who understand the subject matter well and those who are uncertain of the correct answer. When test items are well written, distractors are not meant to trick, confuse, or mislead students.

When creating multiple-choice test items, adhere to the following guidelines:

- Write the stem in the form of a direct question or an incomplete sentence that measures only one learning objective.
- Write a clear, brief stem that contains most of the wording for the test item; doing so helps to avoid placing repeated words in the alternatives.
- Write positive questions as much as possible. Be consistent in labeling negative words if and when negative statements are used.
- Provide at least three plausible, attractive distractors.
- Phrase the choices so that they are parallel and grammatically consistent with the stem.
- Place correct answers in varied positions among the A, B, C, and D choices.
- Place each choice on a separate, indented line, and in a single column.
- Begin responses with capital letters when the stem is a complete question.
- Begin responses with lowercase letters when the stem is an incomplete sentence.
- Do not include choices that are obviously wrong or intended to be humorous.
- Make sure that stems and alternatives do not give students grammatical clues as to the correct response. For example, a stem that calls for a singular response (by using a/an or a singular word ending) would automatically rule out alternatives that are plural, and a stem phrased in the present tense would rule out alternatives phrased in the past tense.
- Make all alternatives close to the same length.
- Avoid using the phrases *all of the above* and *none of the above* as the fourth choice.
- Do not test trivial ideas or information.
- Use correct grammar and punctuation.

Instructors should be aware of the disadvantages of multiple-choice tests, such as:

- They are not well suited to measuring certain cognitive skills, such as organizing and presenting ideas. Essay tests are more effective for this purpose.

- Depending on the test writer's skill, this type of test may not include different difficulty-level test items that measure a variety of cognitive learning levels.
- Creating appropriate and plausible distractors for each stem can require significant time and thought.
- Students who do not know the material may still be able to guess the correct answer.

True-False

The true-false test item is a single statement that the student must determine to be either true or false (**Figure 13.5**). The difficulty in constructing this type of test is creating a statement that is completely true or completely false. True statements should be based on facts, while false statements should be based on common misconceptions of the facts. In addition to the traditional true-false test items, there are also modified true-false test items. Modified true-false items ask the student to explain why an item is false or to rewrite the item to make it true.

One limitation of true-false questions is that students tend to remember the false items on the test as being true, a phenomenon known as the negative suggestion effect. Instructors should review the correct answers to true-false questions with students after scoring the test to help combat this effect.

When creating true-false tests and test items, instructors should consider the following guidelines:

- Write the words *True* and *False* at the left margin if students must mark their answers on the test paper. On computer-scored answer sheets, *True* may be assigned to *A* while *False* is assigned to *B*.
- Provide clear instructions so that students know how to respond to each statement (for example, by writing *True* or *False*, by writing *T* or *F*, by circling either *True* or *False*, or by marking *A* or *B*).
- Create enough test items to provide reliable results. For reliability purposes, more true/false items are needed than the number used for multiple-choice items. A large number of test items minimizes the possibility of guessing the correct answers.
- Distribute *true* and *false* items randomly. Avoid having several questions in a row be *true* or *false*, or any pattern of distribution that students might be able to recognize.
- Avoid determiners (words that indicate a specific answer) that provide unwarranted clues. Words such as *usually, generally, often,* or *sometimes* are most likely to appear in true statements. The words *never, all, always,* or *none* are more likely to be found in false statements.
- Avoid creating items that could trick or mislead students into making a mistake. Ensure only one correct answer is possible.
- Avoid double-negative test items; they are very confusing to students and do not accurately measure knowledge.
- Avoid using personal pronouns such as "you."

True-False Tests

Example 1:
Identify accurate statements about fuel characteristics. Circle the word *TRUE* before each correct statement, and the word *FALSE* before each false statement.

TRUE (FALSE) 1. The shape and size of a fuel affects its ignitability.

Example 2:
Identify accurate statements about fuel characteristics. Place a check mark under the column marked *TRUE* for each correct statement and under the column marked *FALSE* for each false statement.

TRUE   FALSE
_____   ✓   1. The shape and size of a fuel affects its ignitability.

Example 3:
Identify accurate statements about fuel characteristics. Write TRUE for each correct statement and FALSE for each false statement on the blank.

FALSE_____ 1. The shape and size of a fuel affects its ignitability.

Example 4:
Identify accurate statements about fuel characteristics. Fill in circle A for each TRUE statement and circle B for each FALSE statement on scantron.

Ⓐ ●   1. The shape and size of a fuel affects its ignitability.

Figure 13.5 Students can respond to true/false questions by circling the correct choice, placing a check in the correct column, writing out the word ("true" or "false"), or filling in a circle.

- Do not use test items that test trivia or obscure facts.
- Develop test items that require students to think about what they have learned, rather than merely remember it.
- Avoid unusually long or short test items, because the length may be a clue. True items are often longer than false items, because they include a justification.
- Create brief, simply stated test items that deal with a single concept. Avoid lengthy, complex items that address more than one concept.
- Avoid quoting information directly from the textbook.

Matching

Matching test items consist of two parallel columns of words, phrases, images, or a combination of these. In the most common example, students must match a word from the left column with its definition from the right column. The content of a matching test item must consist of similar material, items, or information. For example, a matching section that should evaluate whether students know the component parts of a tech rescue rig should not include information about other topics. Some examples of matching test items are shown in **Figures 13.6 a-c, p. 274**.

A partial list of test items that can be easily made into matching test items includes:

- **Short questions** — With answers
- **Events** — With dates
- **Parts** — With their functions
- **Terms** — With their definitions
- **Objects** — With their names
- **Machines or tools** — With their uses
- **Problems** — With their solutions
- **Causes** — With their effects

Consider the following guidelines for matching tests and matching test items:

- Avoid placing each group of prompts and the list of responses on more than one page.
- Separate matching sections into sets of five problems and responses when using computer or mechanically scored answer sheets.
- Consider preparing one more response than there are prompts. The extra response requires more precise knowledge and prevents students from finding an answer by eliminating all the other possible answers.
- Number the problem statements. Place an answer line to the left of each number unless a separate answer sheet is used.
- Use letters for each response.

Arrange problem statements and responses into two columns: problem statements on the left side of the page and responses on the right. Columns may be titled with appropriate headings, such as Tools and Uses, or Symptoms and Treatments.

NOTE: Instructors should be advised that matching test items may be more effectively and efficiently written as a series of multiple-choice questions.

Matching

Match the correct statement to each term.

_____ 1. Can vary from moderate (unable to see, breathless) to severe (convulsions)

_____ 2. Lowest concentration of a gas or vapor capable of killing a specified species over a specified time

_____ 3. Minimum concentration of an inhaled substance in the gaseous state that will be fatal to the test group (usually within 1 to 4 hours)

_____ 4. Lowest administered dose of a material capable of killing a specified test species

_____ 5. Statistically derived single dose of a substance that can be expected to cause death in 50 percent of animals when administered by the oral route

_____ 6. Minimum amount of solid or liquid that when ingested, absorbed, or injected through the skin will cause death

A. Incapacitating dose (ID)
B. Lethal concentration (LC)
C. Lethal concentration low (LC_{LO} or LCL)
D. Lethal dose (LD)
E. Lethal dose low (LDLO or LDL)
F. Median lethal dose (LD_{50})

a

Matching Test Items

Objective 16:
Match DOT hazards to their placard colors.
Write the correct letters on the blanks.

_____ 1. Nonflammable gas
_____ 2. Health hazard
_____ 3. Water reactive
_____ 4. Explosive
_____ 5. Oxidizer
_____ 6. Flammable

A. Green
B. Orange
C. Blue
D. Yellow
E. White
F. Red
G. Black

b

Pictorial Matching Test Item

Objective 16:
Match DOT placard illustrations to their hazards.
Write the correct letters on the blanks.

Orange _____ 1. 
Yellow _____ 2. 
White _____ 3. 
Red _____ 4. 
Blue _____ 5. 
Green _____ 6.

A. Nonflammable gas
B. Oxidizer
C. Flammable
D. Chemical hazard
E. Explosive
F. Health hazard
G. Water reactive

c

Figures 13.6 a-c Three examples of matching test items. Note that in b and c additional alternatives have been included to discourage students from guessing.

Short-Answer/Completion

A short-answer item is a question for which students must write a correct response. To do so, they must recall previously learned information, apply relevant principles, or understand methods or procedures (**Figures 13.7 a and b**). Short-answer items are often subjective.

Short-Answer Test Items

1. List four types of portable fire extinguishers and then briefly describe what each is designed for.

2. List the class ratings used for portable fire extinguishers and describe the test or qualifications each is based on.

a

Picture Identification

Objective 11:
Identify components of positive-pressure SCBA.

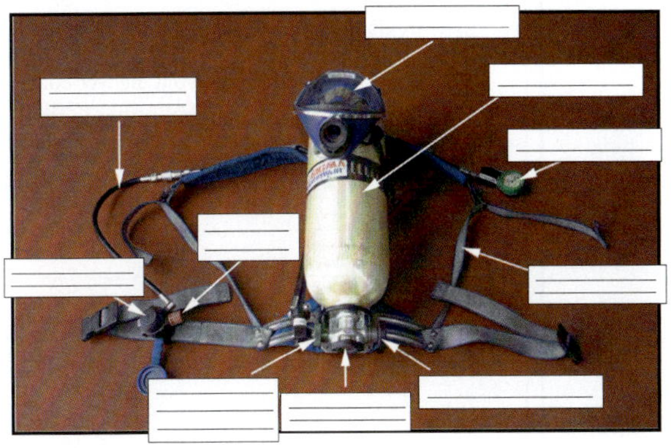

b

Figures 13.7 a and b Short answer test items (a) should be written so that only one correct answer is possible. Completion items (b) should require that students provide a correct answer without being given a list of possible choices.

In contrast to short-answer items, a completion item should be objective. This type of test item is a statement in which key words are replaced with an underlined blank space that students are tasked to fill in.

When creating short-answer/completion test items, instructors should consider the following guidelines:

- On completion test items, create short, direct statements for which only one answer is possible.
- Avoid long statements with a string of blanks to fill.
- Start with a direct question and change it to an incomplete statement.
- Make sure that the desired response is a key point in the lesson.
- Arrange the statement with the blanks at or near the end of the sentence.
- Avoid statements that call for answers with more than one word, phrase, or number.
- Eliminate unnecessary clues, such as answer blanks that vary in length or the use of the words "a" or "an" preceding the blank.
- Write a rubric or detailed answer sheet so that the scorer understands the full extent of possible, acceptable answers to the questions. Refer to **Appendix D** for an example of a scoring rubric.

Essay

Like short-answer test items, essays are subjective. Students must construct an in-depth answer on a topic or question related to a key aspect of the course material (**Figure 13.8, p. 276**). The strength of this item type is that it tests the students' higher level cognitive processes. Students are expected to demonstrate the ability to analyze a topic, create a solution to a problem, or evaluate a system or process. Essay tests also eliminate guessing, because students must know the material thoroughly in order to write an effective essay. Creative students often prefer

> **Chapter 3:**
> **Fire Department Communications**
>
> Answer the following questions completely in the space provided. Each question is worth 10 points. Allow a maximum of 10 minutes for each question.
>
> 1. Explain what information your jurisdiction could track through postincident reports and how this will be used. Be sure your answer contains enough detail to help a new firefighter understand why these reports are important.
>
> _____
> _____
> _____
> _____
> _____
> _____
>
> 2. What information can be gained from reading postincident reports? Give two examples of reported information that might change the way you approach your job.
>
> _____
> _____
> _____
> _____
> _____
> _____
> _____

Figure 13.8 Essay questions require students to display in-depth knowledge.

this type of test item because it allows them a forum to express their perspective of a topic. However, there are disadvantages, such as:

- Essays are time-consuming for students to complete and instructors to score.
- Differences in students' writing ability, penmanship, spelling, and grammar may affect an instructor's ability to easily score the test.
- Students who have difficulty writing or write slowly will be at a disadvantage, especially in a timed test. Refer back to Chapter 7, Student Interaction, for information concerning individual student needs.

Although essay test items are generally easy to construct, instructors should be aware of and practice the following guidelines:

- Choose essay topics that reflect key aspects of the course material.
- Create a rubric that establishes clear scoring guidelines.

- For each essay question, provide clear instructions that define how students should respond, how much time they should spend responding, and how many pages or paragraphs each response should be.
- Provide sufficient time for students to respond to all questions.

Interpretive Exercises

The interpretive exercise is another subjective test item that measures higher level cognitive processes. An exercise consists of introductory material, typically numerical data, a graph, or a paragraph of text, followed by a series of test items. Students read the text or look at the illustrations, then answer the test items, which may be any of the types described in this chapter: multiple-choice, true-false, short-answer/completion, matching, or essay.

Instructors should apply the following rules when creating an interpretive exercise:

- Make sure that all introductory material relates to key learning objectives, and is as concise as possible.
- Apply relevant guidelines for effective item construction for each test item.
- Use test items that require the same type of performance that is listed in the test specifications for the various learning objectives.
- Create original introductory material unfamiliar to students.
- Ensure that the introductory material does not give away the answer to any of the test items.
- Encourage students to read the introductory material to be able to answer test items.
- Provide enough test items, using a variety of item types, to effectively measure students' understanding of the material.

Oral Tests

Oral tests consist of either open or closed questions. When the purpose of the test is to determine knowledge, the questions should be closed, requiring only a single brief answer. When the purpose is to determine how a student responds under pressure, the question should judge both accuracy and presentation. In this second case, the questions should be open, permitting longer answers that may lead to further questions.

Oral tests can be very stressful for students. Some students may find it difficult to present their thoughts in this fashion, regardless of the testing environment. Instructors should provide a relaxed, comfortable atmosphere for the presentation of oral tests. Important aspects of designing and conducting oral tests include the following:

- **Development** — Base oral questions on standard criteria and performance objectives.
- **Validation** — Before presenting oral test items to students, ask for feedback from other instructors, technical experts, or students in other training programs. Revise inappropriate or ineffective questions, and modify questions as needed.
- **Evaluation** — When administering the test, listen carefully to students' responses. Assess responses based on learning/performance objectives and standard criteria.

Oral tests are highly subjective, especially when the questions may be answered a number of ways. To reduce evaluator bias, test developers should provide a scoring rubric that lists all possible correct answers.

Even though scoring is subjective and testing conditions may cause anxiety, an oral test is the most valid and reliable way to test a student's ability to verbally communicate ideas, concepts, and processes. It may also be the best measure of a student's judgment and thought processes.

Performance (Skills) Tests

A performance or skills test measures a student's proficiency in performing any task that involves a psychomotor objective. Examples include demonstrating care of tools and equipment, driving and operating apparatus, and performing emergency care steps and techniques **(Figure 13.9, p. 278)**. Assessment is based on either a speed standard such as timed performance, a quality standard, such as minimum acceptable performance, or both.

Skills associated with developing a performance skills evaluation are shown in **Skill Sheet 13-2**.

Performance tests require students to demonstrate psychomotor proficiency after appropriate practice or drill sessions. Tests must take place under controlled conditions so instructors can make reliable, valid judgments about student performance. Instructors and test developers should consider the following guidelines when creating performance tests:

- **Specify performance objectives to be measured** — All test items should be directly related to the relevant objectives. To save time, each test item should require that students perform several basic skills. For example, a test item that requires ventilating a pitched roof also requires students to demonstrate proper use of ground and roof ladders, cutting tools, safety ropes, and hoselines.

Figure 13.9 Performance tests measure psychomotor skills.

- **Select rating factors on which the test will be judged** — Rate students against a standard, not against the performance of other students **(Figure 13.10, p. 279-280)**. Design a rating form that includes the following:
    — Student's approach to a stated job or procedure
    — Care shown in handling tools, equipment, and materials
    — Demonstration of accuracy
    — Time required to complete a job or procedure safely

- **Provide written instructions that clearly explain the test situation** — Review these instructions orally and give students the opportunity to ask questions if necessary so students understand how to proceed.

- **Confirm a new performance test with other instructors or previous students before administering it to students** — Conduct a trial test to uncover problems that can be corrected.

- **Use more than one test evaluator** — Ask other instructors or officers to be test evaluators. For certification testing, include more than one test evaluator. Ensure that instructors do not evaluate their former students on material they covered together during the course. Provide instructions to evaluators about what they should look for during the test and how to use the rating scales and forms. Testing may be either scored or graded as pass/fail.

- **Follow established procedures when administering the test** — All necessary apparatus and equipment must be ready before beginning the tests. Evaluators must use the same equipment, follow the same task sequence, and rate performance on the same basis for each student. Eliminate all distractions from the testing area so that evaluators can concentrate on observing and evaluating, and students can concentrate on demonstrating their proficiency.

- **Make a score distribution chart after tests have been administered and graded** — Evaluate students with low scores. Focus additional attention on students who demonstrate difficulty in performing manipulative skills.

- **Rotate team members to every position for team evaluation ratings** — Ensure that each student is observed and evaluated in each position in an evolution.

Performance tests have many of the following advantages:

- **Validity** — A performance test is the only valid method of measuring a student's ability to perform manipulative skills.

- **Reliability** — A properly constructed performance test using specific criteria is a reliable measure of performance when coupled with an appropriate rating scale.

DEPARTMENT OF FORESTRY AND FIRE PROTECTION
Office of State Fire Marshal
State Fire Training

Skill Sheet # 4-1

Candidate: _____ Date: _____

ID#: _____

Skill Sheet 4-1
NFPA Standard 1001, 2013 Edition, JPR #6.4.1

Extricate a Victim Entrapped in a Motor Vehicle

Evaluator Instructions: Depending on the number of candidates being tested it may not be practical to have a vehicle for each candidate. It may be necessary to develop props that will simulate each of the different tasks identified.

Equipment Needed: Personal protective equipment, extrication equipment, cribbing, salvage vehicle, backboard/KED, vehicles

Task: Extricate a victim entrapped in a motor vehicle

Performance Outcome: As part of a team, given stabilization and extrication tools, a vehicle, and personal protective equipment, will be able to extricate a victim entrapped in a motor vehicle, stabilize the vehicle, disentangle the victim without further injury, and manage hazards.

Candidate Directive: "Demonstrate proper methods as part of a team to extricate a victim entrapped in a motor vehicle"

No.	Task Steps	First Test		Retest	
		P	F	P	F
	Select two (2) of the following tasks for moving or removing: ☐ Vehicle Roofs ☐ Vehicle Doors ☐ Vehicle windshields and windows ☐ Vehicle steering wheels and/or columns ☐ Vehicle dashboards				
1.	Size up the situation to identify hazards, determine required stabilization, and select appropriate extrication techniques				
2.	Stabilize a vehicle using cribbing and/or shoring material				
3.	Correctly operate both hand and power extrication tools in a safe and efficient manner				
4.	Perform extrication techniques and disentangle victim without causing further injury				
Retest Approved By:		Retest Evaluation:			

Continued on next page

Figure 13.10 Using checklists for evaluation helps to ensure that students are rated against a standard and not against one another. *Courtesy of California State Fire Training and the Oakland Fire Department.*

DEPARTMENT OF FORESTRY AND FIRE PROTECTION Office of State Fire Marshal **State Fire Training**	Skill Sheet # 4-1

Evaluator Comments:	Candidate Comments:

_____ _____ _____ _____
Evaluator   Date   Candidate   Date

_____ _____ _____ _____
Retest Evaluator   Date   Retest Candidate   Date

Figure 13.10 (continued)

- **Skills-based** — Students who may not express themselves well orally or in writing may be able to perform a set of skills as well as or better than other students.
- **Student motivation** — Performance tests motivate students to practice skills, both in and out of class.
- **Sense of accomplishment** — Students who successfully complete well-prepared and carefully administered performance tests will understand their level of achievement and be proud of their accomplishments.
- **Job related** — Students recognize that the psychomotor skills they are being tested for are directly related to the performance of their duties as fire and emergency services personnel.

Affective Domain Test Items

Test items are not normally written to evaluate the affective domain because affective objectives are often difficult to measure. Learning objectives that include a change in affective behavior can be evaluated based on instructor observations during training.

Test Planning

Test planning consists of three major steps:

Step 1: Determine test purpose and classification

Step 2: Identify learning objectives

Step 3: Construct the appropriate test items

Determining Test Purpose and Classification

The purpose of a test refers to the overall expectation of student performance. This purpose includes an indication of whether students will be evaluated in comparison to other students, or based on a set of predetermined qualifications (See the "Review: Criterion-Referenced and Norm-Referenced Tests" box which reviews information from Chapter 9: Testing and Evaluation).

Review: Criterion-Referenced and Norm-Referenced Tests

- **Criterion-referenced tests** — Compare performance against appropriate minimum standards.
- **Norm-referenced tests** — Rate student performance compared to other students.

Norm-referenced tests are very effective for promotional examinations when ranking applicants is desirable. Outside of this particular use, norm-referenced tests are rare in the fire and emergency services. Norm-referenced tests should never be used as end-of-course or certification tests.

The majority of tests in the fire and emergency services use criteria that have been established through the following sources:

- NFPA professional qualifications standards
- Federal, state/provincial, or local requirements
- Other professionally accepted requirements

In addition to the purpose of a test, the classification of a test should also be defined before it is administered. The classification will indicate to students what to expect from the questions as indicated later in this chapter, and also to provide administration a way to group similar tests over time. When planning tests, the instructor should also consider whether the test:

- Is designed to determine readiness for instruction or placement in the appropriate instructional level (prescriptive or placement test)
- Is designed to measure improved progress or identify learning problems that are hampering progress (formative or progress test)
- Is designed to rate terminal performance (summative or comprehensive test)
- Whether the test measures technical knowledge retention and recall in the cognitive domain (written or oral tests)
- Measures manipulative skills in the psychomotor domain (performance or skill tests)
- Measures behavioral changes in attitude, values, or beliefs in the affective domain (written or oral tests)

For review of test classifications as examined in Chapter 9: Testing and Evaluation, see the "Review: Test Classifications" box **on p. 282**.

> **Review: Test Classifications**
>
> Though there are other classifications, most tests in the fire and emergency services fall into one of the following categories:
>
> - **Prescriptive (Pretest)** — Given at the beginning of instruction to establish a student's current level of knowledge
> - **Formative (Progress)** — Quizzes, pop tests, or question/answer periods in class that are given throughout the course or unit of instruction; typically measure improvement and give the instructor and students feedback on learning progress
> - **Summative (Comprehensive)** — Measure student achievement in an entire area on a number of topics covered over a long period of time such as a semester or other major segment of a course
>
> These three classifications have common characteristics and often overlap to some degree. For instance, summative test questions may be included in a formative test to determine whether unit objectives have been learned before advancing to another unit within the course.

Identifying Learning Objectives

After determining the purpose and classification of the test, the instructor must identify the learning objectives the test will evaluate. Learning objectives reflect the course outcomes, which are broad statements explaining what students should have learned at the completion of a course. For example: *Upon completion of the High-Angle Rescue Course, given proper equipment, the student will be able to perform a high-angle rescue.* The test should be based upon the learning or behavioral objectives taken from each part of the course or individual lesson plans because they are specific, detailed, and measurable. Test questions, like learning objectives, are often based upon the cognitive levels of learning originally described in Chapter 11, Lesson Plan Development. For review of student cognitive levels of learning, see the "Review: Cognitive Levels of Learning" box.

> **Review: Cognitive Levels of Learning**
>
> The levels of learning in the cognitive domain, ranked from the simplest to the most complex, are as follows:
>
> - **Remember (Knowledge)** — Students remember, recall, and recognize previously learned facts and theories. They can describe, define, label, list, and match terms and items. Sample Question: *Which of the following is an indication that a victim is in shock?*
> - **Understand (Comprehension)** — Students understand, compare, and contrast information, and estimate future trends. They give examples and explanations, make predictions, and summarize information and ideas. Sample Question: *If one point of the fire tetrahedron is removed during combustion, which of the following will occur?*
> - **Apply (Application)** — Students apply information, rules, and concepts that they have learned to new situations. They compute, demonstrate abilities, solve problems, modify ideas and actions, and operate equipment. Sample Question: *How would you approach a fire scene if smoke had been reported coming from under the eaves of the house?*
> - **Analyze (Analysis)** — Students divide information into its component parts to understand how they relate to one another and to the whole. Sample Question: *Based on the case study included in the chapter, what is the role of Truck Company B at this fire scene?*
> - **Evaluate (Evaluation)** — Students judge the value of materials or actions based on defined criteria using elements from all other levels. They compare, conclude, contrast, discriminate, and justify decisions based on standards and criteria. Sample Question: *Given the information provided about a residential fire in the narrative above, what actions should the responders have taken to prevent the incident from escalating beyond the first residence?*

- **Create (Synthesis)** — Students put parts together to form a new whole. They categorize, create, design, organize, revise, and integrate parts to invent something new. Sample Question: *How would you organize the command structure at a mass casualty incident?*

All instruction should be based on this hierarchy of cognitive levels, first exposing the student to simpler information before moving on to more challenging material. Testing may or may not follow this progression, but should include test items that are written at a cognitive level that corresponds to the learning objectives in the course.

Constructing Appropriate Test Items

After determining the purpose and classification of a test, and identifying the specific learning objectives, test developers and instructors are ready to meet the challenge of matching the specifications with individual test items. Writing effective tests requires that the instructor completes the following tasks:

- Selecting proper level of test item difficulty
- Determining appropriate number of test items
- Eliminating language and comprehension barriers
- Avoiding clues to test answers
- Ensuring test usability
- Ensuring validity and reliability

Selecting Proper Level of Test Item Difficulty

Test item difficulty can be arranged in two significant ways. When a test includes items with a range of difficulties, a student's overall knowledge is tested. Test results in this strategy will favor students who are generally knowledgeable, and most students should be able to answer most questions. In contrast, when a test includes only difficult questions, a student's depth of specific knowledge is tested. Test results in this strategy will favor students who have exceptional knowledge.

Test item difficulty is often a factor of whether the test is criterion-referenced or norm-referenced. Criterion-referenced tests are developed with the intention of evaluating whether students in a class have a working knowledge of a skill or data set. This can be accomplished with test items showing a wide range of item difficulty. On a criterion-referenced test, all test items should be designed so knowledgeable students should be able to pass the test. Criterion-referenced test items should match the cognitive level of the learning objectives in the course.

In contrast, the intention behind norm-referenced tests is a wide distribution of test scores. Test items that can be answered by the majority of students should not be included. The most desirable test items can be easily answered correctly by only half of the class, or less. This level of difficulty ensures that the test results reveal a clear separation between students who know the information and those who do not. This separation allows evaluators to rank students' performance against their peers.

Determining the Appropriate Number of Test Items

The appropriate number of items for any test depends on the following factors:

- Purpose of the test
- Types of test items or performance items
- Desired level of reliability

A test should measure students' abilities in all phases of a course. However, it must also fit within the time constraints for administering the test. Use either an established set of guidelines or a mathematical approach to determine the appropriate number of test items, as follows:

- **Guideline approach** — Write questions to address each difficulty level or multiple levels of learning in the cognitive domain based upon:
  - The length of time allowed for the test
  - The number of learning objectives to be tested
  - The importance (weight) of the learning objectives (learning objectives considered more important should receive more questions)
- **Mathematical approach** — Instructional time for each learning objective determines the percentage of test items that address that objective. For example, an objective that took 30 minutes to teach in a 120-minute course would be addressed by 25 percent of the test items.

Eliminating Language and Comprehension Barriers

Instructors should eliminate, or at least minimize, language and comprehension barriers. One approach is to use words that students would use during training or on the job. Also maintain awareness of the students' learner characteristics and learning styles. Avoid test items that include the following:

- Higher reading level than students possess
- Lengthy, complex, or unclear sentences
- Vague directions
- Unclear graphic materials
- Obsolete words or terms

Avoiding Giving Clues to Test Answers

Test items should not give clues to correct answers. Some areas to avoid include:

- Word associations that give away the answer.
- Plural or singular verbs, or use of the words "a" or "an," that may hint at the answer or eliminate an answer.
- Words that make some answers more likely, such as *sometimes*, or less likely, such as *always* or *never*. *Never* and *always* are appropriate when the student should absolutely know the information, for example, questions that emphasize safety.
- Correct answers that are consistently placed in the same location, such as the Choice B answer in multiple-choice questions.
- Correct answers that are consistently longer than distractors, such as true statements that are always longer than false statements.
- Stereotypical answers.
- Test items that give the answer to other test items.

Ensuring Test Usability

A highly usable test is easy for instructors to give, easy for students to take, and convenient and cost-effective for the training organization **(Figure 13.11)**. Usability is an important factor to consider when developing a test, or when selecting a test that another organization developed. Tests that are difficult to administer or score, or that are difficult for students to understand should be avoided.

Figure 13.11 Construct tests that will be easy to administer, easy for students to take, and easy to score and grade electronically.

A usable test has the following characteristics:

- **Easy to administer** — Tests should include simple, clear instructions for administering the test. All parties, including an instructor with no experience administering a test, should easily understand the instructions.
- **Easy to take** — Tests should include clear directions that tell students how to answer test items, and include sample questions and answers.
- **Appropriate length** — The number of test items included should be sufficient to provide accurate assessment of student learning and obtain valid and reliable results. The number of test items should be limited so all can be answered during the testing period.
- **Cost-effective** — Internally developed tests often require personnel to spend many hours developing them properly. It may be more cost-effective to purchase tests from a professional or specialized publisher.
- **Contains multiple testing instruments** — Some test-preparation firms offer multiple tests with different sets of test items on the same subjects or learning objectives. Different tests over the same subjects may allow instructors to use a variety of textbooks, skills sheets, or other ancillary materials. Multiple tests also make it more difficult for students to cheat.
- **Easy to score** — Answer sheets should be easy to read and easy to use. Ideally, instructors should be able to score and grade tests electronically. This method is not only faster, it is also more accurate.

Ensuring Validity and Reliability

The two most important characteristics of a well-designed test are validity and reliability. Validity is the extent to which a test measures what it is supposed to measure. To validly measure whether students have achieved the desired learning objectives, test items should require students to display specific knowledge of the appropriate learning objectives. One way to ensure validity is to include an ample number of test items for each learning level and content area.

Reliability is the extent to which a test provides consistent, accurate measurements of student achievement. A reliable test should have the following characteristics:

- Clear instructions
- Clear, well-written test items
- Specific scoring criteria

Instructors should take the following steps to help ensure that the tests they write are valid and reliable:

Step 1: Select a representative sample of learning objectives.

Step 2: Select enough test items to represent the skills required in the learning objectives.

Step 3: Select test-item formats that reduce the potential for guessing.

Step 4: Use only the number of test items that an average student can complete in the available time.

Step 5: Determine methods to maintain positive student attitudes toward testing.

At the conclusion of each test, the scores should be analyzed for items that do not meet the stated requirements. Those items should be discarded and rewritten.

Finally, bias is an important factor that can invalidate a test instrument. No tests should be biased toward one group or individual. Bias often includes words or cultural references that are unknown, exclusionary, or difficult to understand for a group or individual because of ethnic, economic, social, gender, or cultural influences and background. Tests should use universally understood and noncontroversial terminology.

Test Scoring Method Selection

The final step in creating a test instrument is determining how to score it, which includes establishing the criteria for passing and failing. Scoring systems vary depending upon the type of test and the importance (weight) of the questions on the exam.

For written tests, a point value should be assigned to each question. All questions of the same type can be given the same value or different questions can be weighted. For example, items that address the most important learning objectives may be worth more (weighted more heavily) than other items of the same item type. Short answer and essay items may also be worth more than other item types because they take more time to complete and require greater understanding of the course material. However an instructor chooses to weight the scoring of different items, the scoring system should always be explained on the test sheet.

Scoring sheets for oral tests should indicate proper and accepted responses and how many points those responses are worth. Oral tests may vary according to the interpretation of the instructor. Scoring sheets should remove this subjectivity to the greatest extent possible.

Similarly to oral tests, performance or skill tests can be very subjective. The training division should guard against subjectivity with the following guidelines:

- Train instructors or test evaluators in the intended steps of the skills to be tested. This training is especially important when instructors represent a variety of jurisdictions.
- Develop checklists for each tested skill, and use them while scoring students' performances. Checklists ensure that the instructors score each step against the same criteria.

As discussed in Chapter 9, some tasks in some skills may be considered mandatory. When creating scoring sheets for mandatory tasks, those tasks should be indicated as pass/fail. Regardless of the student's performance on other tasks, instructions should also indicate that skipping or failing mandatory tasks will result in a failing score.

Finally, criteria for passing the test should be included in the test instructions and on the scoring sheet, so that both the student and the test evaluator are aware of them. Depending upon the type of test and the learning objectives it addresses, the criteria for passing may be different. For example, a written, multiple-choice test may have a passing score of 75 percent, while a performance test of highly hazardous skills may require a score of 100 percent to pass.

Chapter Review

Answer the following questions to review the information provided in this chapter.

1. What are some common considerations for all test instruments?
2. What type of ability does a written test assess?
3. What type of ability does an oral test assess?
4. What type of ability does a performance test assess?
5. What are the three major steps for test planning?
6. How does a Level II Instructor select a test scoring method?

Discussion Questions

The following questions are intended to generate discussion, expand your understanding of the chapter text, and allow you to think critically about what you have learned. Answers to these questions may vary.

1. Why is it important to consider the type of testing instrument to use for fire and emergency services training?
2. What are some advantages and disadvantages to using written tests?
3. What are some advantages and disadvantages to using oral tests?
4. What are some advantages and disadvantages to using performance tests?
5. What issues may arise if one of the major steps for test planning is not used when creating a test?
6. Why is it important to create scoring sheets for any type of test?

Key Terms

Alternatives — Possible answers in a multiple-choice test item.

Distractors — Possible answers in a multiple-choice test item that are incorrect but plausible.

Stem — The question or introductory statement in a multiple-choice test item.

Testing Instrument — Series of test items that are based on learning objectives and collectively measure student learning on a specific topic.

Test Item — Single question on a testing instrument that elicits a student response and can be scored for accuracy.

13-1
Develop test items. [NFPA 1041, 5.5.2]

Task Steps

Step 1: Review departmental policy regarding test development.

Step 2: Determine purpose and type of test item to be developed.

Step 3: Identify specific learning objectives/outcomes to be covered by test item.

Step 4: Determine level of learning to be assessed.

Step 5: Create appropriate cognitive test item **(Figure 13.12)**.

Step 6: Identify correct answer on rubric.

Step 7: Ensure test usability, validity, and reliability.

Step 8: Create answer key.

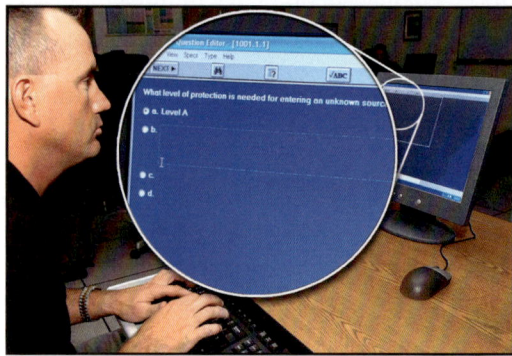

Figure 13.12

13-2
Develop a performance skills evaluation. [NFPA 1041, 5.5.2]

Task Steps

Step 1: Review department policy regarding performance skills evaluation instruments **(Figure 13.13)**.

Step 2: Identify skills to be evaluated.

Step 3: Reference learning objectives of the performance skills.

Step 4: List tasks to be performed **(Figure 13.14)**.

Step 5: Develop a scoring/rating system.

Step 6: Develop a passing score and/or critical failures.

Step 7: Write instructions for the proctor in evaluating performance skills.

Step 8: Identify resources required for the skill being evaluated.

Step 9: Identify the conditions under which the evaluation will take place.

Figure 13.13

	Task Steps	Yes	No
	Tool Cleaning		
1.	Clean tools according to manufacturer's guidelines.		
2.	Dry tools thoroughly.		

	Task Steps	Yes	No
	Tool Inspection		
1.	Inspect tools for damage or wear.		
2.	Inspect parts for tightness and function. a. Ensure all guards are in place and functional. b. Check all electrical components for cuts or other damage.		
3.	Place any tools that require maintenance on a salvage cover or clean surface and tag them out of service.		

	Task Steps	Yes	No
	Tool Maintenance		
1.	Maintain cutting blades and replace blades that are damaged or worn.		
2.	Check fuel level and fill with the correct fuel.		
3.	Check oil level and fill with the correct oil.		
4.	Start all power tools and verify their operation. Turn off power tools after operation is verified.		
5.	Tag tools that must be placed out of service.		
6.	Record cleaning, inspection, and maintenance according to local SOPs.		

Figure 13.14

Supervisory and Administrative Duties

Chapter 14

SECTION B INSTRUCTOR II

Chapter Contents

- **Supervising Other Instructors** 293
  - Establishing and Communicating Goals and Objectives 294
  - Promoting Professional Development. 294
  - Empowering Instructors. 295
- **Scheduling Resources and Instructional Delivery** 297
  - Assessing Factors That Affect Scheduling 297
  - Determining Scheduling Needs 298
  - Determining Requirements. 300
  - Determining Availability 300
  - Coordinating Training. 300
  - Creating a Schedule 301
  - Publishing the Schedule. 301
  - Revising the Schedule 302
- **Recommending Budget Needs** 302
  - Funding Needs Determination 302
- AHJ Budget Policies. 304
- Sources of Funds 304
- Budget Request Justification 306
- **Purchasing Process** 307
  - Determining Funding Sources 308
  - Determining Purchasing Needs 309
  - Contacting Vendors 309
  - Purchase Orders. 309
- **Managing Training Records** 310
  - Training Information. 310
  - Record Management Systems 311
  - Record Auditing Procedures 312
  - Legal Requirements for Training Records 312
- **Chapter Review** 313
- **Discussion Questions** 314
- **Key Terms** 314
- **Skill Sheets** 315

JPRs addressed in this chapter

This chapter provides information that addresses the following job performance requirements of NFPA 1041, *Standard for Fire Service Instructor Professional Qualifications*, 2019 Edition.

5.2.2   5.2.5
5.2.3   5.2.6
5.2.4

Learning Objectives

1. Discuss techniques for supervising other instructors. [5.2.6]
2. Describe the tasks necessary for scheduling instructional delivery and resources. [5.2.2]
3. Explain the process used for recommending budget needs. [5.2.3, 5.2.4]
4. Discuss the components of the purchasing process. [5.2.4]
5. Explain the aspects of managing training records. [5.2.5]
6. Skill Sheet 14-1: Select resources, staff, and facilities for specified instructional sessions. [5.2.2]
7. Skill Sheet 14-2: Perform a needs analysis for training resources. [5.2.3, 5.2.4]
8. Skill Sheet 14-3: Maintain and secure accurate training records so that all agency and legal requirements are met. [5.2.5]

Chapter 14
Supervisory and Administrative Duties

Besides developing and presenting instructional material, the Level II Instructor is assigned duties with the administration and supervision of the training division. This chapter concentrates on the knowledge and skills required to perform instructional resource management duties. They include the following:

- **Supervisory techniques** — Applying sound supervisory skills to create a positive work environment and resolve conflicts between instructors or between instructors and students
- **Resource and instructional delivery scheduling** — Ensuring that courses, instructors, and resources are scheduled for efficient instruction
- **Recommending budget needs** — Creating program and division budgets
- **Purchasing process** — Procuring equipment, materials, apparatus, and facilities for training
- **Training records management** — Maintaining training records in accordance with all legal, agency, and jurisdictional requirements
- **Conducting research** — Applying effective research techniques to both administrative and instructional duties

NOTE: For the purposes of this chapter, a supervisor is anyone responsible for the activities of one or more instructors. The term should be considered synonymous with the phrase *Level II Instructor*.

Supervising Other Instructors

Level II Instructors may be responsible for supervising other instructors. The number of instructors who are directly supervised varies according to the following factors:

- Size of the training division or department
- Supervisor's position and duties
- Abilities of the instructors
- Complexity of the learning environment or learning scenario

In addition to demonstrating effective leadership skills, supervisors should know how to do the following:

- Encourage instructors to participate in the decision-making process.
- Delegate and involve instructors in planning.
- Respect the judgment of instructors.
- Teach, enforce, and follow health and safety rules.
- Coach and mentor instructors.
- Encourage diversity among instructors.
- Acknowledge instructors' accomplishments.

- Treat each instructor fairly and equitably.
- Keep accurate records.
- Keep lines of communication open at all times.
- Build and maintain a professional work environment.
- Resolve conflicts between instructors or between students and instructors.

Above all, the supervisor must consistently apply all these techniques. Lack of consistency undermines a supervisor's authority and ability to accomplish organizational goals.

Level II Instructors must always be leaders, and leadership goes beyond the scope of this manual. Level II Instructors with supervisory duties should seek out other sources to help hone their leadership skills. They should remember that they lead by example. They must adhere to a standard of ethical, moral, and legal behavior that will motivate peers and other instructors to do the same.

The sections that follow detail a supervisor's primary responsibilities within the AHJ. Supervisors should seek out other sources that illustrate good supervisory techniques and leadership skills to supplement the basic information included here.

> **Review: Supervision**
> This section contains information about supervision of instructors as a general skill for Level II Instructors. Supervision information that pertains directly to training can be found in Chapter 8, Skills-Based Training Beyond the Classroom, and Chapter 12, Training Evolution Supervision.

Establishing and Communicating Goals and Objectives

Level II Instructors set the training objectives for instructors under their supervision. The training agenda is only as relevant as the communication between a supervisor and the instructors. Objectives and timelines can be communicated to instructors in two ways:

1. Communication through group meetings where information is shared and discussed
2. Written task sheets, including personalized assignments and deadlines

Supervisors should involve instructors in the decision-making process, including:

- Setting goals and timelines
- Tracking progress
- Providing and accepting feedback
- Establishing a plan for evaluation
- Identifying areas of strengths and weaknesses

Supervisors who communicate goals and objectives clearly and provide periodic progress reports should find that instructors function more efficiently and effectively. In addition, supervisors who involve instructors in establishing the objectives should find that their instructors have more incentive to fulfill the objectives.

Promoting Professional Development

Instructors must maintain and develop their professional skills. They benefit from having a supervisor who actively recommends or provides opportunities for development. In some cases, a well-informed supervisor can provide this training directly. By emphasizing instructors' personal professional development, supervisors demonstrate their commitment to developing well-trained, highly qualified instructors. This one-on-one instructional coaching also helps the continual fostering of instructional methods and techniques in a team setting. Supervisors can help develop instructors by utilizing professional development opportunities such as:

- Higher learning educational facilities
- Mentorships
- Conferences and conventions
- Fire instructional organizations

Empowering Instructors

Level II Instructors should empower the instructors they supervise. The empowerment shows confidence in their skills, judgment, and abilities. It also helps to motivate instructors and improve morale. To effectively empower instructors, supervisors must relinquish some authority **(Figure 14.1)**. The supervisor should focus on:

- Celebrating instructors' accomplishments
- Offering recognition for quality performance
- Resolving conflicts
- Setting a good example
- Maintaining a positive attitude

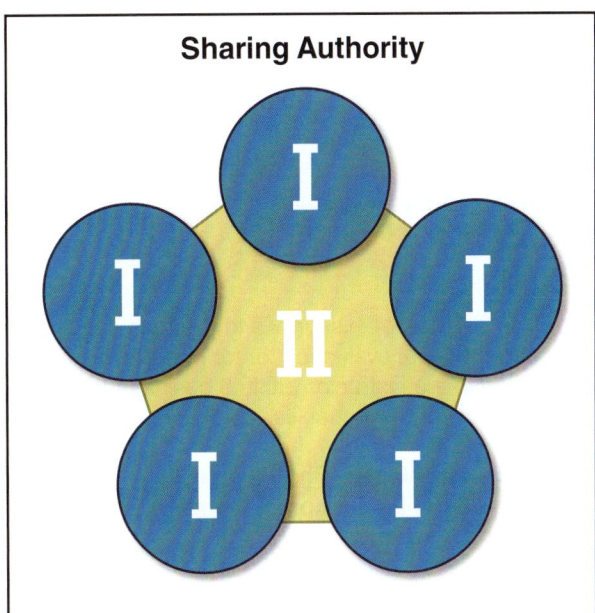

Figure 14.1 To achieve greater effectiveness, supervisors must share authority with instructors.

Celebrating Instructor Accomplishments

When personnel meet objectives or achieve significant accomplishments, a supervisor should celebrate the achievement as soon as possible. This acknowledgement shows that individual and team contributions are important to the success of the organization. A supervisor should announce the accomplishment to the rest of the department/organization and congratulate the participants on the results.

Offering Recognition for Quality Performance

Recognition can be earned through effort or participation. Level II Instructors can use it as a motivational tool. In addition to making public acknowledgements of accomplishments, the following actions are examples of ways to recognize instructors:

- Hold group gatherings or parties to encourage unit cohesiveness and spirit.
- Make positive statements about the skills and abilities of instructors, whether individually or as a team.
- Make appropriate comments on the instructor's job performance evaluation, which can result in raises or promotion **(Figure 14.2, p. 296)**.

Resolving Conflicts

One of a supervisor's most important roles is to resolve conflicts between instructors or between instructors and students. In any conflict, the supervisor has to function as an intermediary. Ideally, a conflict should be resolved to the satisfaction of all parties involved, although this may not always be possible. Resolving conflicts takes practice, but the following guidelines may help **(Figure 14.3, p. 297)**:

- Focus on the conflict, not the individuals involved.
- Depending on the situation, speak to the parties of a dispute individually and privately.
- Allow the parties of the dispute to express themselves freely.
- Get the whole story from all involved, gather the facts, and don't jump to conclusions.
- Make decisions based on policies and procedures, and avoid personal bias.
- Show how the solution to the conflict supports the mission of the training and professional development.

Field Instructor Performance Evaluation Form

The purpose of this form is to evaluate the performance of an OSU FST field instructor. It may be used for an annual evaluation, or for performance during a specific training event. A signed copy must be provided to the instructor within 5 business days following the evaluation, and a copy must be placed in the instructor's personnel file.

Instructor Name: Sharon Myers   **Date of Evaluation:** July 10, 2018
Purpose of Evaluation: Annual ☑ Other ☐
Period Covered: 2017-2018   **Evaluator:** Thomas Payne

Strengths
Sharon is an outstanding instructor. She takes the time to learn the needs of the department and then delivers the training needed to meet those needs. She is well prepared for every class. Sharon actively engages the students in the content and ensures all learning objectives are achieved. Comments from fire chiefs and students are always positive. Sharon is an exemplary ambassador for OSU FST.

Areas for Improvement
Patience with low-motivation students. Sharon, at times, displays a lack of patience for students who are not committed to training being conducted. She has dismissed several students from a training session who were not actively participating and following instructors. Sharon should work with the fire chiefs and supervisor of those students to correct any behavior or performance which does not meet her expectation.

Improvement Plan/Expectations
Attend the OSU professional development workshop "Generations in the Classroom". Sharon is to attend the next available offering, which should be in August before the start of the fall semester. A workshop certificate is to be provided to me on completion of the training.

Comments of Field Instructor (Optional)
I understand the issue with my patience. I will work on my response to students who lack motivation.

Thomas Payne   8/9/18
Evaluator Signature   Date

Sharon Myers   8-9-18
Field Instructor Signature   Date

Figure 14.2 Comments on instructor performance evaluation forms should address specific points and include strengths and areas for improvement. *Courtesy of Oklahoma Fire Service Training.*

Maintaining Positive Examples/Attitudes

A supervisor should express a positive attitude toward all personnel. This attitude is reflected through personal actions that set a good example for others, which fosters mutual respect in other working relationships. Even if a supervisor's evaluation recommends changes to an instructor's teaching style or communication methods, the supervisor's positive attitude will also encourage one in the instructor.

Scheduling Resources and Instructional Delivery

Another basic task for a Level II Instructor is scheduling resources and instructional delivery. Training resources include the elements needed to present a course or curriculum:

- Personnel
- Funding
- Facilities
- Materials
- Information
- Time

For any course or curriculum, at any level of the organization, a supervisor should follow these steps when creating a schedule:

Step 1: Assess factors that affect scheduling.

Step 2: Determine scheduling needs.

Step 3: Determine requirements.

Step 4: Determine availability.

Step 5: Coordinate training.

Step 6: Create a schedule.

Step 7: Publish the schedule.

Step 8: Revise the schedule (as needed).

NOTE: Skills associated with selecting resources, staff, and facilities for specified instructional sessions are shown in **Skill Sheet 14-1**.

Figure 14.3 Supervisors must strive to resolve conflicts professionally, focusing on results instead of personalities.

Assessing Factors That Affect Scheduling

A supervisor has to prioritize several factors when creating a schedule, including the following:

- **Training requirements** — These include government mandates for all covered topics, and the minimum amount of time to be spent on each. Local and state/provincial mandates may apply.
- **Physical resources** — Supervisors must always reserve classroom and training space, and may need to reserve props and apparatus. They may also have to order student workbooks and other instructional texts.

- **Instructor availability** — Instructors may serve multiple roles within the AHJ in addition to providing training. Supervisors should arrange the schedule to minimize conflict with their other duties and with any holidays or scheduled leave **(Figure 14.4)**.
- **Student availability** — It is best to schedule training when the largest possible number of students are available. Work schedules, personal obligations, and other conflicts to training must be considered. If personnel cannot attend because of an approved scheduling conflict, the organization may provide alternative sessions.
- **Minimum staffing levels** — Training should be planned so that it will not compromise any emergency response needs.
- **Budgetary considerations** — Funds must be available to support the cost of instruction. If funds are not sufficient for all priorities, some courses may need to be postponed or cancelled.
- **Environment** — Adverse environmental conditions may create safety hazards for students, so specific types of training should not be scheduled in unprotected areas during those conditions. Similarly, training that is weather-related, such as ice rescue, should take advantage of appropriate weather conditions **(Figure 14.5, p. 300)**.
- **Contractual language** — Often employment contracts may have specific language that limits the days of the week, times of the day, and high/low temperatures at which scheduled training may take place.

Training Priority Levels

Any training that the supervisor considers should be classified based on the following priority levels:

- **Priority 1: Training that is mandated by federal, state, or provincial agencies** — Examples may include the following:
  — Certification and recertification of fire service personnel
  — Certification and recertification of emergency medical service providers
  — Hazardous materials responder competency requirements
  — Emergency responder respiratory protection recertification
  — Continuing education that is required by the federal, state, or provincial government
- **Priority 2: Training that is required to meet goals determined by the organization or jurisdiction** — For example, an organization may want to expand its ability to respond to technical rescue incidents. In a given year a priority could be set to certify all emergency responders to the Operations Level of structural collapse and a select group of emergency responders to the Technician Level.
- **Priority 3: Training that is not mandated or required but would benefit the AHJ or service** — While an organization may have required all emergency responders to meet the Operations Level of structural collapse, resources may also allow it to certify everyone at the Technician Level. Other types of Priority 3 training may include the following:
  — Attending state or national conferences
  — Additional training that does not address specific organizational goals

Determining Scheduling Needs

The next step in establishing a training schedule is to determine the organization's training needs. They will change over time, based on the following factors:

- Equipment or technology
- Personnel
- Legal concerns
- Observations at emergency incidents
- Requests from the public

Firefighter I & II Course
Fire Fighting Operations Block

DAY: __THURSDAY__   I/S: __Captain Rogers__   DATE: __10 MAY 18__

CLASS #: 0301	CLASS #: 0304	CLASS #: 0309	CLASS #: 0314
Day: 13	Day: 10	Day: 7	Day: 4
OBJ: CAR FIRES	OBJ: UNIT III	OBJ: LADDERS	OBJ: Hose Ops
PRIMARY INSTR(s)	PRIMARY INSTR(s)	PRIMARY INSTR(s)	PRIMARY INSTR(s)
Parker	Monroe	Gordon	Richards
Kent	Danvers	Stark	Wayne
MULTI-INSTRS	MULTI-INSTRS	MULTI-INSTRS	MULTI-INSTRS
Mason		Lane	Hall
Kord		Barton	Lance
Rhodes			
STUDENTS: 16	STUDENTS: 14	STUDENTS: 15	STUDENTS: 17
CLASS #: 0317	CLASS #:	LEAVE / TRAINING	APPOINTMENTS
Day: 1	Day:	Curry – SL – 5/11	0930 – Blake – Dr Appt
OBJ: UNIT I	OBJ:	Dibney – Veh Extr Trng	1330 – Kord – Cert Test
PRIMARY INSTR(s)	PRIMARY INSTR(s)	Grey – AL – 5/7 to 5/11	
Maximoff		Logan – Instr II Trng	
Blake		Prince – AL – 5/7 to 5/18	
MULTI-INSTRS	MULTI-INSTRS		
STUDENTS: 18			

INSTRUCTOR ROSTER: 24 ASSIGNED			
0 Barton	0 Hall	0 Monroe	AVAILABLE
0 Blake	0 Kent	0 Parker	Foster
0 Curry	0 Kord	0 Prince	
0 Danvers	0 Lance	0 Rhodes	
0 Dibney	0 Lane	0 Richards	
0 Foster	0 Logan	0 Rogers	
0 Gordon	0 Mason	0 Stark	
0 Grey	0 Maximoff	0 Wayne	
NOTES:			
Foster to fill in for Blake and Kord, as needed.			

Figure 14.4 Careful scheduling will enable instructors to fully prepare for their teaching responsibilities.

The supervisor should create a list of training courses or programs required to provide the minimum level of training. The schedule should reflect the short- and long-term needs of the organization in order to establish recurring and projected training needs. Recurring needs include recertification or annually mandated training. Projected training needs are based on increases in service levels, expansion of coverage areas, or changes in staffing.

Determining Requirements

Local jurisdictions and other governmental authorities usually mandate specific training. However, Level II Instructors may have to determine additional requirements rather than rely on government mandates. Required training includes the following elements:

Figure 14.5 Seasonal weather variations create unique training opportunities.

- Specific topics to be taught
- Preparation for certification or testing
- Minimum number of contact or teaching hours
- AHJ mandates

The supervisor should determine the type of required training, the amount of time to be allocated to each course, and the sequence in which the courses must be taught. Because training mandates have a tendency to change over time, the supervisor should research the requirements annually.

Determining Availability

Supervisors should determine the availability of instructors and facilities before creating a training schedule. This determination can be made by using surveys that list the required training types and expected training time. Instructors and facility managers complete the surveys, indicating the best times for training.

The availability of students is the next determination that a Level II Instructor should make. Training should be scheduled when the participants are most likely to be able to attend. Instruction may be delivered over the Internet or closed-circuit television at a time that is convenient for the organization. However the instruction will be delivered, supervisors should consider scheduled leave patterns in each jurisdiction.

Volunteer organizations usually schedule training in the evenings and on weekends, and they make attendance mandatory. However, specialized or individual training may require that volunteers attend training sessions during the workday that other agencies provide. This scheduling places a burden on students and requires them and their organizations to be flexible.

Coordinating Training

Larger classes are generally more cost-effective, so supervisors should contact other fire and emergency service organizations to see if joint training sessions can be scheduled **(Figure 14.6)**. In addition, training for nonemergency services organizations may be coordinated with the planned training. For example, the American Red Cross can provide CPR training for emergency services responders while they provide training for the general public. The training division could also offer similar training for civilians during internal CPR classes.

While more cost-effective, training that consists of multiple agencies or organizations expands the scope of the training supervisor's duties. To meet safety and training objectives, the Level II Instructor should assign supervisory roles to instructors. This technique of supervision is much like following the Incident Command System (ICS) organizational chart.

Figure 14.6 Joint training exercises can build teamwork across jurisdictions.

National, state/provincial, and regional training schedules should be consulted to take advantage of their offerings. This consultation is especially appropriate when a small number of students require training in a specialized topic not normally offered by the local training division.

Creating a Schedule

After determining needs, requirements, availability, and alternate sources for training, supervisors can create a training schedule on a 12-month cycle or a more long-term plan. They can get help organizing with tools such as computer software, dry erase boards, or easel pads. Before publishing a final schedule, supervisors should distribute a draft with times, topics, names of assigned instructors, and locations to members of the AHJ. Providing this information to instructors before setting a final schedule allows them to inform the supervisor of any conflicts.

Alternative dates should also be included when possible. Supervisors should consider providing makeup sessions in case of inclement weather, instructor or student absences, emergency incidents, and other situations. They should remain as flexible as possible to ensure that courses can be provided efficiently.

A table or chart listing courses, dates, and instructor assignments should also be distributed to the teaching staff **(Figure 14.7)**. One effective format is to list instructors' names in the left-hand column, with dates across the top row, and course names in the appropriate intersecting box.

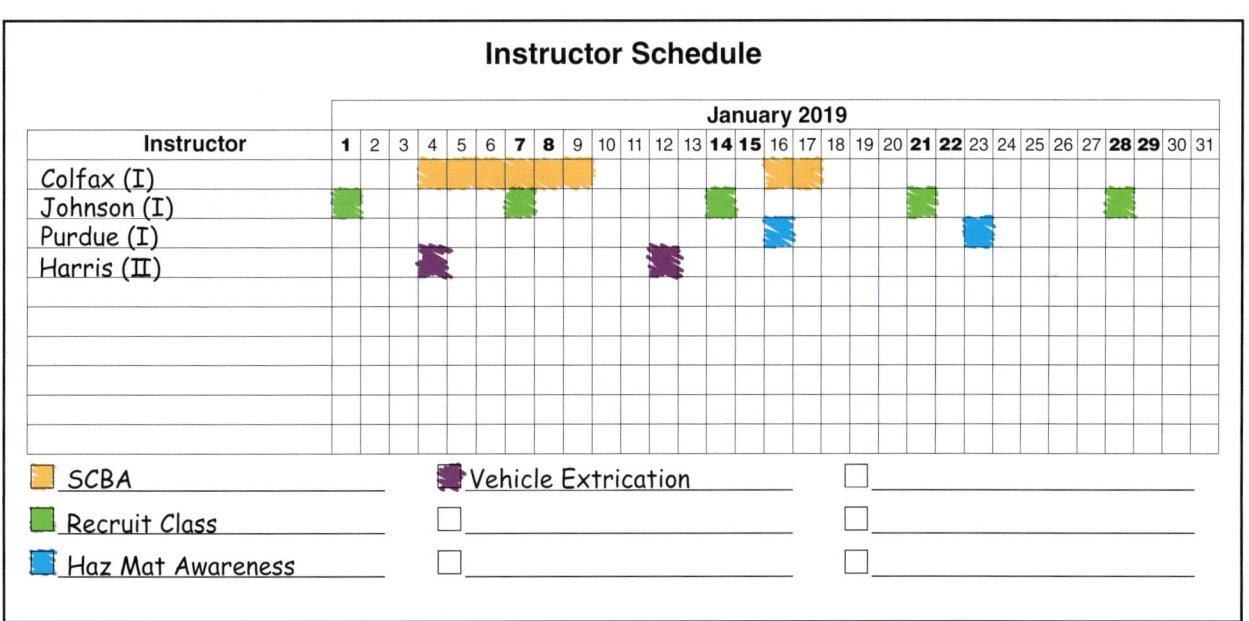

Figure 14.7 A chart listing instructors and their sessions (such as this general monthly chart) will provide a good overview of the upcoming teaching schedule.

Publishing the Schedule

Once a schedule is created, prospective students can access the completed schedule in a number of ways. Printed versions of the schedule and catalog can be distributed at all worksites and facilities. Supervisors may wish to provide an entire year's training schedule. Registration details, course prerequisites, and other information should be provided in a course catalog. The schedule can be sent to an e-mail list, posted to the training organization's website, and/or managed using computer calendar programs **(Figure 14.8)**. Combining these features on a single website may be the most effective way to allow students to register for courses.

Revising the Schedule

Both short- and long-term training schedules may require revision, for any of the following reasons:

- Instructor availability conflicts
- Inclement weather
- Lack of funds
- Lack of equipment or materials
- Lack of facilities
- Unforeseen situations
- Changes in the amount of time required to present courses
- Creation of newly mandated courses

The supervisor and training staff should review the schedule periodically to make necessary changes. Schedule changes should then be communicated to the teaching staff and organization members.

Recommending Budget Needs

In their role as supervisors, Level II Instructors may determine the budgetary needs for training in their AHJ. Typically, instructors submit budget requests to the supervisor, who then submits a budgetary form to an organization. The line items on a budget form may include:

- Purchase of materials and equipment
- Salary and benefit costs
- Miscellaneous costs
- Justification for each item

If any line item contains an unusual request or proposes a funding increase, the justification should include a summary of the training need along with relevant background research. The sections that follow introduce Level II Instructors to the process of recommending budget needs.

Funding Needs Determination

Supervisors must accurately estimate training cycle costs. Research may be necessary to accurately present these costs. For any course, there are typical operating expenses or needs, including the following:

- **Instructor pay** — Salary and benefits for all instructors
- **Equipment** — Includes audiovisual equipment, computers, and training gear
- **Course materials** — Purchase price of externally created lesson plans, or the labor cost of internally created plans
- **Student manuals** — Textbooks, workbooks, and student guides for each student
- **Audiovisuals** — Purchase price of computer-generated slides, video, and audio, or the labor cost of their internal production
- **Training ground expenses** — Fees for using a training ground or remote site, and the labor and construction cost of overhauling acquired structures
- **Travel expenses** — Costs incurred if instructors have to travel to training sites

2018 COMPANY TRAINING SCHEDULE		
	Instr.	**Subject**
JANUARY		
Monthly Subject:	C.O.	Infection Control SOP
Other:	A.T.O.	Annual Respirator Fit-Testing
HazMat:	A.T.O.	Hazmat TECH Annual Refresher Training
Officer's Class:	T.O.	Suspicious White Powder SOP
MultiCompany Drill:	T.O.	Suspicious White Powder Incident
Rescue Drill:	A.T.O.	Ice Rescue
EMS CEU:	T.O.	All Shifts: Infection Control Refresher
FEBRUARY		
Monthly Subject:	C.O.	Ch. 17, Loss Control (Essentials 6th ed.)
	C.O.E.	Annual Respirator Refresher Training
Driver Training:	A.T.O.	Autry Tech Driving Simulator [Refresher and New Driver Training]
Officer's Class:	T.O.	Administrative Policies & Procedures SOP
Rescue Drill:	T.O.	Confined Space Rescue
EMS CEU:	E.M.S.	A-Shift: Ryder - The Obese Patient
	E.M.S.	B-Shift: Compton - The Obese Patient
	E.M.S.	C-Shift: McDevitt - The Obese Patient
MARCH		
Monthly Subject:	C.O.	Ch. 18, Protecting Fire Scene Evidence (Essentials 6th ed.)
Officer's Class:	T.O.	FF Safety & Health SOP
Other:	T.O.	Emergency Services Instructor Course (Prerequisite for Lieutenant) Dates: March 7-11, 2018
	T.O.	ERG & Hazardous Communication Annual Refresher
	P.F.I.	Quarterly Physical Fitness Review
HazMat:	A.T.O.	Quarterly HazMat Drill
EMS CEU:	E.M.S.	A-Shift: Palmer - Head Injuries
	E.M.S.	B-Shift: Gates - Head Injuries
	E.M.S.	C-Shift: Craig - Head Injuries
APRIL		
Monthly Subject:	C.O.	EFD Rules and Regulations Annual Refresher
Other:	C.O.	Hydrant Flowing
Driver Training:	A.T.O.	EVDT Course [Refresher and New Driver Training]
Officer's Class:	T.O.	Special Situations SOP
Rescue Drill:	T.O.	Grain Bin Rescue
EMS CEU:	E.M.S.	A-Shift: Copeland - Chest Trauma
	E.M.S.	B-Shift: Williams - Chest Trauma
	E.M.S.	C-Shift: Jenkins - Chest Trauma

Figure 14.8 All instructors should be informed about the upcoming training offered within the organization. *Courtesy of Oklahoma State Fire Service Training.*

> **Unexpected Cost of Training**
>
> Departments with paid personnel can have expensive unintended consequences. That may be due to minimum staffing requirements or other commitments driving overtime or back-filling requirements to enable courses to be taught. A "free" class could actually cost thousands of dollars not anticipated in the budgeting process.

Supervisors should estimate the cost of each resource need in their **operational budget** request, and do research to make sure estimates are accurate. With manuals for example, supervisors should consult the publisher's catalogue to determine the price of a book, and multiply that amount by the number of expected students. The number of students and the title of the manual would also be included in the budget request.

Sometimes a course or curriculum requires that organizations make a large, one-time purchase like those found in **capital budgets**. These requests may require greater justification, or other sources of funding outside of the operational budget.

Most of the time, the funding items that instructors identify are meant to be part of the AHJ's operational budget. Completing an operational budget often involves updating the requests from the previous year's budget to reflect the current needs. A percentage is usually added that represents the rate of inflation based on the federal government's cost-of-living estimate.

AHJ Budget Policies

Supervisors must become familiar with their AHJ's budgetary policies. As previously stated, AHJ budget policies typically involve completing budget request forms **(Figure 14.9)**. Supervisors should also include justifications for the line items in their budget requests as policies dictate.

AHJ policy may require supervisors to compare their budget requests to the previous year's budget. Similarly, the AHJ may create a projected budget and ask the supervisors to compare what they see as their training needs against what the AHJ is prepared to budget, then justify any differences. In some budget cycles, supervisors are asked to determine training needs based on a fixed allocation. When there is not enough funding to cover all training needs, supervisors must justify requests and may have to search for additional funding.

Sources of Funds

Governments require revenue to provide services. Typically, this money comes from a combination of property taxes and sales taxes, but other sources include:

- Trust funds
- Enterprise funds
- Bond sales
- Grants/gifts
- Fundraising **(Figure 14.10, p. 306)**

Although instructors may not control most of these sources, they can be directly involved in grants/gifts and fundraising. Instructors who write grants or are responsible for accepting gifts must be aware of the budgetary and reporting rules of the AHJ.

In some municipal departments, even if the supervisor raises additional revenue, they cannot spend beyond the budget. However, smaller departments may have contracts with local governments that allow them to spend whatever money they can raise as part of their contracts with local governments.

Many fire and emergency services organizations supplement budgets with grants and charitable contributions from either corporations or private citizens. This type of funding is usually intended to address a specific need. In

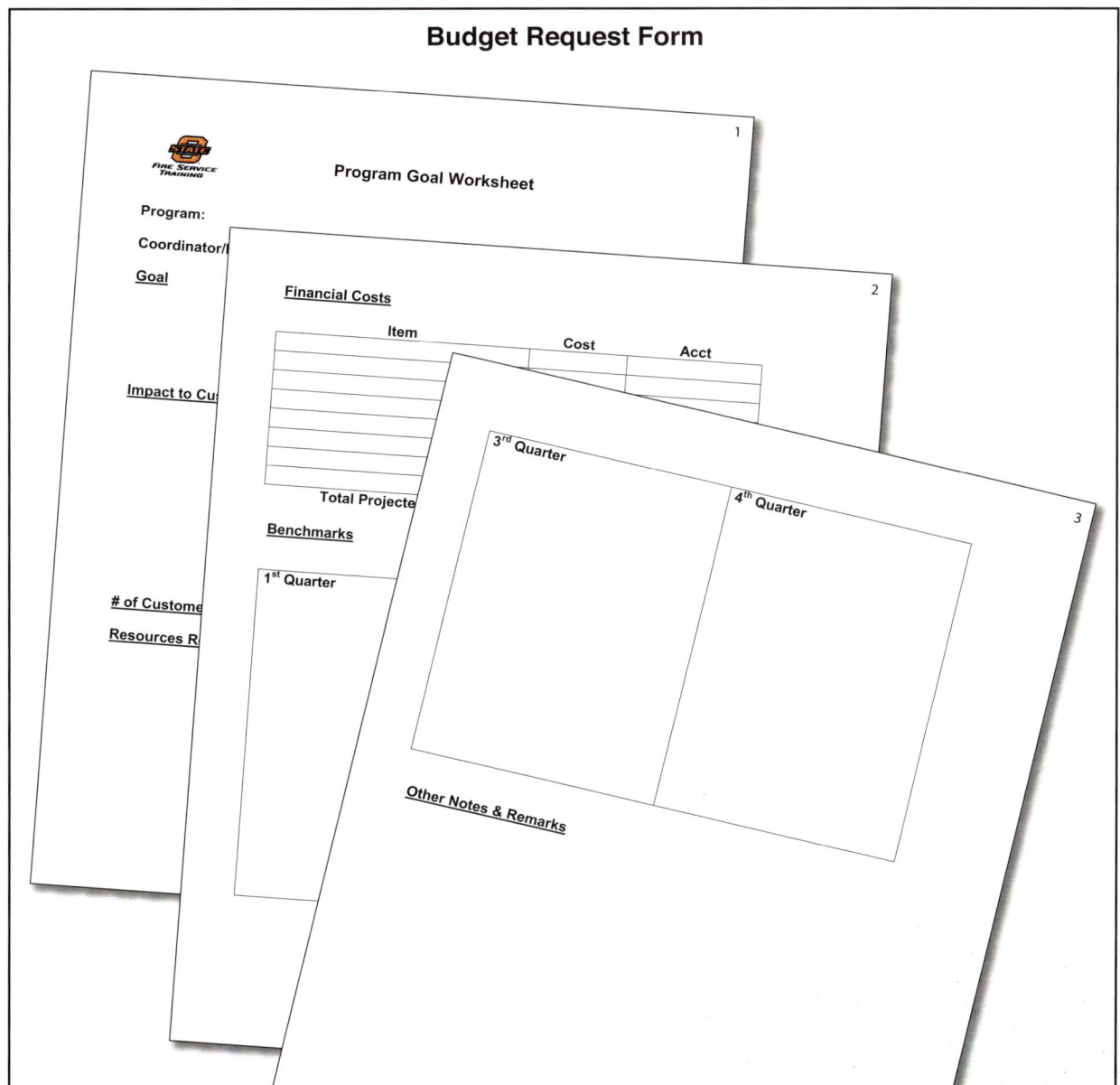

Figure 14.9 Instructors must understand the procedures and forms necessary for making budget requests. *Courtesy of Oklahoma State Fire Service Training.*

some states a portion of all fire insurance premiums paid into the insurance industry is returned to local fire and emergency services organizations to pay for training and instructional materials.

Charitable Contributions

In many jurisdictions, service clubs and other civic organizations have donated funds to purchase specialized equipment such as hydraulic rescue tools or automatic external defibrillators (AEDs). It is important that funds donated for capital purchases are used for that purpose only — not for operating expenses.

Grants

Grants to address specific organizational needs are available from governmental and nongovernmental organizations (NGOs). In the U.S., grants from the USFA, DHS, and Department of Transportation (DOT) provide local emergency responders with training and equipment to deal with a variety of incidents. The U.S. Department of

Figure 14.10 Public funds for emergency services come from a variety of sources.

Agriculture also provides grants to rural fire departments. NGOs or nonprofit organizations also give money to fund programs such as fire prevention training open to the public through the fire and emergency services organizations **(Figure 14.11)**.

The application process for securing grants can be challenging, especially for small organizations. Grant writing can require a skilled writer to be successful; however, any individual well versed in the process may be able to obtain funding. Supervisors should assign the task to staff members with the necessary skills and time or contract it to a professional service.

While many grants are based on specific needs, such as hazardous materials training, some government programs provide funding for defined purposes but with few restrictions. These consolidated funding streams are known as **block grants**. They provide organizations with flexibility in running their programs, and minimize the bureaucratic aspects of the budgeting process. An example of a block grant funded program might be a safety awareness program that provides smoke detectors, fire extinguishers, and fire safety education to older adults.

Block grants are being used more frequently in the fire and emergency services, because the fire service has done an excellent job of being accountable and responsible for the funds. Block grants also bolster community involvement; the funds obtained through the grant directly benefit the community.

NOTE: Figure 14.12 gives examples of how different funding sources function.

Budget Request Justification

Justifying a budget request requires thoroughly documented research and supporting evidence through a needs analysis to prove to the AHJ that the request is valid. Often the documentation used in the justification is the same research that was initially used to prepare the budget. Sources for this justification include the following:

Figure 14.11 The emergency services may sponsor training that is accessible to the public.

- **Organization's financial history** — Primary source of data to support the budget request, based on the actual cost of providing the services in previous cycles. Some of this history includes the cost of fuel, maintenance, utilities, parts, training, and operating supplies. It can be used to justify the operating budget or a capital request such as the replacement of an apparatus.

- **Actual equipment, material, or service costs** — Average product or item costs according to vendors' catalogs or price lists. This can also be determined by examining the jurisdiction's existing contracts for materials and services.

- **Third-party evaluations** — Information available through the state/territorial/ provincial insurance commissions or fire service accrediting agencies.

- **Training mandates** — This is particularly important when higher levels of government place training requirements on local services and then fail to provide adequate funding.

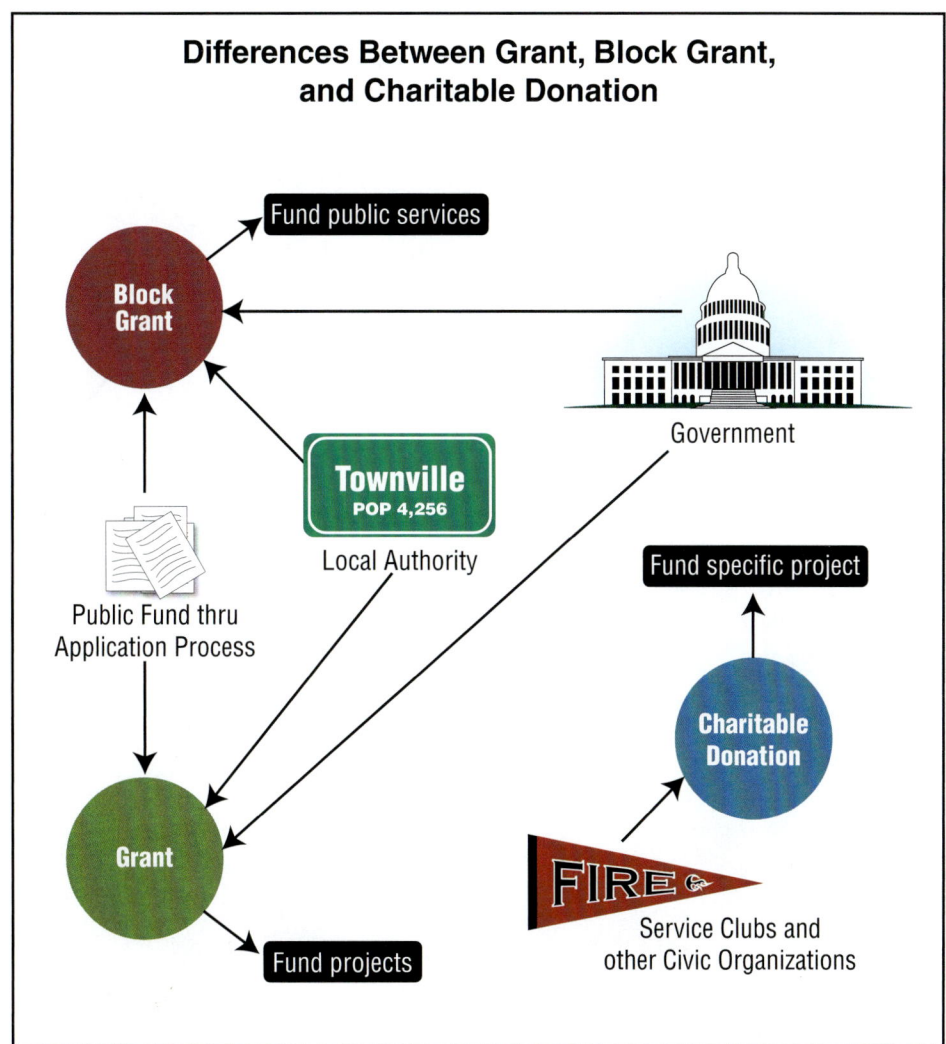

Figure 14.12 It is important to know the regulations for securing and using grants and donations.

- **Contractual requirements** — Labor/management contracts and contracts for services that an organization is obligated to provide, such as mutual or automatic aid response.
- **Injury reports and fire losses** — Incidents that resulted in firefighter injuries or fatalities may be indicators of areas where training within the organization needs to be reviewed.
- **New programs or services** — Companies that produce training materials or courses may introduce programs or services with which the organization wants to become involved.
- **Training refinements** — Deficiencies in the training regime may make additional funding necessary to refine curricula or courses.

  NOTE: Skills associated with performing a needs analysis for training resources are shown in **Skill Sheet 14-2**.

Purchasing Process

To perform its mission, a fire and emergency service organization must gather training resources such as course materials, equipment, and apparatus. Purchasing may be the responsibility of a supply, apparatus, or logistics chief; a county or city clerk; or a member of the jurisdiction's central purchasing department. A Level II Instructor may also be responsible for purchasing training materials or equipment **(Figure 14.13, p. 308)**. Regardless of their jurisdiction's size, all supervisors should understand and follow approved purchasing procedures.

Figure 14.13 In order to meet the AHJ's training needs, the Level II Instructor must understand the purchasing process for the acquisition of equipment, materials, and apparatus.

Supervisors must ensure that the organization receives the materials that were budgeted for and ordered, and keep purchasing expenses within the approved budget. In doing so, they must only spend funds on the items for which the money was allotted. For example, money budgeted for manuals should not be spent on training equipment.

The following sections outline aspects of the purchasing process. Some of the information included in these sections may be completed during the budgeting process or afterward.

Determining Funding Sources

The first step in the purchasing process is to determine sources of funding. The most common sources were discussed earlier in this chapter, in the section on budgets. But another option that local laws may permit is a lease or lease/purchase arrangement. Primary funding sources that could affect instructor purchases include the following:

- **Operating funds** — Designated in the annual training budget; includes personnel and small-resource costs.
- **Capital funds** — Designated in an annual budget for purchasing capital items, which are items that cost more than an allowable fixed value. These items are specifically requested during budget preparation, and if approved they are purchased through a bidding process.
- **Grants** — Awarded by governmental agencies and NGOs, often to fund a specific program. Grants do not have to be paid back, but the jurisdiction receiving the grant must be accountable for how the funds are spent.

- **Direct donation** — This is a method for obtaining funds from prospective donors. The donation may be targeted for a specific purpose or may be a general donation.
- **Leases or lease/purchases** — Another method of obtaining capital items, although they are not funding sources. Purchasing ordinances and jurisdictional laws determine their use. In some cases, it may be illegal to obligate funds in the subsequent budget, thereby preventing a lease. A supervisor must research this type of acquisition before including it in bid specifications. A cost/benefit analysis must be made to compare the direct purchase of equipment with the lease/purchase process cost. Definitions:
  — A lease can be used when equipment is needed for a short duration or for extended periods.
  — A lease/purchase arrangement allows the purchase cost to be spread over several years.

Determining Purchasing Needs

Determining purchasing needs should have been completed during the budget process in accordance with AHJ policies. If equipment integral to the continuation of training unexpectedly breaks, then adjustments may have to be made in the budget to purchase replacement equipment.

Needs determined during the budgeting process may be left vague. When the time comes to purchase the items in the budget, the following actions should be performed:

- **Review the standards and regulations that mandate the purchase of specific types of training equipment** — Review the legal mandates that the AHJ creates for the operation of fire and emergency services training organizations.
- **Review the current resources** — Assess how well equipment and materials meet the organization's training requirements, and determine whether instructors have quick access to them in adequate quantities.
- **Determine the amount of funds available** — Determine whether the training organization has the necessary funds for a selected purchase. It may be necessary to locate additional funds, transfer funds from unused accounts, or cancel the associated training.

Contacting Vendors

Supervisors should contact available vendors for the resources they need to purchase. In some cases, supervisors may be limited in the number of vendors from whom they can make purchases. Jurisdictions may have a publisher from whom they purchase manuals as a matter of policy. Even in these cases, the manufacturer should be contacted. They may be offering special pricing because of agreements with the jurisdiction.

If multiple manufacturers are available, the supervisor in charge of purchasing should find out which manufacturer offers the best product at the best price. Resources may be available through a state training agency, university, community college, or other third party either for a small fee, on loan, or as a donation.

Before purchasing from unfamiliar vendors, supervisors should conduct basic research. They can review the business histories of the vendors and manufacturers they represent. Supervisors should ask recent purchasers about their experiences with the vendor.

NOTE: In all cases the instructor should consult with the AHJ to ensure they comply with the purchasing policies of the jurisdiction.

Purchase Orders

To actually release funds for purchases, the supervisor first completes a purchase order, which usually includes the following information:

- Resources being purchased
- Cost of resources being purchased
- Entity from which resources are being purchased

- Any authorization or account numbers needed for processing
- Appropriate signatures to approve the request

Supervisors must become familiar with their jurisdiction's procedures for processing purchase orders, and with the accounting personnel who oversee this process. Larger departments may have a chief financial officer. In smaller departments, the city or county clerk's office may handle purchase orders.

Supervisors should strictly follow the AHJ's procedures. If they do not, authorities may question their use of funds because there is no documentation of how purchases were funded. In addition, following procedures helps ensure purchases are completed on time and resources are gathered when they are needed.

Managing Training Records

Level II Instructors may be involved in developing and retaining their organization's training records or in developing record-keeping systems. Each jurisdiction will have certain information that the AHJ wishes to keep as part of a training records system. Supervisors should always follow the AHJ's policies when retaining or discarding information. The following sections describe important aspects of record-keeping. Skills associated with maintaining and securing accurate training records are shown in **Skill Sheet 14-3**.

Training Information

The type and format of training records may vary widely, depending upon the needs of the AHJ. NFPA 1401, *Recommended Practice for Fire Service Training Reports and Records*, provides examples of different training forms as well as other helpful information on their design and procedures for effective management. In addition, NFPA 1403, *Standard on Live Fire Training Evolutions* provides guidance on the use of acquired structures and fixed facilities for live-fire training.

Important information to be gathered for any training records system usually includes, but is not limited to, the following:

- Course name
- Dates and hours of each training session
- Names of instructors for each training session
- Student attendance rosters
- Topics taught at each session
- Lesson plans, workbooks and texts, tests, videotapes, and other course material, curriculum, and program event documentation and processes
- Evaluation/testing scores of students when applicable, as well as practical skills sheets used as evaluation criteria
- Course evaluations that students provide
- Required training and certifications that the student has completed
- Assigned or required training that was missed
- Performance deficiencies and aptitudes noted during evaluations
- Recommended remedial training to correct deficiencies identified by testing
- Reports of skill deficiencies during routine or emergency operations
- Any other information that the organization deems appropriate, including privileged information such as student identification numbers and locations of training sessions

> **Review: Types of Training Records**
>
> The following are different types of training records:
> - **Attendance records** — Evidence that a student has completed a specified number of hours of training in a topic, such as respiratory protection or hazardous materials incident response.
> - **Certification records** — Which certifications individuals have completed.
> - **Incident/injury record** — Documentation of student injuries during training.
> - **Test records** — Documentation of a student's test results and testing history; these must remain secure after becoming part of a student's personal training records and/or personnel files.
> - **Training schedules** — History of drills and classes that were offered and facilities used.
>
> Supervisors may also have contact with medical records, personnel records, and maintenance records. If these records fall under the supervisor's responsibility, he or she should follow policies and procedures for completing and maintaining them securely and confidentially.

Training records are usually maintained at the company, district/battalion, and administrative levels of the fire and emergency services organization. Each level supplies the level above with information until the records become part of an organization's information management section. Instructors and supervisors typically collect information for some or all of the following types of training:

- Daily training performed by the designated instructor, such as entry-level or recruit training. Details:
  — These records confirm the hours of training that each student receives.
  — In the U.S., the Occupational Safety and Health Administration (OSHA) mandates these records, which must be made available upon request to the U.S. Department of Labor at the federal or state level.
- Company-level training, such as basic skills refresher training in ropes and knots.
- Organizational training delivered to all members of an organization, such as time-management or cultural-sensitivity training.
- Individual self-study, such as preparation required to develop a training course.
- Individual training, such as courses in public information or media events courses for a public information officer.
- Special training that a source outside the organization provides, such as courses attended at the National Fire Academy or the state/provincial fire academy.
- Degrees, certificates, or levels of training that members of the organization achieve.

In order to properly maintain training records, supervisors need to consider other record types they will need to retain. They should know what information may be included in a training record system and how records will be managed within the system.

Record Management Systems

Level II Instructors collect records from instructors, then store them in a secure location. The benefits of securing accurate training records include the following:

- Retains documents so that authorized parties may review them when necessary
- Helps administrators determine which training areas have been overemphasized or underemphasized
- Documents the fact that required training has been completed, or that required mandates are being met
- Provides information to help administrators plan and schedule future training programs
- Provides a degree of protection against lawsuits

Most types of records discussed in this manual, even test scores, can be stored in a computer database. Computer databases vary by organization, the information they contain, and the digital security management they use. Supervisors should become familiar with the organizations' databases. In addition, they should pay special attention to its security measures and maintain database security.

Record Auditing Procedures

Record auditing is the process by which a document in a record-keeping system can be traced to its creator. To ease this process, include the following information on any record or report:

- Name of the person who completed a record or report
- Names of contributors
- Dates and times the records were completed

Naming conventions should be established for computer files and folders, and they should be organized. There may be software that organizes the files and makes them easily searchable.

The record-keeping system should be evaluated regularly. Doing so should ensure that records are easily retrieved.

Legal Requirements for Training Records

An accurate record-keeping system can provide an organization with documentation for legal proceedings, management reviews, and accreditation programs, such as those conducted or administered by the following organizations:

- Insurance Services Office
- Federal and state requirements
- State boards/bureaus of EMS and firefighter certification and accreditation

State/provincial and federal governments have laws that affect what information must be gathered and how it must be stored. These laws determine how long records must be retained, how privacy must be safeguarded, and which records must be available to the public. However, exemptions exist to these requirements.

Retention Length of Records and Reports

The length of time that records must be retained by the organization depends on state/provincial and local laws and the type of record. For instance, in some states employment application forms must be retained for a specified number of years, while exposure report forms and documentation must be retained for the same or longer number of years following the retirement or termination of the employee. Instructors should consult the AHJ's human resources or legal departments for retention guidelines on all types of records **(Figure 14.14)**.

Privacy of Records and Reports

Confidential records include personnel files, individual training records, and medical files. Other private information includes Social Security numbers and test scores. Specific information regarding privacy requirements includes:

- **Social Security number** — Many organizations no longer use Social Security numbers for records identification. The practice of using other identification methods reduces the potential for improper use or identity theft.

Figure 14.14 When necessary, the AHJ should provide instructors with guidelines for retaining organizational records and reports.

- **Personnel files** — Training records may be considered part of a private employment file, a fact that requires an organizational system to limit access to them. Organizations should develop and adopt policies that limit access to training records to only those personnel with a legal need to know.
- **Test scores** — Scores are considered privileged information. They are available only to authorized personnel who have a specific need to know. Examples:
  — In the U.S., the Family Educational Rights and Privacy Act (FERPA) prohibits the release of this type of information, with state and local restrictions. FERPA only applies to organizations receiving money from the U.S. Department of Education. As a general rule, a local fire department does not have to follow the FERPA requirements; however, it is recommended that they follow FERPA as a best practice. As a result, the practice of posting test scores and other personal information on bulletin boards is no longer allowed.
  — In Canada, the Office of the Privacy Commissioner of Canada manages privacy laws. Additionally, all provinces have enacted privacy legislation, such as Ontario's Municipal Freedom of Information and Protection of Privacy Act (MFIPPA). Canadian instructors must be aware of their duties and responsibilities under the applicable legislation.

Public Access to Records and Reports

While individual personnel records are confidential, many other organizational records are not. Organizations should take care when recording any information, because it may eventually become public. Instructors should follow the policies of the AHJ to determine which records and reports are publically accessible.

NOTE: If an outside party approaches instructors for information by an outside party, the instructors should notify and consult their AHJ counsel before releasing the information.

Open Records Act Exemptions

Many states exempt some information from the provisions of the Open Records Act. Federal law presumes that all records are open and places the burden on the jurisdiction to demonstrate that any requested materials are exempt. When a public record contains exempt and nonexempt material, the exempt portion must be removed and the remaining nonexempt material disclosed. Federal employees must be aware of the application of the Freedom of Information Act on government agencies. Some examples of exemptions include the following:

- Medical and other records involving personal privacy
- Records relating to pending investigations
- Records that the federal government requires be kept confidential
- Trade secrets and certain information of a proprietary nature
- Research data that has not been published, patented, or otherwise publicly disseminated
- Confidential evaluations submitted to a public agency in connection with the hiring of a public employee

Chapter Review

Answer the following questions to review the information provided in this chapter.

1. What are some common supervision techniques used when supervising other instructors?
2. What tasks are necessary for scheduling instructional delivery and resources?
3. How is the process of recommending budget needs managed by a Level II Instructor?
4. What components are involved in the purchasing process?
5. How are training records managed by a Level II Instructor?

Discussion Questions

The following questions are intended to generate discussion, expand your understanding of the chapter text, and allow you to think critically about what you have learned. Answers to these questions may vary.

1. What is one technique you would employ as a supervisor for other instructors? Why?
2. Why is prioritizing scheduling factors important for an instructor?
3. What kind of information is critical in justifying a budget need? Why?
4. Why is it important for Level II Instructors to understand and follow approved purchasing procedures?
5. What experience have you had with incorrectly managed records as a student or Level I Instructor?

Key Terms

Block Grant — Government grant allocated to help local authorities provide general services, with few restrictions.

Capital Budget — Budget intended to fund large, onetime expenditures, such as those for fire stations, fire apparatus, or major pieces of equipment.

Grant — Donated funding from a government or private source, typically secured through a competitive application process; funds may be separate from an organization's operational or capital budget.

Operational Budget — Document that outlines operating expenses for any course, curriculum, or training program.

14-1

Select resources, staff, and facilities for specified instructional sessions.
[NFPA 1041, 5.2.2]

Task Steps

Step 1: Assess factors that affect scheduling of the instructional session(s) **(Figure 14.15)**.
Step 2: Determine staffing and resource requirements.
Step 3: Determine availability of resources, staff, and facilities.
Step 4: Coordinate training for the instructional session(s) **(Figure 14.16)**.
Step 5: Create a schedule for the instructional session(s).
Step 6: Publish the instructional session schedule.
Step 7: Revise the instructional session schedule, as needed.

Figure 14.15

Figure 14.16

14-2
Perform a needs analysis for training resources.
[NFPA 1041, 5.2.3, 5.2.4]

Task Steps

Step 1: Identify training goals **(Figure 14.17)**.
    a. Priority one – training that is mandated
    b. Priority two – training required to meet organizational goals
    c. Priority three – training that would benefit the organization

Step 2: Determine resources needed to meet goals **(Figure 14.18)**.

Step 3: Obtain and review department budget processes and equipment acquisition policies.

Step 4: Create timeline for training resources acquisition.

Step 5: Review budgetary requirements regarding training resources and budgeting processes.

Step 6: Complete training resources acquisitions based on department policy.

Step 7: Submit completed forms to supervisor.

Figure 14.17

Figure 14.18

14-3
Maintain and secure accurate training records so that all agency and legal requirements are met. [NFPA 1041, 5.2.5]

Task Steps

- **Step 1:** Complete attendance form.
- **Step 2:** Document course name.
- **Step 3:** Document dates and hours of instruction.
- **Step 4:** Document names of instructor(s).
- **Step 5:** Document method of instruction.
- **Step 6:** Document test scores of students.
- **Step 7:** Document certification results of students (if applicable).
- **Step 8:** Ensure records include name of person creating the record, as well as time and date of completion.
- **Step 9:** Ensure confidential information is secured **(Figure 14.19)**.
- **Step 10:** Ensure records are retained for length of time required by the AHJ.

Figure 14.19

Instructor and Class Evaluations

Chapter 15

SECTION B INSTRUCTOR II

Chapter Contents

Supervisory Instructor Evaluations....... 321
    Supervisory Evaluation Tools......... 321
    Performance Evaluation Processes 322
    Course Evaluations................. 324
Findings from Evaluations 328

Instructor Strengths................ 328
Instructor Weaknesses.............. 328
Chapter Review 328
Discussion Questions 328
Skill Sheets 329

JPRs addressed in this chapter

This chapter provides information that addresses the following job performance requirements of NFPA 1041, *Standard for Fire Service Instructor Professional Qualifications*, 2019 Edition.

5.2.6
5.5.3

Learning Objectives

1. Describe the process for evaluating instructors. [5.2.6]
2. Describe the process for developing class evaluation instruments. [5.5.3]
3. Explain the benefits of evaluation findings. [5.2.6]
4. Skill Sheet 15-1: Administer a Level I Instructor performance evaluation. [5.2.6]
5. Skill Sheet 15-2: Develop class evaluation forms. [5.5.3]

Chapter 15
Instructor and Class Evaluations

One of the responsibilities of the Level II Instructor identified in NFPA 1041 is to evaluate Level I Instructors and learning environments. These evaluations measure effectiveness and efficiency based on criteria established or adopted by the AHJ. This chapter discusses supervisory and student evaluations and the findings that Level II Instructors can glean from them, such as instructor strengths and weaknesses.

Supervisory Instructor Evaluations

The evaluation of Level I Instructors is based on the observations of supervisors and students, shared via an instrument. Student evaluations provide instructors with information about the effectiveness of a course and their performance as instructors. Student feedback also helps to form the supervisor's formal evaluation of an instructor, which addresses classroom conduct and teaching ability.

Many fire and emergency services organizations have a personnel evaluation program that establishes guidelines, requirements, and timelines for evaluations. Waiting for the formal evaluation timeline should not deter supervisors from providing ongoing, informal performance feedback.

Level II Instructors who serve as supervisors may periodically evaluate the instructors who work for them. These supervisors must be familiar with the organization's evaluation policies. In general, the instructor should know the following elements of a personnel evaluation program:

- Forms
- Guidelines
- Processes
- Findings

Supervisory Evaluation Tools

A supervisory evaluation tool should address criteria for the skills and behaviors an instructor should have; it may include the following points:

- Classroom management considerations
- Effective interaction with students
- Proper use of verbal and nonverbal communication techniques
- Proper use of audiovisual training aids or other equipment
- Communication of all learning objectives to students enrolled in the course
- Appropriate use of questioning techniques
- Use of appropriate instructional methods
- Knowledge of subject matter

The Level II Instructor should document these criteria on an instructor evaluation rubric. This rubric is used to evaluate the effectiveness of the Level I Instructor and can also provide a means for qualitative feedback. See **Appendix F** for a Level I Instructor evaluation rubric.

Evaluation is an important and ongoing process throughout curriculum development and instruction. The supervisor and/or evaluator can gather valuable information by observing instructors as they teach and interact with students. The evaluator must remember that an instructor is more than a subject matter expert. The instructor needs to be capable of relating sometimes difficult topics to students. Instructors need to be adaptive to student learning styles as they change along with their generation. As the evaluator observes the instructor/student interaction some areas of importance include:

- Level of student participation
- Student reaction to exercises and activities
- Student questions and comments
- Student knowledge of the subject
- Student learning and interest levels

With feedback from observation, instructors can change or modify instructional methods. Observations may indicate needed changes to a course or its materials. The supervisor should remember the instructor evaluation may be a tool to assist in preparing an employee evaluation. However, this is only one part of the overall supervisory evaluation. Consult the IFSTA manual, **Fire and Emergency Services Company Officer,** to address how to conduct supervisory evaluations.

Performance Evaluation Processes

The Level II Instructor has informal and formal means to evaluate Level I Instructors. Informal evaluations occur often and are based on the instructors' observations at work (**Figure 15.1**).

Figure 15.1 Informal evaluations include watching classes in progress.

The formal evaluation may occur on a regular schedule or as needed. This type of evaluation uses the results of the informal observations as well as students' instructor evaluations to generate comments and suggestions. Formal evaluations are always held privately between the evaluator and the evaluated instructor. The format of evaluations differs among jurisdictions. Guidelines are as follows:

- **Communicate the evaluation process** — Subordinate instructors should be made aware of what to expect at each stage of the process.
- **Apply in a timely manner** — Conduct evaluations at appropriate intervals, which is especially important for performance evaluations linked to a specific incidence of unsatisfactory performance.
- **State criteria clearly** — Maintain written job-performance criteria for review at each successive evaluation.
- **Ensure uniform standards** — Apply the same job-performance standards regardless of gender, race, ethnicity, age, or other classifications.
- **Maintain thorough records** — Place each evaluation in the employee's personnel file. Although these records are confidential, give a copy to the instructor being evaluated.
- **Ensure objectivity** — Base the evaluation on established criteria. Objectivity is essential in all of the organization's personnel evaluations.

During a formal or informal critique, a supervisor should give positive feedback before addressing areas that need improvement. Supervisors should give specific, constructive suggestions. Comments should refer to observed or credibly reported behaviors.

Informal evaluations have the advantage of immediacy. For example, an instructor-supervisor may wish to review course outcomes and student evaluations with instructors after training courses. This debriefing allows the supervisor to immediately address performance weaknesses and to identify potential problems with course materials.

Effective instructors should be praised in front of students and peers. But if instructors perform below expectations, comments and suggestions for improvement should be offered privately. A written record should also be kept of informal evaluations, to be used for reference during formal evaluations. Maintain these records according to AHJ policy.

Instructor evaluations can be divided into the following stages:

- **Before the Evaluation** — Steps:
  1. Communicate the date, time, and location of the evaluation to the instructor being evaluated.
  2. Review the evaluation materials, including student evaluation surveys, previous formal evaluations, and documented personal observations.
  3. Discuss the evaluation process with the instructor being evaluated.
  4. Be familiar with organizational policies or criteria.
- **Observation** — Steps:
  1. Have the organizational policies or criteria available to review.
  2. Take notes while observing in order to be able to give specific feedback to the instructor **(Figure 15.2)**.
  3. Use the evaluation tool that the organization provides to reduce the potential for subjectivity and ensure consistency between evaluations.
  4. Remember the following points while observing:
     — Evaluate instructors based on their classroom presentations.
     — Do not prejudge instructors.
     — Assign evaluation points to teaching skills and topic knowledge as a way to objectively judge the instructor's ability.
     — Look for instructional qualities in all presentation, demonstration, and performance areas of teaching.
- **Performance Review** — Steps:
  1. Discuss the evaluation in private **(Figure 15.3)**.
  2. Reinforce that one purpose of the evaluation is to recognize areas where an instructor can improve his or her instructional style and/or communication methods.
  3. Complement the instructor's strengths.

Figure 15.2 Taking notes during observations can help the evaluator give specific, useful feedback.

Figure 15.3 Supervisors should hold performance reviews in private.

4. Ask for the instructor's opinion on his or her performance.
5. Make note of the instructor's comments to compare with the notes made during the formal evaluation.

After the formal evaluation, the supervisor must hold the instructor accountable for making necessary improvements. The supervisor should conduct further observations to determine whether an instructor has adequately addressed specific deficiencies identified during the evaluation. If not, the supervisor must hold formal meetings with the instructor to determine whether supplementary training would help.

Inexperienced or underperforming instructors may benefit from observing or working with more experienced instructors, who act as mentors and model appropriate teaching methods. Working with other instructors gives inexperienced instructors opportunities to observe models of the desired instructional methods and teaching behaviors. Mentors must be carefully selected to ensure that they model appropriate methods and behaviors **(Figure 15.4)**. Skills associated with administering a Level I Instructor performance evaluation are shown in **Skill Sheet 15-1**.

Figure 15.4 Instructors may be assigned to work with more experienced instructors.

Course Evaluations

In addition to evaluating Level I Instructors, Level II Instructors typically create and distribute course evaluations to students so they can provide feedback about courses they have taken. The course evaluation provides valuable information on the curriculum and materials, the course environment, and the instructional staff that delivered it. The information allows the program staff to evaluate how the course is performing and being received, and will give direction for course improvement and development. Skills associated with developing class evaluation forms are shown in **Skill Sheet 15-2**.

Student evaluations of instructor performance are highly subjective and may vary greatly **(Figure 15.5)**. The following factors may influence student responses on evaluations:

- Student's background knowledge of the subject
- Student's reason for attending the class (voluntary or mandatory)
- Personality of the instructor or student
- Preconceived ideas held by the student
- The way the student feels as he or she completes the form

The three areas of course evaluation are:

- Environment where the program was delivered, which may include the physical environment such as a classroom or training facility, or an online environment for a blended or electronically delivered course

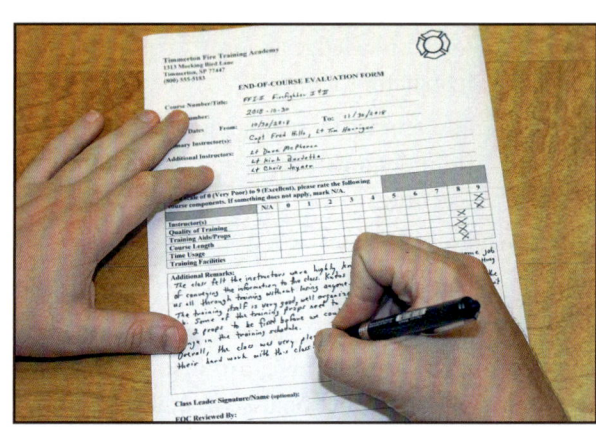

Figure 15.5 Student evaluations provide a significant amount of feedback about a course.

- Course itself, including the subject matter, course materials, and delivery method
- Instructional staff who delivered the course

Evaluation of the learning environment helps quantify whether external factors may have detracted from the course. The evaluation may be expanded if the course was delivered electronically.

Course content and materials evaluations are subjective and can vary widely. Evaluations should ask students about the benefits they feel they received from a class, focusing on perceptions of their learning experience and the learning environment. A thorough evaluation should address the following essential areas of the instructional process:

- **Reaction** — Were students satisfied with the course? What reason do they give for their opinion?
- **Knowledge** — What knowledge did students acquire and demonstrate?
- **Skills** — What skills did students acquire and demonstrate?
- **Behavioral changes** — Did the course change how the student would address a situation?
- **Learning Environment** — Was the course format appropriate and did it provide a distraction-free, safe place to learn?
- **Safety** — Were appropriate safety precautions followed throughout the class?
- **Materials** — Were the provided teaching aids and equipment sufficient to help students meet objectives?
- **Results** — Did the course meet the student expectations? Was the course directly related to their jobs and duties?

Evaluation tools should include questions on the following instructor characteristics:

- Presentation skills
- Interest and enthusiasm for the subject
- Ability to answer students' questions
- Ability to engage in course content to promote student learning
- Interaction with students
- Time-management skills
- Whether or not students would choose to take another class with this instructor

Questions on most instructor/course evaluations ask students to respond on a continuum that ranges from very satisfied to very dissatisfied. Assigning numerical values to these responses – for example, from 1 (poor) to 5 (best) – allows supervisors to statistically average the results from one or more classes (**Figure 15.6, p. 326-327**).

Evaluations should always leave space for open-ended answers or comments. Doing so allows students to express more complete thoughts that may not be represented by a numerical value or specific question. Students should not be required to sign the evaluation form, but an optional signature line may be provided for students who volunteer to be contacted in order to provide additional information.

Course evaluations and the decision of how and when to complete them varies depending on the AHJ. In some cases, evaluations are completed directly at the end of the course. If adequate time and facilities are not provided, the validity of the evaluation may be compromised due to the student rushing to complete the evaluation. Other courses use an evaluation instrument completed online after the course has finished. This allows students to consider their thoughts on the course and ideally will allow them an adequate amount of time for completion. However, if there is no requirement for the evaluations to be completed as part of the course process, the response level for this type of evaluation tends to be lower. Some programs will hold completion certificates until an evaluation is returned. The decision as to what method of evaluation to use will vary between AHJs and may vary depending on the course.

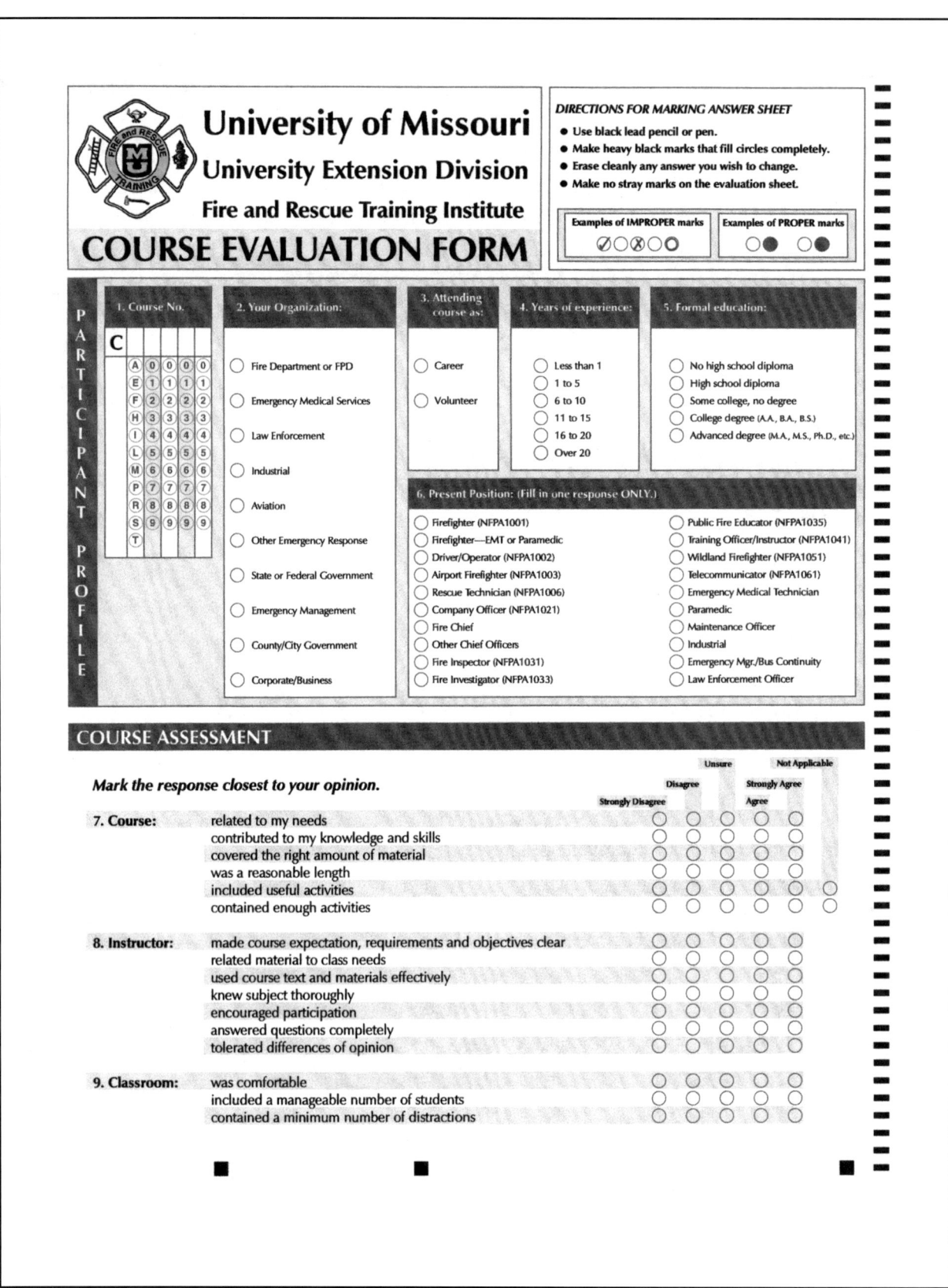

Figure 15.6 A thorough evaluation form should permit input on all aspects of training. *Courtesy of Fire and Rescue Training Institute, University of Missouri.*

COURSE ASSESSMENT (continued)

Mark the response closest to your opinion.

Scale: Strongly Disagree | Disagree | Unsure | Agree | Strongly Agree | Not Applicable

10. Outside Activities:
- included a manageable number of students
- adequate/enough equipment available
- activities performed were realistic
- activities performed were relevant to the course objectives
- adequate opportunity given to perform the activities
- no outside activities

11. Visual materials were:
- related to the course
- good quality
- in appropriate number
- easy to see

12. Printed materials were:
- well organized
- complete
- readable (printed well)

SUGGESTIONS

13. How could the course content or structure be improved?

14. How could the instructor improve the class delivery?

15. How could the classroom be improved?

16. How could outside activities be improved?

17. How could the audiovisual materials be improved to increase learning?

18. How could the printed materials be improved to increase learning?

Please feel free to use additional blank paper, if additional comments are needed.

CHECK FRONT PAGE TO MAKE SURE YOU HAVE COMPLETED ALL QUESTIONS.

SCANTRON CUSTOM FORM NO. F-15591-UOMFRT   **Thank you for your cooperation.**   ©SCANTRON CORPORATION 2001 ALL RIGHTS RESERVED.   PC2 1901-254-5 4 3 2 1

Figure 15.6 *Continued*

A supervisor should discuss and emphasize the importance of course evaluations with instructors. Instructors must understand the importance of turning in all evaluations for a course and not editing or altering them. These are valuable opportunities for improvement for the course and the instructor, and should be regarded as such.

CAUTION: Altering, deleting, or editing course evaluations in any way is unethical and may be a violation of the law.

Findings from Evaluations

Evaluations aim to identify an instructor's strengths and weaknesses. After making that determination, the supervisor should craft a response to those strengths and weaknesses. The instructor should always be included in the decision-making process used to apply strengths constructively or correct weakness.

Instructor Strengths

Strengths should be cultivated and used to the benefit of the individual instructor, as well as the division, organization, and community. Through coaching and mentoring, a supervisor can assist the instructor in determining the best use of strengths. This assistance may involve further training opportunities, career-path decisions, advanced degrees, or potential promotions. Additional authority and responsibilities can reward the instructor and further utilize strong attributes. From a supervisory standpoint, helping an instructor build upon strengths is one of the best ways to build confidence. Playing to an individual's strengths helps to ensure high-quality instruction over the long term.

Instructor Weaknesses

The supervisor has numerous tools available to address instructors' weaknesses. Counseling, coaching, and mentoring can all be useful when trying to help an instructor improve performance.

The supervisor should work with the instructor to determine the steps necessary to overcome a weakness. This action gives the instructor greater ownership and a stake in the corrective process and final outcome. To effectively correct a weakness, an instructor must be able to take responsibility for both the weakness and the method for correcting it. When these tools prove ineffective, further steps must be taken in accordance with AHJ policies.

Chapter Review

Answer the following questions to review the information provided in this chapter.

1. What guidelines should the Level II Instructor follow during the formal evaluation process of instructors?
2. What guidelines should the Level II Instructor follow when developing class evaluation instruments?
3. How do supervisors address an instructor's strengths and weaknesses revealed during instructor evaluations?

Discussion Questions

The following questions are intended to generate discussion, expand your understanding of the chapter text, and allow you to think critically about what you have learned. Answers to these questions may vary.

1. Why are instructors evaluated?
2. What components of a class evaluation instrument provide the most information for course improvement?

15-1
Administer a Level I Instructor performance evaluation. [NFPA 1041, 5.2.6]

Task Steps

Step 1: Identify departmental policies for evaluating Level I Instructors.

Step 2: Identify evaluation method(s) for evaluating Level I Instructors.

Step 3: Observe and evaluate teaching presentation of the Level I Instructor.

Step 4: Complete evaluation form, identifying the Level I Instructor's strengths and weaknesses.

Step 5: Coach the evaluated Level I Instructor by recommending changes in instructional style or communication methods **(Figure 15.7)**.

Step 6: Allow for feedback from the Level I Instructor **(Figure 15.8)**.

Step 7: Document evaluation records per AHJ policies.

Figure 15.7

Figure 15.8

15-2
Develop class evaluation forms. [NFPA 1041, 5.5.3]

Task Steps

Step 1: Review department policy regarding class evaluation instruments.

Step 2: Develop a class evaluation form that allows for student feedback regarding **(Figure 15.9)**:
 a. Instructor performance
 b. Instructional methods
 c. Communication techniques
 d. Learning environment
 e. Course content
 f. Student materials

Step 3: Determine an appropriate review method.

Figure 15.9

330 Chapter 15 • Instructor and Class Evaluations

Chapter 16

Course and Curriculum Development

SECTION C — INSTRUCTOR III

Chapter Contents

Four-Step Development Model 334	Evaluating the Curriculum 341
Identifying Training Needs 334	Evaluating Testing Instruments 342
Needs Analysis 335	Test Validity and Reliability 342
Job Performance Requirements....... 336	Test-Result Analysis................ 342
Gap Analysis..................... 337	Test-Item Analysis 343
Cost/Benefit Analysis 338	Chapter Review 345
Designing a Program or Curriculum...... 338	Discussion Questions 345
Curriculum Design 338	Key Terms 345
Course Design 339	Skill Sheets 346
Implementing the Curriculum 339	

JPRs addressed in this chapter

This chapter provides information that addresses the following job performance requirements of NFPA 1041, *Standard for Fire Service Instructor Professional Qualifications*, 2019 Edition.

6.2.5	6.3.3	6.3.6
6.2.6	6.3.4	6.5.3
6.3.2	6.3.5	6.5.5

Learning Objectives

1. Describe the four-step curriculum development model. [6.3.2, 6.3.3]
2. Explain the processes used to identify training needs. [6.2.6, 6.3.2, 6.3.3]
3. Describe the parts of course or curriculum design. [6.3.3, 6.3.4, 6.3.5, 6.3.6]
4. Discuss the steps included when implementing a course or curriculum. [6.3.2, 6.3.3]
5. Identify data used to evaluate a course or curriculum. [6.2.5, 6.5.3]
6. Explain the processes used to evaluate test instruments. [6.3.2, 6.5.5]
7. Skill Sheet 16-1: Perform a needs analysis for a training agency. [6.3.2]
8. Skill Sheet 16-2: Develop program goals based on a needs analysis. [6.3.3]
9. Skill Sheet 16-3: Develop course objectives based on a standard. [6.3.5]
10. Skill Sheet 16-4: Design and implement curriculum based on course outcomes. [6.3.3, 6.3.4, 6.3.6]
11. Skill Sheet 16-5: Evaluate student evaluation instruments. [6.5.5]

Chapter 16
Course and Curriculum Development

Section C. Instructor III

This chapter and the ones that follow discuss the JPRs affecting Level III Instructors in NFPA 1041, *Standard for Fire Service Instructor Professional Qualifications*. The previous chapters, which present information for Levels I and II, are prerequisite knowledge for these remaining chapters. Where appropriate, review sections will be included in the upcoming chapters, but with a shift in emphasis toward JPRs for Level III.

The Level III Instructor manages the organization's training program, designs and modifies courses and curricula, and creates course goals **(Figure 16.1)**. Requisite knowledge includes instructional design and technical writing. These responsibilities expand the Level II Instructor's task of creating and modifying lesson plans.

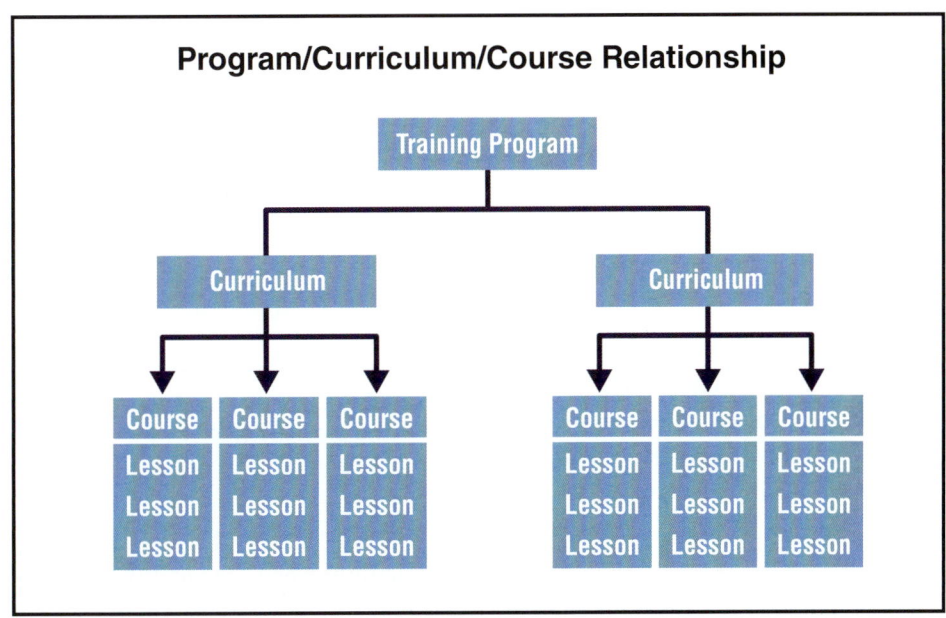

Figure 16.1 A planning model will illustrate how each curriculum component fits within the overall training program.

Review: Level III Instructor Prerequisites

NFPA 1041 includes prerequisite skills for performing many of the duties in this chapter. If you feel that a review of prerequisites is necessary, they are addressed in the following locations:

- **Lesson planning** — Chapter 11, Lesson Plan Development
- **Instructional methods** — Chapter 6, Classroom Instruction; Chapter 8, Skills-Based Training Beyond the Classroom
- **Characteristics of adult learners** — Chapter 2, Principles of Learning; Chapter 6, Classroom Instruction

- **Instructional media** — Chapter 4, Instructional Materials and Equipment
- **Development of evaluation instruments** — Chapter 13, Test Item Construction; Chapter 15, Instructor and Class Evaluations
- **Conducting research** — Chapter 11, Lesson Plan Development
- **Performance-based education** — Chapter 6, Classroom Instruction; Chapter 8, Skills-Based Training Beyond the Classroom

Four-Step Development Model

In order to create and develop a **course** or **curriculum** that can be used as part of an overall training **program**, the Level III Instructor should use a planning model to guide his or her decisions. For the purpose of this chapter, the outlined four-step model should be an effective method for most Level III Instructors and most courses. This model includes the following steps (**Figure 16.2**):

- **Step 1: Identify training needs** — Perform a needs analysis to determine the course or curriculum required to meet the organization's needs and jurisdiction's mandates. Use this step to determine the need for alterations to courses or curricula when deficiencies are found in the *evaluate* step.
- **Step 2: Design a course or curriculum** — Design a course or curriculum to meet the established requirements and AHJ goals.
- **Step 3: Implement the course or curriculum** — Perform a pilot presentation of the course or curriculum. Add the course to the training schedule when it meets the identified needs.
- **Step 4: Evaluate the course or curriculum** — Determine the effectiveness of the course or curriculum in meeting requirements. Conduct evaluations following the initial pilot test and following each presentation of the course.

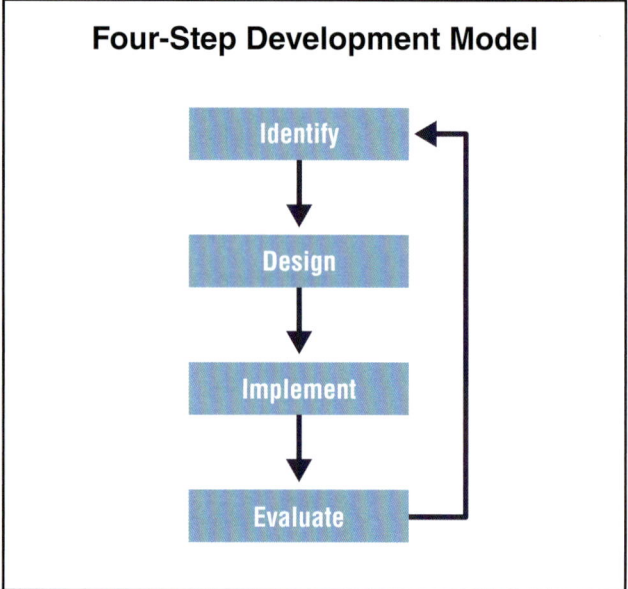

Figure 16.2 Each part of the four-step planning model builds on the previous step.

Courses may be revised, replaced, or removed from the curriculum when the *identify* and *evaluation* steps no longer provide desired outcomes. At the same time, curricula may be altered, replaced, or abandoned when program evaluations indicate such an action is necessary. The remaining sections in this chapter elaborate on these steps and offer suggestions for revising a course or curriculum after completing the steps.

Identifying Training Needs

The identification step begins with the realization that a change in the organization's operating environment has occurred, requiring a change in the organization's service delivery. The realization that changes need to be made to either courses, curricula, or programs may come in the following forms:

- State/provincial or federal mandate that may require fire department training
- Request from a stakeholder group such as a governing body

- Operational deficiencies found via postincident critiques
- Changes in department mission scope
- Changes in nationally recognized standards
- Program evaluation and review
- Emergency services and industry based research

When it is apparent that a change is required or that different or additional services are required, the organization must perform a **needs analysis**. A needs analysis should determine how the service level and capability differ from new requirements. It may also indicate how much and what type of training ensures the organization can provide the new level of service. This analysis is a function of the administration of the organization and may include representatives from all divisions of the organization. The processes that are available to the Level III Instructor for identifying needs include the following:

- Needs analysis
- Job performance requirements
- Gap analysis
- Cost/benefit analysis

In addition, the Level III Instructor should understand how work should be assigned within the organization to ensure the efficient creation of quality courses and curriculum.

Needs Analysis

With thoughtful questioning, carefully designed surveys, and in-depth research, a needs analysis accomplishes several functions such as the following:

- Defines levels of knowledge, skills, and attitudes of emergency service personnel
- Supports the department mission and does not contradict the department strategic plan
- Determines whether a need exists and indicates whether the need would receive the most benefit from training, equipment to perform a task, administrative policies or commitment, or a change in procedures **(Figure 16.3)**
- Identifies specific individuals, work groups, or organizations that need training, equipment, or procedure change
- Identifies a method to achieve desired levels of knowledge, skills, or attitudes

Figure 16.3 Many records, reports, and personnel materials are assessed while conducting needs analysis.

To determine needs in the areas of knowledge, skills, or attitudes, the Level III Instructor analyzes the following items:

- Operational reports
- Injury records
- Accident reports
- Results of promotional exams
- Personnel, operation, and training records
- NFPA standards

Various methods determine individual training needs or readiness for a training course or curriculum such as the following:

- **Survey** — Process that the organizations use to determine the level of training or skill of potential students.
- **AHJ training history** — Reviewing the training history may indicate whether prerequisite training has been previously offered.
- **Prerequisite training or skill level requirement** — Minimum level of skill required (prerequisite) for prospective students before they enter certain training courses. Some AHJs use regularly scheduled skill evaluations as records of the skill level of their firefighters and EMS responders.
- **Training records** — The following elements contained in training records may be used to determine whether a student has the knowledge or skills to be admitted to a course:
  — Completion of specific courses
  — Training certifications
  — Professional development activity
  — Verification of skills or required knowledge on a pretest

Regardless of the technique used to determine training needs, the Level III Instructor needs to know the knowledge or skill level of the organization's members before proceeding with course or curriculum development. This guideline prevents the level or complexity of the instruction from being inappropriate for some or even all of the course or curriculum. The Level III Instructor should use the analysis process to determine training requirements that will be the basis for the required curriculum.

Records and reports are the raw information for making decisions about training needs. To justify the creation of courses, Level III instructors should be able to organize this information into a proposal for the course. For example, the instructor could calculate a percentage of how many personnel have completed EMS training. Or the instructor may create a statistical analysis of firefighter injuries in the jurisdiction based upon individual reports over the previous year. Collating data and showing patterns are essential to completing a needs analysis. Skills associated with performing a needs analysis for a training agency are shown in **Skill Sheet 16-1**.

Job Performance Requirements

When evaluating training materials, Level III instructors should establish the skills/knowledge that are prerequisite for students to take a course, and the skills/knowledge that students must master over the duration of a course. These parameters must be established in order to effectively and efficiently develop and present a course or curriculum to the intended audience. Determination of the proper instructional methodologies will coincide with this process.

Starting and ending points are often determined by the skills that personnel must learn and perform on jobs. Level III Instructors often start their needs analysis by identifying and reviewing national standards and determining minimum job performance requirements (JPRs) or performances required for a specific job based on those standards. In addition, Level III instructors need to understand how JPRs apply in their jurisdictions. Jurisdiction-specific JPRs may need to be developed in order to meet training needs.

JPRs are grouped according to job duties. After determining JPRs, planners evaluate resources and select learning objectives that will be used in designing and implementing the revision of existing courses and curricula or developing new programs.

Professional qualifications standards can be used for the following purposes:

- Designing and evaluating training
- Certifying personnel
- Measuring and critiquing on-the-job performances
- Defining hiring practices

- Setting organizational policies, procedures, and goals

Professional qualifications standards are often written in the JPR format and organized by areas of responsibility (duties). The standard containing each duty and its JPRs defines and describes the job, giving Level III Instructors and planners an end point around which to design and teach a training curriculum.

Gap Analysis

A **gap analysis** is another template the Level III Instructor can follow. In this approach, the Level III Instructor can utilize a standard, regulation, or best practice as a basis for analysis **(Figure 16.4)**. The Level III Instructor can then compare the standard, regulation, or best practice to the policies, practices, and behaviors in the organization and see where the gaps are. A gap analysis can help modify the AHJ's programs with immediate corrections and long-term planning.

In order to perform a gap analysis, the Level III Instructor should collect information on specific job tasks. He or she can use the following formal or informal methods to collect information, and then use that information to determine the steps that compose the task:

- Formal methods:
  — Carefully designed and executed surveys
  — Checklists
  — Observations
  — Structured interviews
  — Research analyses
  — Tests
- Informal methods:
  — Conversations
  — Casual observations of activities and habits
  — Other unobtrusive measures

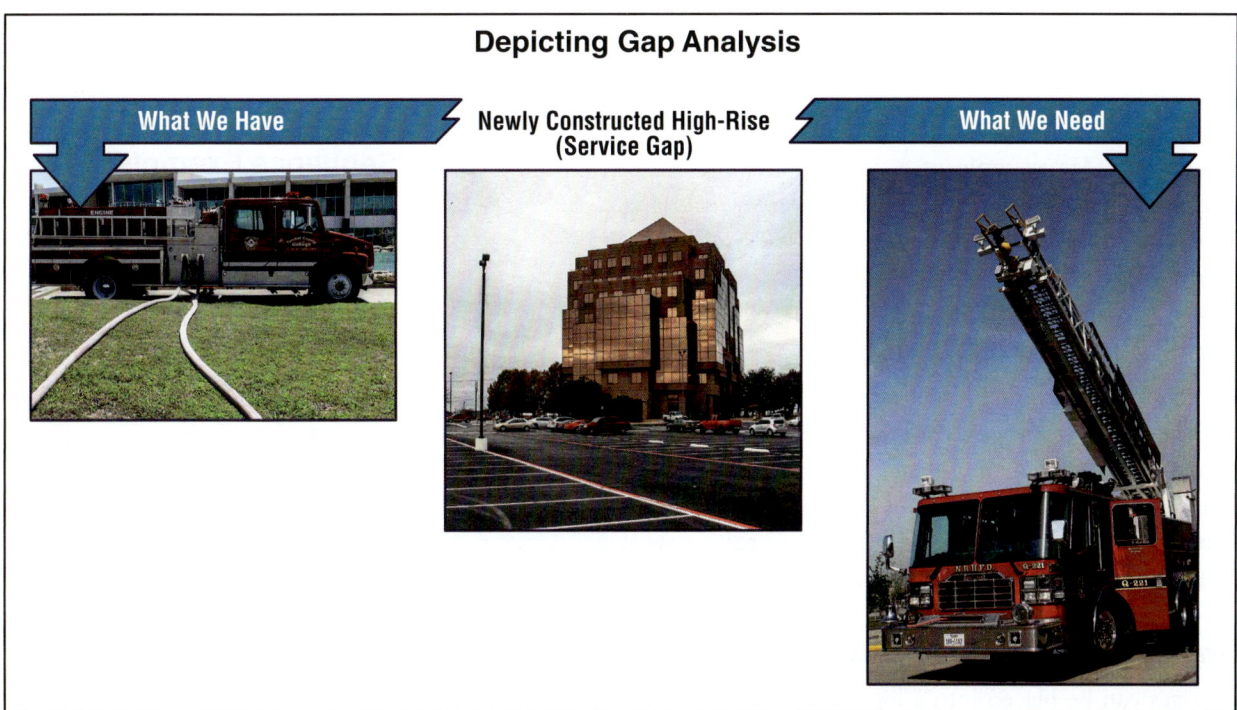

Figure 16.4 Gap analysis compares what a department has versus what it needs.

Cost/Benefit Analysis

As part of the needs analysis, the Level III Instructor should perform a **cost/benefit analysis**. For example, the cost of creating a curriculum or providing the training internally should be estimated and compared to the cost of purchasing an existing curriculum or contracting with an external agency to provide the training. It may be more cost-effective to purchase curriculum or use training available from state/provincial or national agencies or other sources. For instance, instead of developing a pumper driver/operator course, Level III Instructors could use curriculum from an outside publisher.

Each hour of actual training time is generally the result of many more hours of development. Development hours also translate to funds expended for salaries and resources. Many organizations may have neither the personnel hours nor the funds to develop courses and curriculum.

Designing a Program or Curriculum

Once the type and quantity of required training have been identified, a Level III Instructor designs a program or curriculum to meet those requirements. Tasks at this level of the process include:

- Curriculum design
- Course design

Curriculum Design

Curriculum design is oriented toward determining which courses should be grouped together to meet a larger outcome. The Level III Instructor must understand the components and characteristics of curriculum outcomes and how the outcomes and objectives support the knowledge and skills for which the outcome was written. The Level III Instructor must also know how to correlate those outcomes to AHJ goals. To design a curriculum, the Level III Instructor must perform the following tasks:

- Identify the curriculum outcomes.
- Identify courses for the curriculum.
- Sequence courses into a curriculum.

Identify a Curriculum Outcome

Curriculum outcomes resemble course outcomes except that they apply to a number of courses collectively. An example of a curriculum outcome may be: *The Technical Rescue Curriculum will provide students with the knowledge and skills to meet NFPA certification requirements.*

Identify Courses for the Curriculum

A curriculum comprises a number of courses. When developing a curriculum, courses should be analyzed to determine which might be included in the curriculum. Any areas that are not covered after adding existing courses may indicate a need to create new courses to meet the curriculum goal.

Sequence Courses into a Curriculum

Courses can be placed into a logical and progressive sequence using a general outline format **(Figure 16.5)**.

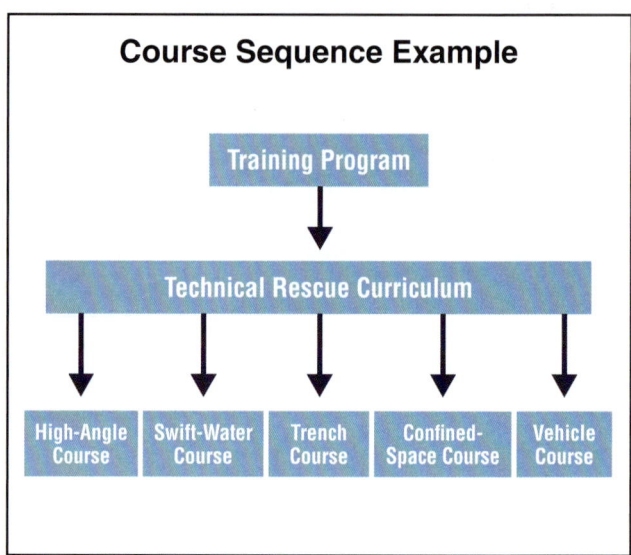

Figure 16.5 Courses in a curriculum may follow a rigid or flexible order, depending on many variables including prerequisite skills.

The sequence of courses may be more flexible than the sequence of lessons. In addition, courses may be completed over time with only a few prerequisites. However, courses must be completed satisfactorily before the following occurs:

- Curriculum is considered complete.
- Student performance is validated.
- Certificates are awarded.

Course Design

To design a course, the Level III Instructor must perform the following tasks:

- Identify program goals (**Figure 16.6**).
- Develop learning objectives.
- Group similar learning objectives.
- Develop lesson plans.
- Sequence lessons into a course outline.
- Create appropriate student, course, and instructor evaluation instruments, keeping learner characteristics in mind.

NOTE: This information is covered in detail in Chapter 11, Lesson Plan Development.

NOTE: Skills associated with developing program goals based on a needs analysis are shown in **Skill Sheet 16-2**. Skills associated with developing course objectives based on a standard are shown in **Skill Sheet 16-3**.

Figure 16.6 Here is an example of a program or course goal: *As part of a team, students safely enter a live-fire prop and extinguish a fire.*

Fire Service Terminology and Technical Writing

Knowledge of fire terminology and technical writing are listed as prerequisite knowledge for Level III Instructors in NFPA 1041. By the time instructors are considering becoming a Level III Instructor, they are generally very familiar with fire service terminology. When in doubt about terminology, instructors should consult other references such as IFSTA's **Fire Service Orientation and Terminology**.

Technical writing is very broad and outside the scope of this manual. If Level III Instructors feel their writing needs work, they should seek out professional development opportunities to become better writers. Instructors should also consider giving anything they write to someone else for review.

In general, the Level III Instructor should remember three guidelines for producing good technical writing:

- Be concise.
- Write to the intended audience of the document.
- Ensure that grammar and spelling are correct.

Implementing the Curriculum

During instructional design, program developers must remember the importance of implementing the training curriculum. When Level III Instructors have completed the initial design steps through the final writing of lesson plans and development of evaluation instruments, their next logical step is to implement the curriculum following the principles of student-centered learning.

Implementation requires the Level III Instructor to perform or delegate the following functions:

- Acquire funding and facilities.
- Determine instructor and student time requirements.
- Find or train qualified instructors.

The Level III Instructor must confirm that the course or curriculum can be presented in the facility. The location, design, equipment, and instructional technologies of a facility directly affect the activities that instructors can direct and students can perform to achieve course objectives.

Before implementation begins, Level III Instructors perform a detailed evaluation. Ideally, evaluations have happened throughout the design phase. These evaluations include the following items:

- Reviews by peers or supervisors
- Consultations with subject-matter experts
- Reviews by prospective instructors
- Any other steps that perform the following functions:
    — Help to ensure the quality of the course or curriculum
    — Monitor the course or curriculum continually to ensure its focus on meeting performance objectives

Once everyone is satisfied that the evaluation and corrections indicate that the course is ready to move forward, Level III Instructors can implement the training course or curriculum. Implementation typically includes the following steps:

Step 1: Obtain final course approval.
Step 2: Assemble, create, research, or select appropriate teaching aids, instructional technologies, and course materials.
Step 3: Schedule facilities and equipment.
Step 4: Select qualified instructors.
Step 5: Establish appropriate training records systems or databases.
Step 6: Schedule and announce the course or curriculum.
Step 7: Approve student registration.
Step 8: Present a pilot version of the course.

Each organization or jurisdiction may have different requirements for obtaining course or curriculum approval. Level III Instructors may give presentations to authorities about how a course supports the learning goals and objectives. The presentation gives information on the financial and time commitments that the organization must invest to implement the course. Level III Instructors should provide details on required facilities and equipment and the necessary instructor and student hours.

Once a course has been approved, it can be implemented. The first implementation is known as a **pilot course**. Its purpose is to evaluate course effectiveness as follows:

- Verify that its content meets course and curriculum goals.
- Evaluate the appropriateness of facilities, props, materials, logistics support, and audiovisual training aids.
- Evaluate its design and time allotment.
- Evaluate the effectiveness of its learning activities.

Pilot courses should be taught by experienced instructors who are able to adapt sections of a course that are untried **(Figure 16.7)**. These instructors should also be qualified to teach the subject matter.

Figure 16.7 Any time a pilot course is presented to students, the assigned instructor must be able to teach the course effectively and efficiently.

Evaluation of the pilot delivery is based on feedback received from instructors, observers, and students as well as student performance on evaluations. The evaluation forms and surveys for a pilot course may be more detailed than what would normally be used to evaluate a course that has been taught before. Evaluation may also occur during a pilot course on a lesson-by-lesson basis with more specific feedback from students. Pilot evaluations also present opportunities to determine the effectiveness of the course evaluation process and instruments.

Alterations correct any problems discovered in course evaluations. After making changes to the course, it should be pilot tested again to determine whether the changes helped.

NOTE: Skills associated with designing and implementing curriculum based on course outcomes are shown in **Skill Sheet 16-4**.

Evaluating the Curriculum

Level III Instructors and students can perform a summative evaluation at the end of each course to determine whether the curriculum met the educational goals of the organization. This evaluation answers the following questions:

1. Did students meet learning objectives and course and curriculum goals?
2. Was training conducted as designed and within the resources allocated by the AHJ?

   Any of the following data can be analyzed as part of a summative evaluation:

- Students' scores on written or performance tests
- Students' behaviors observed in the field
- Feedback from students and instructors
- Feedback from field supervisors

A summative evaluation is performed after each course. In part, the trends identified in student performance help evaluators judge whether the course met course and curriculum goals. These evaluations also determine whether course delivery continues as is, continues with adjustments, or is discontinued. To accurately measure the improvement in student performance, course or curriculum evaluators need to compare the post-course performance to the performance level observed in the needs analysis and pretests (if given).

In addition to evaluating students' performances and whether they meet course learning objectives and JPRs, Level III Instructors must take into account an evaluation system for course materials, testing instruments, and instructor performance. Chapter 15, Instructor and Class Evaluations, presents a detailed discussion on student evaluations.

Figure 16.8 Level III Instructors must evaluate whether curriculum materials reinforced the course objectives and were appropriate for the AHJ's goals.

Instructors and students should have input into the assessment of teaching and learning materials. The results should indicate whether the materials and resources supported the objectives and whether they were relevant or applicable to job requirements (**Figure 16.8**).

One or more instructors with inadequate teaching skills can jeopardize an excellent course design with appropriate training materials. When establishing instructor evaluation instruments, ensure that appropriately prepared instructors answer the following questions:

- How well does the instructor know the subject being taught?

- How enthusiastic is the instructor for the topic and teaching?
- How well does the instructor provide the knowledge and resources necessary to meet the lesson and course objectives?
- How well is the instructor prepared?
- How well is the instructor organized?
- How well does the instructor relate to and/or interact with students?
- How well does the instructor respond to and answer student questions?

During the presentation of a curriculum, an instructor should be periodically evaluated by more experienced instructors with subject-matter expertise. Based upon staff schedules or AHJ policies, Level III Instructors can also choose an evaluation schedule of specific or regular intervals throughout a course. Instructor evaluations should follow all necessary state/provincial and federal employment laws, such as Equal Employment Opportunity (EEO) job requirements. Be sure to comply with all anti-discrimination and privacy laws, and be aware that any breaches of confidentiality could bring legal action. Instructor evaluations can influence promotions and terminations; therefore, the evaluation document should be reinforced with observation statements and proper language that justify the action that will be taken.

Student input is also an important component in instructor evaluation; they can complete carefully designed course-feedback instruments that elicit objective responses. The results of instructor evaluations may show the need for enhanced instructor development or only the need for appropriate instructor orientation, particularly for new courses.

Evaluating Testing Instruments

Giving a test is only part of the learning process. Scoring and analyzing the results according to AHJ policies also serve necessary functions. The primary purpose of analyzing test results is to improve the teaching/learning process. Performing simple test result analyses provides instructors, test developers, and organizations with information on test validity and reliability.

The results of a test analysis identify items that may need altering, rewording, or restructuring to make them more easily understood and significant to the objectives being tested. Level III Instructors can perform some simple analyses to assess the effectiveness of test items. A list of corrective actions is also provided in the sections below to help an instructor determine how to alter test items. Skills associated with evaluating student evaluation instruments are shown in **Skill Sheet 16-5**.

Test Validity and Reliability

As defined previously, validity is the degree to which a test measures what it was designed to measure. Reliability is the consistency of test scores from one measurement to another and a condition of validity.

Validity has a specific meaning when interpreting criterion-referenced tests. For these tests, validity refers to the measurement of mastery or nonmastery by a student compared against the established learning objectives.

Similarly, **reliability** has a specific meaning when interpreting criterion-referenced tests. For these tests, reliability indicates the consistency of results in classifying mastery or nonmastery of an individual.

The length of a test has some correlation to reliability and validity. In general, longer tests are more reliable than short tests but not necessarily more valid. However, long tests may lack reliability if the test is overly long. When assessing reliability and validity, test length should be evaluated.

Test-Result Analysis

Most test-result analysis uses computer software. The software can analyze the average scores for any given exam. The software will indicate the questions most frequently missed and how many students missed them. Using these numerical values, a test's overall difficulty can be established. Tests that most students failed may need revisions to ensure that more students pass the next one. A high failure rate may show the test does not accurately reflect the learning objectives taught during the class. Similarly, tests with a very high passing rate may not indicate a recall of learning objectives by the student.

> **What This Means to You**
>
> You are evaluating student scores on an exam. The average grade on the exam according to your scoring software is 60 percent, 10 percent lower than the criterion passing level for the test, which is 70 percent. You are concerned that the test may be too difficult. You begin examining the individual test items to find out how many students got each right and which answers were selected.
>
> On question No. 10, five out of the twenty students who took the test selected the correct answer, response C. The question only had 25 percent correct, well below the criterion for the test. Upon further examination, those who got the question wrong all selected the same distractor instead of the correct answer. You discover that there are similar questions. You have to make a determination as to whether the distractors offer the students a reasonable choice or whether the stems are leading students toward the chosen distractors.
>
> In this situation, you would make the recommendation that the test is unreliable and requires revision. The revisions could be performed yourself or someone in your agency. If the test was provided by a third-party, you may consider examining the other materials you have received from them and/or consider using a different publisher in the future.

Test-Item Analysis

Test item analysis allows instructors to use systematic methods to assess the quality of an item on an examination. Three simple measures of quality can be computed when using a multiple-choice test format: difficulty index, discrimination index, and distractor analysis. Analyses of other test formats such as short-answer and essay rely on proper test item construction.

Most of the time, instructors are interested in measuring minimal competency in an area, and thus they rely on criterion-referenced tests. Ideally, an item's difficulty should be similar to the test's criterion level. Difficulty is a simple proportion of how many students in a class or training group selected the correct answer. If the percentage of students who answered the item correctly is close to the criterion (passing) level for the test, then the item is likely appropriate for the test **(Figure 16.9)**.

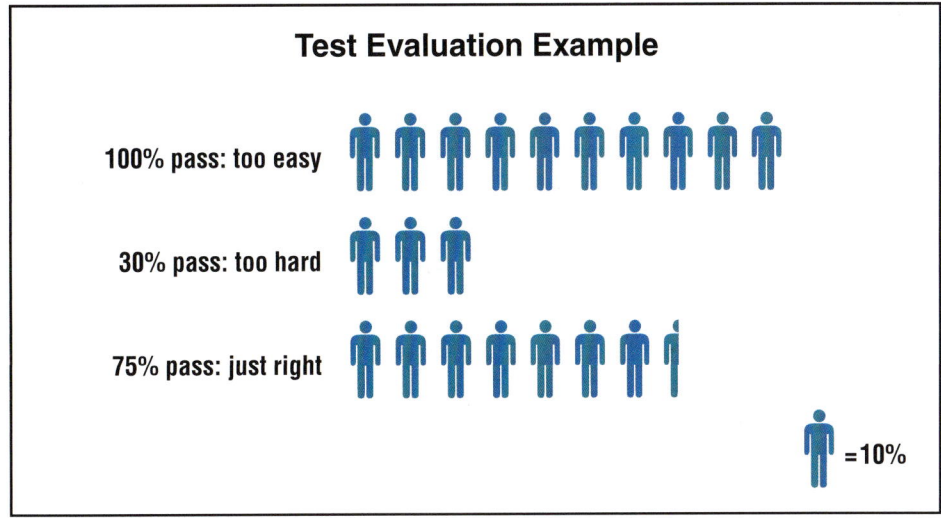

Figure 16.9 A test question that proves to be too easy or too difficult should be removed.

Part of assessing the difficulty of a test item is evaluating the item's distractors. This evaluation applies specifically to multiple-choice questions. Distractors are wrong answers that should be written to make them enticing options for a student to choose. Distractor analysis is simply examining responses for each test item to determine the distribution of responses across the range of possibilities. For example, the correct answer is not likely to be any specific choice out of four options. Item quality refers to the effectiveness of the distractors. Distractors that are not selected by any test taker need revision.

Some procedures that Level III Instructors can use to revise a test are as follows:

- **Remove poor test items and recalculate the score** — Take this action after carefully evaluating the test item and determining that its content or structure misled students.
- **Review the test analysis, adjust the test items, and give the test again** — Take this action when distractors or test items need rewording.
- **Teach the lesson again and retest** — Take this action when it appears certain critical information was not taught or was not clear to students.

Other options include a review of the instruction, circumstances of the instruction, and testing situations. Any of these factors may have confused, misled, or distracted students. Adjust the instruction, teach the lesson again, and administer the test as necessary when critical changes need to be made. When two or more identical classes are held at the same time, another corrective analysis technique is to compare test results of all classes, then determine the appropriate steps. Always adhere to AHJ policies and applicable laws when performing any test-item analyses and revisions of evaluation materials.

Course or Curriculum Modification

The evaluation of the training course or curriculum may determine its need for revision. The instructor or training manager should regularly assess the need for curriculum revision to improve learning or provide students with new learning opportunities. Revisions may become necessary because of changes in operational standards, department protocols, new technologies, or the standards of appropriate accrediting bodies.

When planning revisions, the instructor should take the following actions:

- Determine whether the revision is mission-appropriate for the department/organization, training division, or agency. Does the proposed revision help and support the organization to meet its mission statement or goal? If not, then the revision should not be made or the course or curriculum should be altered so that it does meet the mission statement or goal.
- Involve other branches or divisions of the department/organization that may be affected by proposed changes.
- Involve the entire training division or agency membership in the development process.
- Evaluate the resources and potential needs regarding staffing, funding, time, technology, equipment, props, and facilities.
- Communicate clearly to the organization the reason for the proposed revision as well as the anticipated results of the change.
- Create clear, concise revision proposals that meet the criteria for any new course or curriculum.
- Apply the four-step planning model to the revision of the course or curriculum. Periodic evaluations of all courses and curricula should be scheduled to determine whether they still meet the original needs of the organization because the current course outcomes may not resemble those of the original course design.

Chapter Review

Answer the following questions to review the information provided in this chapter.

1. What are the four steps of the curriculum development model?
2. What processes are used to identify training needs?
3. What is the difference between curriculum design and course design?
4. What are the steps to implement a curriculum or course?
5. What types of data are used to evaluate a curriculum or course?
6. What processes are used to evaluate test instruments?

Discussion Questions

The following questions are intended to generate discussion, expand your understanding of the chapter text, and allow you to think critically about what you have learned. Answers to these questions may vary.

1. Why is it helpful for Level III Instructors to follow the four-step curriculum development model?
2. What situation have you experienced as an instructor that necessitated a needs analysis?
3. What courses or curriculum have you created as an instructor?
4. What functions can a Level III Instructor perform or delegate when implementing a curriculum or course?
5. What are some summative evaluation collection challenges that Level III Instructors may encounter?
6. What are some reasons a Level III Instructor would want to revise curriculum or course materials?

Key Terms

Cost/Benefit Analysis — Systematic methodology to compare costs and benefits to make cost-effective funding decisions on projects.

Course — Series of lessons that lead to the completion of a discipline or certification.

Curriculum — Series of courses in which students are introduced to skills and knowledge required for a specific discipline.

Gap Analysis — Comparison between standards/regulations/best practices and actual behaviors within the organization to determine where real world performance differs from best practice.

Needs Analysis — Assessment of the gap between what exists and what should exist, with regard to an organization's staffing, services, equipment, or training.

Pilot Course — First implementation of a newly developed course; intended to allow instructors to evaluate a new course and make changes for improvement.

Program — Collection of curricula and the resources necessary to deliver the instruction for those curricula.

Reliability — A condition of validity; the extent to which a test or test item consistently and accurately produces the same results or scores when given to a set of learners on different occasions, marked by different assessors, or marked by the same assessors on different occasions.

Validity — Extent to which a test or other assessment technique measures the learner qualities (knowledge or skills) that it is meant to measure.

Perform a needs analysis for a training agency. [NFPA 1041, 6.3.2] 16-1

Task Steps

Step 1: Review the AHJ's mission and strategic plan to determine what need(s) exists.

Step 2: Identify specific individuals, work groups, or organizations that need training, equipment, or procedure change(s).

Step 3: Identify method(s) to achieve desired levels of knowledge, skills, or attitudes by analyzing:

    a. Operational reports

    b. Injury records

    c. Accident reports

    d. Results of promotional exams

    e. Personnel, operation, and training records

    f. NFPA standards

Step 4: Determine training requirements for the basis of the required curriculum using:

    a. Surveys

    b. AHJ training history

    c. Prerequisite training or skill level requirements

    d. Training records

Step 5: Perform needs analysis **(Figure 16.10)**.

Figure 16.10

16-2
Develop program goals based on a needs analysis. [NFPA 1041, 6.3.3]

Task Steps

Step 1: Review needs analysis for training program.
Step 2: Identify gaps in training program based on JPR requirements.
Step 3: Write/develop program goals.
Step 4: Determine the specific reasons, purpose, or benefits of each goal **(Figure 16.11)**.
Step 5: Determine how each goal will be measured.
Step 6: Determine how each goal will be attained.
Step 7: Determine the time frame of each goal.
Step 8: Write each goal in a consistent format, such as (action verb) (object) (modifiers). Example: "*To train 90 percent of the AHJ's Level II Instructors to become Level III Instructors by July, 2021.*"

Figure 16.11

16-3
Develop course objectives based on a standard. [NFPA 1041, 6.3.5]

Task Steps

Step 1: Using goals developed for training program, identify course/program standards **(Figure 16.12)**.

Step 2: Identify what students will be able to do when they have completed instruction:

  a. Identify the appropriate level of learning. Example: *Knowledge, Comprehension, Application, Analysis, Synthesis, Evaluation*

  b. Select a verb that is definite and measurable and that signifies a demonstrable learning outcome. Example: *Describe the four-step curriculum development model.*

  c. Add criteria to indicate how or when the objective will be observable. Example: *By the end of this course, the student will be able to apply the four-step curriculum model to develop a new training course.*

Step 3: Technically write/complete course objectives based on desired JPR standards.

Figure 16.12

16-4
Design and implement curriculum based on course outcomes.
[NFPA 1041, 6.3.3, 6.3.4, 6.3.6]

Task Steps

- **Step 1:** Identify the curriculum goals and outcomes **(Figure 16.13)**.
- **Step 2:** Identify courses for the curriculum.
- **Step 3:** Develop course learning objectives, grouping similar objectives as needed.
- **Step 4:** Develop lesson plans.
- **Step 5:** Sequence the courses into a course outline.
- **Step 6:** Identify best method of delivery for course outcomes.
- **Step 7:** Determine time and learning environment in which to deliver lesson.
- **Step 8:** Create appropriate student, course, and instructor evaluation instruments.
- **Step 9:** Perform pilot presentation of the curriculum.
- **Step 10:** Evaluate curriculum following each presentation for revision, replacement, or removal.

Figure 16.13

16-5
Evaluate student evaluation instruments. [NFPA 1041, 6.5.5]

Task Steps

Step 1: Review item analysis of evaluation instrument **(Figure 16.14)**.

Step 2: If needed, alter, reword, or restructure evaluation instrument to make it more easily understood and significant to the objectives being tested.

Step 3: Verify validity and reliability of evaluation instrument.

Step 4: Adjust the instruction as needed, per AHJ policies and procedures.

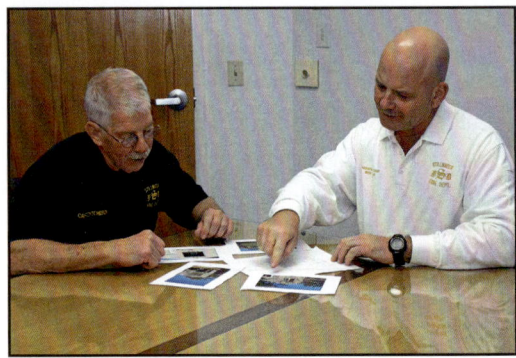

Figure 16.14

Chapter 17

Training Program Evaluation

SECTION C INSTRUCTOR III

Chapter Contents

Methodologies 353	Course and Instructional Design Evaluations 359
Evaluation Types and Categories 354	**Evaluation Results and Recommendations** 360
Evaluation Plans 356	**Chapter Review** 362
Observation of Instructors 357	**Discussion Questions** 363
Supervisor Surveys 357	**Key Terms** 363
Organizational Evaluations 358	**Skill Sheets** 364
Performance Measurement 358	

JPRs addressed in this chapter

This chapter provides information that addresses the following job performance requirements of NFPA 1041, *Standard for Fire Service Instructor Professional Qualifications*, 2019 Edition.

6.2.8
6.5.2
6.5.3
6.5.4

Learning Objectives

1. Describe various methodologies used to evaluate a fire and emergency services training program. [6.5.4]

2. Discuss the types and categories of evaluations used in fire and emergency services programs. [6.5.4]

3. Discuss the ways that instructors gather behavioral data as part of an evaluation plan. [6.5.3, 6.5.4]

4. Describe the steps to be taken after reviewing evaluations in order to make program recommendations. [6.2.8, 6.5.2]

5. Skill Sheet 17-1: Create a plan for evaluating courses within a training agency. [6.5.3]

6. Skill Sheet 17-2: Create a plan for evaluating a training agency program. [6.5.4]

7. Skill Sheet 17-3: Develop a system for evaluation results. [6.5.2]

8. Skill Sheet 17-4: Present evaluation findings, conclusions, and recommendations for training programs to agency administration officials. [6.2.8]

Chapter 17
Training Program Evaluation

From time to time, training programs need to be evaluated for relevance and effectiveness. Level III Instructors should be familiar with evaluation design and the ways that a thorough program evaluation can benefit the organization.

Program evaluation requires that Level III Instructors consider:

- Methodologies
- Evaluation types and categories
- Evaluation plans
- Evaluation results and recommendations

Methodologies

Designing a program evaluation under a specific methodology will depend on the program's purpose and level. In a broad sense, the purpose of program evaluation is to determine value. In order to determine the value of a program, departments must have clearly defined goals that are aligned with the AHJ's mission. These are strategic level, mission-driven items expected of the training division.

On a different level, evaluation of a program will involve examining measureable outcomes to determine whether the program is meeting its goal. These evaluations are defined as formative or summative assessments and are conducted at the course level. Formative assessments promote in-class adjustments to meet the needs of a particular group of students as they work toward the learning objectives. Summative assessments involve documenting learning outcomes and determining overall effectiveness of the course.

Once the methodology and purpose are defined, the Level III Instructor will need a specific framework or design to conduct the evaluation. The purpose and level of the evaluation must be clarified, as programmatic and course level evaluations answer different questions.

Program and course-level evaluations often include instructor and student input. Supervisors and administrators may also share their perspectives. The primary source of information comes from instructors and students because they are directly involved with the delivery of a course and have firsthand knowledge at the point of instruction. Instructor and student input is obtained through course surveys, observation by instructors, instructor feedback, test analyses, and skill evaluations. Supervisors and administrators are indirectly involved through interaction with students after the course.

Qualitative methodology is a nonnumeric form of research, while quantitative methodology is numeric, and mixed methods research is a combination of both. These research approaches serve different purposes:

- **Qualitative methodology** — Based on nonnumeric analysis, although data may have numeric values. This methodology determines whether the program meets the values established for the program. This type of data is difficult to tabulate into precise categories, and it is usually gathered through the following methods:

— Open-ended survey questions

— Interviews of students

— Course and curriculum evaluations

— Instructor and supervisor observations **(Figure 17.1)**

- **Quantitative methodology** — Based on a numeric or statistical analysis; the program is evaluated against numeric criteria. Information gathered locally may be judged against statistics gathered on a local or national level. Some detractors of this methodology claim the content limit placed on the respondent's answers may discourage fresh concepts, and survey responses are typically limited. Quantitative methodology works best when information is gathered from a large number of respondents. Data is gathered from questionnaires that contain:

    — Yes/no questions

    — Checklists

    — Preference scales **(Figure 17.2)**

- **Mixed methods research** — A combination of the qualitative and quantitative methods in the same survey or questionnaire. It allows the user to examine questions in a more comprehensive manner.

An example of qualitative versus quantitative research may be represented by the construction of course evaluation forms. A form on which students respond to open-ended questions in an essay or short answer format yields qualitative results. A form that requires students to rate certain aspects of the course on a 1 to 5 scale yields quantitative results.

Figure 17.1 Qualitative evaluation includes information gathered directly from students.

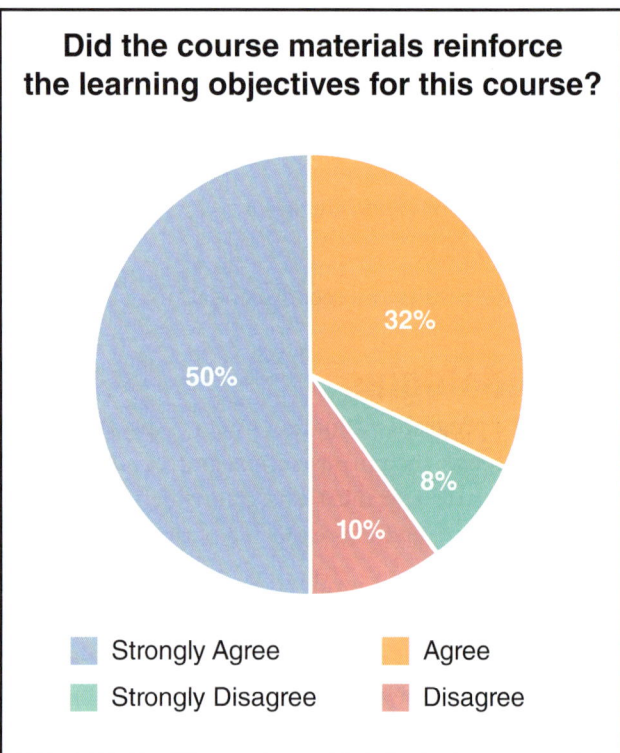

Figure 17.2 Data from questionnaires can be graphed to give a clearer picture of course evaluations.

Evaluation Types and Categories

When developing an evaluation plan, the Level III Instructor can use various types of evaluations to gather information on topics such as:

- Instructor effectiveness
- Value of course components
- Effectiveness of program goals
- Use of facilities

When tailored to meet specific AHJ criteria, evaluations can help jurisdictions determine whether they are meeting their organizational goals. Two generally accepted categories of evaluations used for programs are:

- **Formative evaluation** — Gathers information to help improve the program while in progress. This evaluation type is intended to isolate any evident weaknesses or understand the program's strengths and build on them.

- **Summative evaluation** — Assesses the achievements and/or outcome of the program. The evaluation measures achievement of goals and the effect on the student.

Three generally accepted types of evaluations used for programs are (**Figure 17.3**):

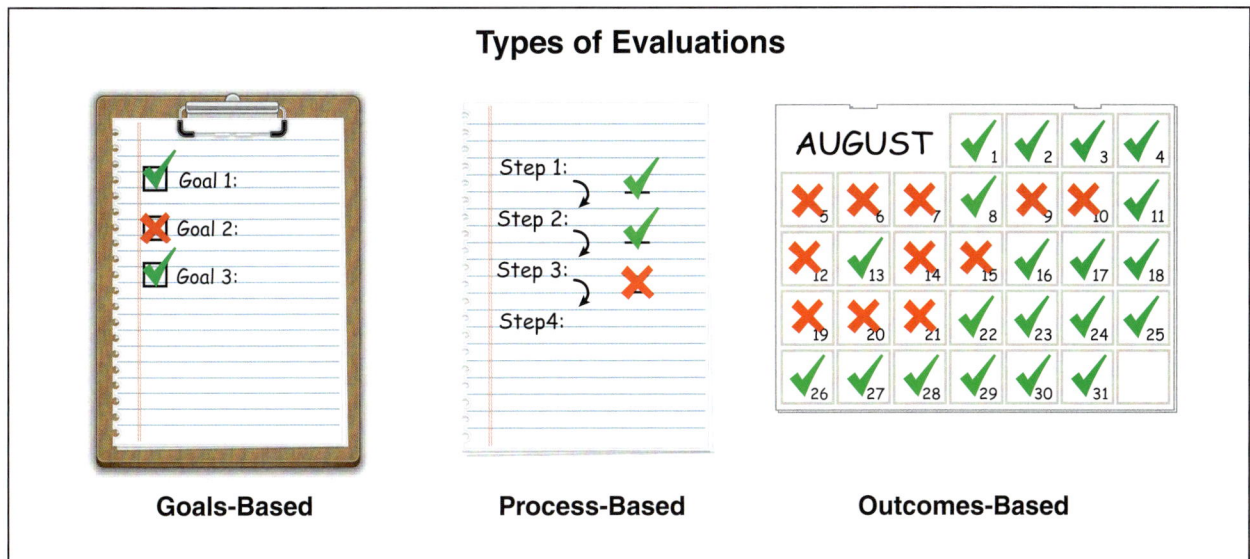

Figure 17.3 The type of evaluation conducted will influence the information collected.

- **Goals-based evaluation** — Determines how well a program meets its goals or objectives through summative assessment.
- **Process-based evaluation** — Determines how a program works in practice and highlights its strengths and weaknesses through formative assessment. It focuses on a lesson, course, or curriculum. In organizations that provide training, process-based evaluations may refer to the formal plan for developing or revising curricula.
- **Outcomes-based evaluation** — Provides measurable benefits resulting from the program; for example, after the department offered an incident safety course to their line officers, in the following twelve months the department observed a 20 percent reduction in workplace injuries.

The following actions should be included in a program evaluation:

- Set criteria by identifying the problems or issues that must be addressed.
- Determine what data is available for evaluation.
- Determine what new data will have to be collected for evaluation.
- Identify the most effective evaluation model.
- Determine the amount of time necessary to complete the evaluation.
- Identify the personnel responsible for making the evaluation.
- Conduct the evaluation.
- Analyze the results and develop conclusions.
- Establish multiple sets of reasonable actions to take.
- Prepare a final report.
- Develop revision strategies and priorities.

An evaluation plan should be a formal element in the process. Selecting and performing the appropriate evaluation type will generate the most useful information about each specific stage of program development, delivery, or revision.

Evaluation Plans

The framework for all types of evaluations should be maintained in a written plan that contains the step-by-step process for the jurisdiction's evaluations. The plan should be constructed to follow AHJ policies and procedures, which in turn ensures the process is conducted in a consistent manner. The plan must contain enough information for an authorized party to understand and follow it. Records must be kept on all formal evaluations.

When an evaluation plan is applied to a program, the results can be used as the basis for the evaluation report and serve as the outline for organizing collected results. A complete evaluation plan includes a review of behavioral objectives and test results to allow an informed decision about its accurate measurement of objectives. This review process enables Level III Instructors to meet the primary purpose of evaluation, which is to improve the teaching and learning process. Many evaluation techniques may be used to assess learning levels and determine how instruction can be modified to enhance learning.

One model for evaluating training in an organization is Kirkpatrick's Four-Level Training Evaluation Model, as follows:

- **Reaction of students** — What they thought and felt about the training; corresponds to student evaluations of courses and instructors
- **Learning** — Resulting increase in knowledge or capability; corresponds to cognitive tests given to students
- **Behavior** — Extent of improvement in student capabilities and application of learned skills; corresponds to psychomotor skills tests for students and instructor evaluations for instructors
- **Results** — Effects on the department or environment resulting from trainee performance; corresponds to training program evaluations

NFA Evaluation Plan

The evaluation process used by the National Fire Academy (NFA) is based on the systems approach and includes the following elements **(Figure 17.4)**:

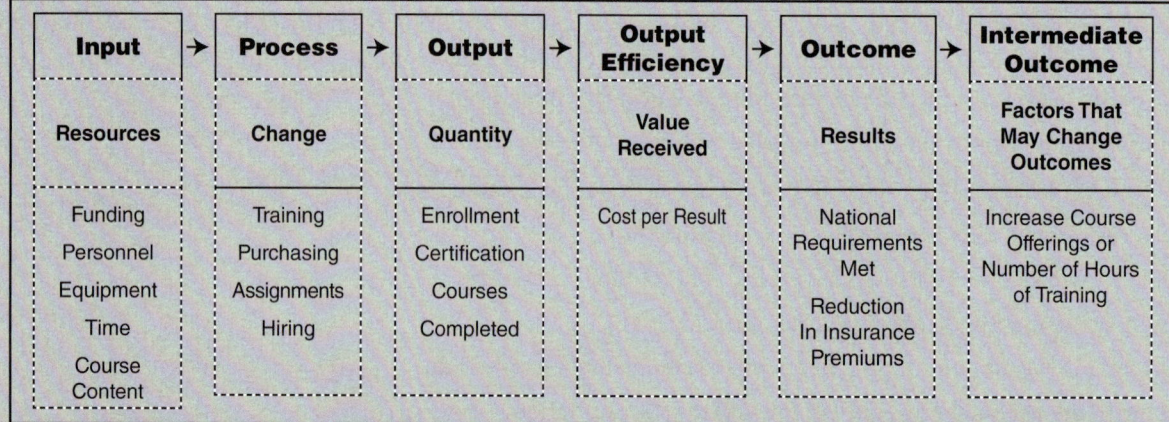

Figure 17.4 The systems approach to evaluation analyzes set training elements to determine outcomes.

- **Input** — Resources such as personnel, equipment, and finances devoted to the activity being measured; resources are stated in quantities such as:
  — Amount of money allotted for training
  — Number of instructors
  — Number of courses taught
  — Number of hours per class
- **Process** (*may also be referred to as* **transformation** *or* **thru-put**) — Change to resources; any activity such as training, purchasing, or hiring

- **Output** — Quantity of services that are the result of input and process; quantity may be stated in the following terms:
  - Number of students enrolled
  - Hours of training provided
  - Number of certification requirements met
- **Output efficiency** — Input divided by the output; for example, the amount of money allotted divided by the number of hours of training delivered results in a cost per instructional hour of training
- **Outcome** — Results of the process
- **Intermediate outcome** — Factors that are expected to directly change an outcome; may include an increased number of enrollments or hours of training

Evaluation occurs when the intent is to assess the achievements or outcomes of the program. Details about these evaluations include:

- Encompasses the effectiveness of the program and how it fits into the department mission
- Identifies the need for new programs
- Provides a comparison with past outcomes
- Documents the achievements of the program
- Documents job-related changes
- Relies on evidence-based documentation

There are several ways in which Level III Instructors can gather behavioral data about their organizations' programs that support the above process. Behavioral data should be collected over several categories to reflect an overview of an entire program.

Observation of Instructors

As mentioned above, Level III Instructors can gather evidence through various evaluation methods. Most of this evidence is observational during classes throughout a course **(Figure 17.5)**. With feedback received during instruction, instructors can change or modify instructional methods to meet the goals of the class. In some cases, it will be clear that a course needs to be changed on a permanent basis. The input from instructors is essential when making decisions about a program under development.

Figure 17.5 Level III Instructors can observe instructors during class set-up and instruction to form their own evidence of class efficiency and effectiveness.

Supervisor Surveys

The training course or lesson evaluation process includes surveys completed by supervisory personnel. Supervisors witness how students trained in the AHJ perform on the job after training. Similarly, they are aware of any changes in skills, behaviors, and attitudes that students display following course completion. Supervisory personnel provide invaluable information about the successes and results of training courses.

To gain student input, survey forms can be distributed or made available to supervisory personnel after the completion of the course or lesson. These surveys can also be repeated at certain intervals such as three months after a course and again six months after a course. Survey items should be constructed to assess how a student has applied the learned knowledge, skills, or behavioral habits while on duty.

Some organizations use similar course or lesson evaluation forms for students to determine how they are using the information or skills learned in a course. These forms may be distributed and completed at the same time as the supervisor survey. By having the student and supervisor complete their individual surveys, a more complete and honest critique of the material presented to the student can be obtained.

Evaluating a training course includes determining student satisfaction (**Figure 17.6**). Students who participate in effective training sessions are more likely to appreciate the time spent on the course. Likewise, students are the first to recognize an ineffective training session.

Figure 17.6 Conducting written evaluations will enable instructors and other decision makers to gauge the course's effectiveness.

Organizational Evaluations

An organizational evaluation is intended to compare an organization's mission statement with its output. The mission statement of a training division may be based upon an established standard (such as an NFPA standard) or may have been developed locally.

Overall performance of the organization, which includes the divisions or branches of the department, is judged against the criteria established for the organization. Organizational evaluations may be conducted internally, or by a third party including AHJs and other organizations. Internal evaluation is a quality-control measure that is part of the planning process. Outside agencies that may evaluate the organization include the local jurisdiction, state/territory/province, third-party accrediting agencies, or national government. These organizations provide a process that allows jurisdictions or organizations to be evaluated on the quality of fire and emergency services that are available.

The results of organizational performance evaluations, whether internal or external, help the leadership of the AHJ improve delivered services and adjust to changes in the expected outcomes Organizational evaluations are viewed as management tools that result in continued improvement of the course program.

Performance Measurement

The training division may develop performance measurements to evaluate success. These performance measurements can be used to justify existing budgets, plan for expectations or needed improvements, and support the efficiency of the training division. These performance measures can be tracked on a quarterly basis or on a regular schedule in order to meet the goals of the AHJ. Examples of items that could be part of a training division's performance measurement include (**Figure 17.7**):

- Percentage of firefighters certified to levels of the NFPA Professional Qualifications standards
- Percentage of individuals who attend special team training-
- Percentage of students who successfully complete recruit academies
- Number of training injuries per year
- Customer approval ratings taken from course evaluations
- Output numbers to include number of courses, students, and instructional hours
- Cost per instructional hour of training

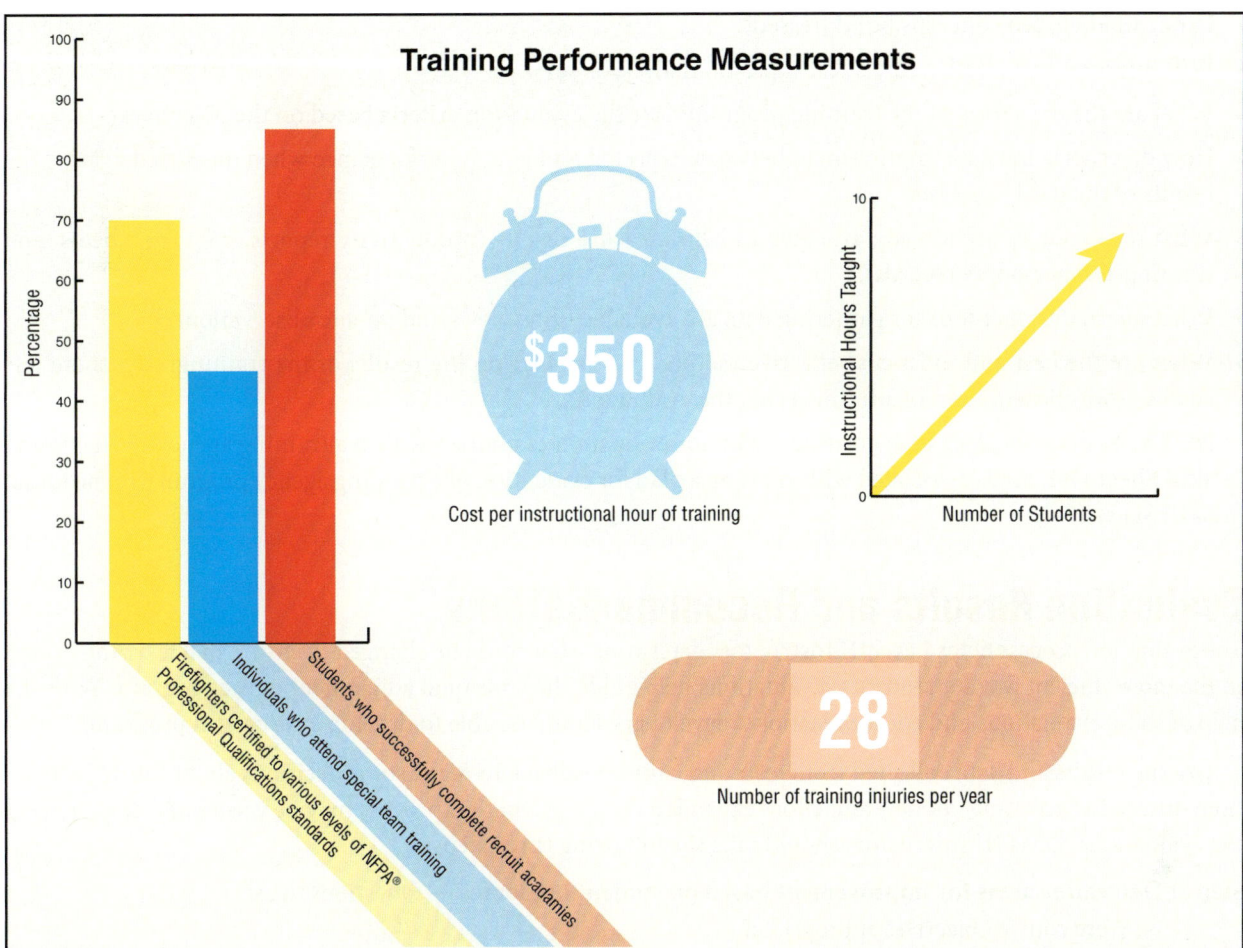

Figure 17.7 Multiple items should be used to track overall training performance to give a well-rounded idea of the program's effectiveness.

Course and Instructional Design Evaluations

As discussed in Chapter 16, Course and Curriculum Development, instructional design is composed of the analysis of training needs, the systematic design of learning activities, and the assessment of the learning process. As training requirements change or when activities are no longer effective, the instructional process changes.

Level III Instructors must always be alert to ways of improving or updating instruction or even eliminating instruction. Instructional content, methods, and techniques must remain flexible in order to serve their intended purpose.

Test scores are only one form of evidence to assess the success of the learning process. For example, students' behavior in training environments may indicate a higher awareness of safety and skill-related knowledge than their scores indicate.

Regardless of form or design, the evaluation process must be useful to its users; thus, data collection must be accurate. Evaluation instruments, as mentioned previously, should be practical to administer, and results should be presented in a manner that is easily understood by all audiences. Planning the approach to evaluation is also important. To get the most value from a course evaluation and meet course objectives, the Level III Instructor must plan the approach by thinking through and answering the following questions:

- How did training affect students' attitudes and behavior? What did they learn? How was the organization affected by the results? How are these factors quantified?

- How can identified concerns be addressed?
- Will information be gathered via tests, questionnaires, or surveys?
- What are the objectives of the training program? Are the evaluation criteria based on the objectives?
- How do criteria indicate improvement between expected and actual performance when measured against the results of the needs analysis?
- What data sources are already available to help measure results (productivity reports, daily log sheets, and training and personnel records)?
- What alternative methods for gathering data are available (interviews and on-site observations)?
- What are the best and most cost-effective methods for measuring the results of the training? Are there less costly, more efficient ways of administering the evaluation?

NOTE: Skills associated with creating a plan for evaluation of courses within a training agency can be found in **Skill Sheet 17-1**. Skills associated with creating a plan for evaluation of a training agency program can be found in **Skill Sheet 17-2.**

Evaluation Results and Recommendations

The evaluation process helps Level III Instructors determine what could be changed to ensure that learning occurs in the most efficient way for students. Evaluations help guide the continual adjustment of the program. With the help of these evaluations and course reactions, developers should be able to develop a successful program.

Use the results and data collected from tests and course feedback to make some decisions about future instruction; otherwise, any time, money, and effort expended on the evaluation process may be wasted. After reviewing the evaluations, Level III Instructors should take the following three steps:

Step 1: Determine areas for improvement based on student outcomes — Questions to ask:
 — Were course objectives appropriate?
 — Did the course outline and materials match the learning objectives?
 — Did the instructional methods and aids facilitate learning?
 — Did the instructor's presentation style impede learning?
 — Was the learning environment conducive to learning?
 — Were testing instruments and test administration valid and reliable?
 — Were there personal or logistical problems that inhibited learning?
 — What student learning weaknesses were revealed?

Step 2: Identify actions to correct deficiencies — Actions to follow:
 — Revise course objectives.
 — Modify the course outline and materials to match objectives.
 — Change instructional methods/aids.
 — Improve presentation style.
 — Enhance the learning environment.
 — Alter evaluation techniques to produce valid and reliable instruments.
 — Provide support for students and offer referral services.
 — Provide extra learning opportunities such as learning activity packets, individual or pair assignments, group study, or one-on-one tutoring.

Step 3: Document and report results as appropriate — Actions to follow:
 — Keep daily records of training activities.
 — Maintain individual training records using the most effective, AHJ-approved record-keeping system.
 — Retain class test results and results of analyses.

NOTE: Provide written progress reports to management that outline findings in Steps 1 and 2.

Evaluations can also be used in the review of material and overall evaluation of the training program. Some agencies refer to this as their **Customer Approval Rating (CAR)** or Customer Satisfaction Rating (CSR). This data is easy to manage and evaluate when the information is provided in a computer database. Small departments may manage this process manually.

The examples in **Table 17.1** show the use of the overall CAR. From this data an organization can look at areas of the course that may need improving, such as instructor competency and classroom usability. Each category has been assigned a value. Some categories, such as the outdoor factor in the Case Law course, may not apply.

Table 17.1
Course Performance Ratings Examples

Farm Accident Rescue
Total Courses: 1

CAR	3.53
Instructor	3.59
Classroom	3.40
Outdoor	3.50
Visuals	3.63
Print	3.60

Farm Machinery Rescue: Technician
Total Courses: 1

CAR	3.36
Instructor	3.67
Classroom	2.90
Outdoor	3.07
Visuals	3.63
Print	3.53

Completing the NFIRS* Report and Writing Incident Reports
Total Courses: 1

CAR	2.60
Instructor	2.70
Classroom	2.47
Outdoor	2.22
Visuals	2.75
Print	2.73

Fire and Emergency Case Law
Total Courses: 1

CAR	3.40
Instructor	3.41
Classroom	3.70
Outdoor	NA
Visuals	3.18
Print	3.57

*NFIRS = National Fire Incident Reporting System
Source: *Courtesy of Fire and Rescue Training Institute, University of Missouri.*

Overall, the purpose of program evaluation is to influence the decision-making process at the departmental level. For the Level III Instructor, that includes recommendations to the AHJ administrator on improvements to trainings offered and possible training development. Level III Instructors should prepare to present their findings using various methods for their target audiences. A few considerations will improve the overall impression of the work, such as:

- Demonstrate that a qualitative and quantitative methodological approach was used in a manner that promotes good decision-making.
- Present evidence, in the form of outcome measurements, in a non-biased manner.
- Include a final correlation that shows the relationship between AHJ goals and training program outcomes.

- Present findings and conclusions using statistical analysis and summaries of the findings to present an argument for proposed changes.
- Base recommendations on gathered evidence and statistical analysis.
- Provide an indication that evidence was evaluated based on its relevance. Any relevant information should be accounted for as part of the recommendation. Including such evidence helps to minimize bias in the recommendation.
- Ensure that any recommendation reflects AHJ goals and follows all applicable laws.

If the Level III Instructor finds that new courses, policies, and procedures are needed, explaining how changes would support AHJ goals can help recommendations gain acceptance.

NOTE: Skills associated with developing a system for evaluation results can be found in **Skill Sheet 17-3**. Skills associated with presenting evaluation findings, conclusions, and recommendations for training programs to agency administration officials can be found in **Skill Sheet 17-4**.

Discussion Exercises

Example 1:

In recent years, there has been an increase in driving-related incidents with fire apparatus. The fire chief has asked you (the training officer) to determine if the training division may be able to address this. As the training officer, evaluate the following:

- Driving records of members
- National fire service driving trends and data

Upon further review, you note that the driver training program is ten years old, and has not been taught to department members over the past two years.

- How would you evaluate the course for currency and relevance?
- How would you evaluate off-the-shelf courses for use in your department?
- If you use an off-the-shelf course, how would you evaluate its effectiveness?

Example 2:

Recently your fire chief asked the training division to consider the increased complaints from the police department and state troopers related to incidents on the local interstate (or local four-lane state highway). These complaints focused around fire department apparatus placement and fire department interaction with law enforcement. Since the training division does not offer a traffic incident management course, the fire chief has requested that you (the training officer) develop and offer some type of training program for all members. With several nationally recognized Traffic Incident Management training courses available, how would you do the following:

- Evaluate these existing courses to determine which best fits in your fire department?
- Implement one of these courses in the upcoming training schedule?
- Evaluate the effectiveness of this course in improving the interaction with law enforcement, and in improving apparatus placement?
- What data would you need to assist in your evaluation of program effectiveness?

Chapter Review

Answer the following questions to review the information provided in this chapter.

1. What are some methodologies used to evaluate a fire and emergency services training program?
2. What types and categories of evaluations are used in fire and emergency services programs?
3. How can instructors gather behavioral data as part of an evaluation plan?
4. What steps should a Level III Instructor take, after reviewing evaluations, in order to make program recommendations?

Discussion Questions

The following questions are intended to generate discussion, expand your understanding of the chapter text, and allow you to think critically about what you have learned. Answers to these questions may vary.

1. What are the pros and cons of the various program evaluation methodologies?
2. How do evaluation types and categories help jurisdictions determine whether they are meeting their organizational goals?
3. Why is it important for Level III Instructors to improve the teaching and learning process?

Key Terms

Customer Approval Rating (CAR) — Organizational rating that adds a quantifiable ranking to qualitative evaluations. *Also known as* Customer Satisfaction Rating (CSR).

Formative Evaluation — Evaluation of a new or revised program in order to form opinions about its effects and effectiveness as it is in the process of being developed and tested (piloted). Its purpose is to gather information to help improve the program while in progress.

Goals-Based Evaluation — Summative assessment of whether or not a given program has met its intended goals.

Mixed Methods Research — Approach to research that is a combination of the qualitative and quantitative methods.

Outcomes-Based Evaluation — Summative assessment of what benefit was gained or enhanced as a result of a particular program.

Process-Based Evaluation — Formative assessment of how well a program functions after it has been implemented and is currently in progress; measures strengths, weaknesses, and efficiency.

Qualitative Methodology — Evaluation based on non-numerical analysis and intended to assess the quality of something.

Quantitative Methodology — Evaluation based on numeric or statistical analysis and intended to discover quantifiable data.

Summative Evaluation — Evaluation of a program after all of its various components have been implemented and established; the evaluation is intended to measure achievement of intended goals and the effect of the program on the student (outcomes).

17-1
Create a plan for evaluating courses within a training agency. [NFPA, 1041, 6.5.3]

Task Steps

Step 1: Review AHJ policies and procedures **(Figure 17.8)**.

Step 2: Review course objectives to verify they are measureable.

Step 3: Identify course evaluation instruments and techniques to collect data.

Step 4: Identify process for providing feedback.

Figure 17.8

17-2
Create a plan for evaluating a training agency program. [NFPA, 1041 6.5.4]

Task Steps

Step 1: Review AHJ policies and procedures.

Step 2: Review AHJ goals and objectives.

Step 3: Identify training agency program evaluation instruments and methods to collect data **(Figure 17.9)**.

Step 4: Identify process for providing feedback.

Figure 17.9

17-3
Develop a system for evaluation results. [NFPA 1041, 6.5.2]

Task Steps

Step 1: Review AHJ policies and procedures.

Step 2: Review AHJ goals and objectives.

Step 3: Identify applicable laws that would affect evaluation results system development.

Step 4: Identify methods of providing feedback **(Figure 17.10)**.

Step 5: Determine areas for improvement based on evaluation results.

Step 6: Identify actions to correct deficiencies.

Step 7: Document and report results, according to AHJ policies and procedures.

Figure 17.10

17-4

Present evaluation findings, conclusions, and recommendations for training programs to agency administration officials. [NFPA 1041, 6.2.8]

Task Steps

Step 1: Prepare written report for administration officials that includes evaluation findings, conclusions, and recommendations reflecting AHJ goals, policies, and procedures.

Step 2: Conduct oral presentation for administration officials that includes evaluation findings, conclusions, and recommendations reflecting AHJ goals, policies, and procedures **(Figure 17.11)**.

Step 3: Present unbiased, supported recommendations that reflect AHJ goals, policies, and procedures.

Step 4: Follow up as needed.

Figure 17.11

Chapter 18

Training Program Administration

SECTION C INSTRUCTOR III

Chapter Contents

Record-Keeping Systems 371	**Formulating Budget Needs** 381
Disclosure of Information 374	Equipment 381
AHJ-Level Needs Analysis 374	Sources of Instructional Materials 382
Committee Meetings 374	Maintenance and Repair of a
Data Organization and Interpretation ... 375	Training Center 382
Development of Training Policies,	**Equipment Purchasing Policies** 383
Procedures, and Guidelines 375	Developing a Request for
Policies 375	Proposal (RFP) 383
Procedures and Guidelines 376	Creating Purchasing Specifications 385
Identifying a Need for a New Policy,	Bid Evaluation and Awarding of
Procedure, or Guideline 377	the Contract 387
Developing a Draft Document 378	**Human Resource Management** 388
Submitting a Draft for Review 378	Instructional Staff Selection 388
Adopting a Policy, Procedure, or	Instructor Qualifications 388
Guideline 379	Position Advertising 389
Publishing a Document 379	**Personnel Evaluations** 389
Implementing a Document's	**Chapter Review** 389
Contents 379	**Discussion Questions** 390
Evaluating Effectiveness 379	**Key Terms** 390
Standards that Influence Training 380	**Skill Sheets** 391

JPRs addressed in this chapter

This chapter provides information that addresses the following job performance requirements of NFPA 1041, *Standard for Fire Service Instructor Professional Qualifications*, 2019 Edition.

6.2.2   6.2.5   6.3.2
6.2.3   6.2.6   6.5.2
6.2.4   6.2.7

Learning Objectives

1. Discuss the functions that a record-keeping system should perform. [6.2.2]
2. Explain disclosure of information. [6.2.2, 6.5.2]
3. Describe the Level III Instructor's role in conducting an AHJ-level needs analysis. [6.3.2]
4. Discuss the process of developing organizational training policies, procedures, and guidelines. [6.2.3]
5. Explain the influence that standards have on training policies. [6.2.2]
6. Describe the process to formulate budget needs. [6.2.6]
7. Explain the equipment purchasing process. [6.2.7]
8. Identify the Level III Instructor's human resource responsibilities. [6.2.4]
9. Discuss the process of personnel evaluations. [6.2.5]
10. Skill Sheet 18-1: Organize a training record system that can be readily accessed. [6.2.2]
11. Skill Sheet 18-2: Develop policy recommendations for meeting agency training goals. [6.2.3]
12. Skill Sheet 18-3: Formulate budget needs. [6.2.6]
13. Skill Sheet 18-4: Create agency specifications for purchasing training equipment. [6.2.7]
14. Skill Sheet 18-5: Select and evaluate instructional staff applicants. [6.2.4]
15. Skill Sheet 18-6: Create a performance-based evaluation plan for instructional staff. [6.2.5]

Chapter 18
Training Program Administration

Level III Instructors have the highest level of administrative duties in a training agency or division. They maintain the training record-keeping system for the AHJ and are the starting point for training policies and procedures. Level III Instructors create forms the AHJ uses and are responsible for training-related purchases. This chapter describes the administrative duties for Level III Instructors.

NOTE: In smaller departments, Level I or II Instructors may be responsible for some of these duties.

Record-Keeping Systems

Record-keeping encompasses a range of information organizing skills and results in the proper documentation of the policies and transactions of the organization. Information stored in records can be used for many purposes, including the development or alteration of policies, procedures, and guidelines. Record-keeping also includes the following:

- Statistical analysis
- Strategic handling of data pulled directly from tests
- Other evaluations to establish trends and relationships

This information can help formulate training policies, mandate safe practices, or change procedures. An organization must create and preserve adequate and proper documentation of its record-keeping activities system that supports operational needs, protects individual rights, and promotes accountability. The Level III Instructor should be aware of all legal requirements affecting the AHJ's record-keeping procedures.

Record-keeping systems may be manual or automated. These systems involve the collection, organization, storage, preservation, and disposal of hard copy records and materials. These records may include:

- Reports
- Forms
- Maps
- Graphs
- Illustrations
- Audio/video recordings
- Photographs

An automated system uses the electronic storage of data in databases. These allow instructors direct access to the files for uploading or updating records. Record-keeping systems (also known as records management systems) should require password-protected computer security to limit access to authorized personnel. Firewalls and Internet security should be established to prevent outside intrusion of servers that instructors can access remotely. Level III Instructors should work closely with their information technology departments to establish security measures.

Either type of system must be able to perform the following functions:

- Organize records
- Index records
- Allow only authorized staff members to access the records
- Ensure that records are easily found

Standardized forms ensure that information mandated by law and stored in the record-keeping system is consistent and complete. Forms are developed based on the types of information outlined in the organization's policies and legal requirements.

When the Level III Instructor creates a form to aid in gathering a type of data, the relevant policies and legal requirements surrounding each type of data should be researched. At the same time, the instructor should investigate whether forms that gather the same information exist or are provided elsewhere in government.

The widespread use of computers makes form creation easy. Forms such as course attendance sheets may be simple documents that get printed and filled in manually. Information from completed forms may be compiled into databases **(Figure 18.1)**.

Figure 18.1 Written information can be entered into electronic databases.

Regardless of other formatting and style features, forms should have fields that are consistent with other training forms used by the organization. For instance, Field 1 may always contain the date; Field 2 may always contain the instructor's name; Field 3 may always contain the student's name, and so on. The more consistent the assignment of data to specific fields, the easier the form will be to use and the easier the information will be to locate, analyze, and store.

Record-keeping systems consist of more than the software applications or filing systems designed to manage the records. The system is also composed of the following resources:

- **People** — Personnel trained in the operation of the system.
- **Policies** — Codified statements that define the system, the data to be gathered, and how data are stored, accessed, analyzed, and disposed of.
- **Procedures** — Methods used to meet policy requirements.
- **Tools** — Record-keeping instruments designed to manage the records over time; these instruments include disposal schedules and access and security classification systems.
- **Technology** — Software, hardware, physical storage, and disposal equipment.
- **Training and Education** — Ongoing personnel training and refresher courses required to manage the system.
- **Maintenance** — System support that ensures the system operates correctly and efficiently and is protected in the event the system experiences a malfunction. This may include system updates when available.

Level III Instructors have the task of ensuring that these components work together and that the system functions according to AHJ policies. In addition, Level III Instructors should consult applicable standards for record-keeping in the fire and emergency services as a model for their record-keeping policies. NFPA 1401, *Recommended Practice for Fire Service Training Reports and Records*, provides some examples of training forms as well as information on the design and procedures for effective record-keeping. The format of records may also depend on the type of training for which information is being gathered **(Figure 18.2)**. Skills associated with organizing a training record system that can be readily accessed can be found in **Skill Sheet 18-1**.

BROCKTON FIRE DEPARTMENT

Standard Roll Call Form

Hours and Subject Codes

COURSE ______________________________

LOCATION ______________________________

SCHOOL NUMBER ______________________________

STUDENT NAME (Print Clearly)	STUDENT ID	Date of Birth	FDID#					CAREER	VOLUNTEER

Figure 18.2 A roll-call form should provide spaces for relevant student and classroom information.

Review: Instructor III Prerequisites

Level III Instructor skills build upon Levels I and II. NFPA 1041 has certain information for record-keeping that was described in other chapters of this manual. If the instructor or student feels a need to review this material, it can be found in the following chapters:

- **Types of training records** — Chapter 10, Records, Reports, and Scheduling; Chapter 14, Supervisory and Administrative Duties
- **Legal requirements affecting record-keeping** — Chapter 14, Supervisory and Administrative Duties
- **Public access to records** — Chapter 14, Supervisory and Administrative Duties
- **Privacy and Open Records Act exemptions** — Chapter 14, Supervisory and Administrative Duties

Disclosure of Information

As described in Chapter 14, Supervisory and Administrative Duties, Level II Instructors must be able to identify what records are public and what records are protected as private. Level III Instructors will often be asked to apply this knowledge to facilitate a disclosure of information. **Disclosure** is a legal term referring to the act of giving out information either voluntarily or to meet legal requirements or AHJ policy requirements. A disclosure is often a formal process, such as when a court order requests training records.

Some records can be released without authorization. For example, if planning documents instructors used to create a course are requested as part of an investigation, they should be freely available. Similarly, instructional materials can be requested without compromising privacy.

Disclosures get complicated when the records requested have privacy requirements attached to them. For example, under the **Family Educational Rights and Privacy Act (FERPA)** test records cannot be disclosed without the written student consent. A chief or commanding officer must tell a student of any request to see their test records. Similar requirements exist for other personnel records and any medical records.

AHJ-Level Needs Analysis

Level III Instructors apply needs analyses to policies at all levels of their AHJ. Level III Instructors identify and address needs projected for the future rather than reacting to immediate situations. For example, Level III Instructors examine the personnel training and educational needs of their organizations and make decisions about selecting staff or adjusting the instructional duties of existing personnel. Needs analysis may include identifying other jurisdictions that can provide training facilities, mutual aid, and equipment when funding does not exist to acquire these resources.

AHJ-wide needs and task analysis are generally not the responsibility of one individual, nor is it performed without research and data analysis. The sections that follow discuss these issues as they relate to task and needs analysis.

> **Review: Prerequisites for Needs and Gap Analyses**
>
> Prerequisite information for performing needs and gap analyses has been discussed in this manual. Students or instructors can review this information in the following chapters:
>
> - **Development of JPRs** — Chapter 11, Lesson Plan Development
> - **Lesson planning** — Chapter 11, Lesson Plan Development
> - **Instructional methods** — Chapter 6, Classroom Instruction; Chapter 8, Skills-Based Training Beyond the Classroom
> - **Characteristics of adult learners** — Chapter 2, Principles of Learning; Chapter 6, Classroom Instruction
> - **Instructional media** — Chapter 4, Instructional Materials and Equipment
> - **Development of evaluation instruments** — Chapter 13, Test Item Construction; Chapter 15, Instructor and Class Evaluations
> - **Conducting research** — Chapter 11, Lesson Plan Development

Committee Meetings

Involving employees in the analysis process is one strategy that Level III Instructors use to include instructors in the decision-making process. Involving employees creates an opportunity for the Level III Instructor to delegate research tasks. Delegating tasks and involving employees in decision-making are often accomplished during a meeting.

Level III Instructors should schedule meetings at appropriate intervals to allow those attending the meetings the time necessary to perform their duties. Meetings should have an agenda so that attendees know what to expect and can plan to spend the time efficiently.

Whenever possible, meetings should be limited to relatively few people, especially when input is going to be requested from those in attendance. The participants can then communicate needed information to others **(Figure 18.3)**.

Finally, Level III Instructors may attend committee meetings to report findings to higher-ranking members. Any presentations they make should be concise and informed by evidence from research.

Figure 18.3 Committee meetings should include attendees who can give input to the entire panel.

Data Organization and Interpretation

All data collected must be organized and interpreted. There are many ways of doing this, and Level III Instructors will discover what methods work best. The most important outcome of interpretation is that the information tells a story, supports a proposal, and/or answers an analysis question. For example, research stating that training injuries were lower than the previous year is good news. The Level III Instructor can justify continuing those policies if that information can be correlated with changes in training policies or courses from the previous year.

Development of Training Policies, Procedures, and Guidelines

Written policies, procedures, and guidelines are essential for the effective operation of any fire and emergency services organization. Place into writing the expectations of the organization based on its values and strategic and operational plans.

The documents must contain current and appropriate information. Therefore, the organization should have a process to perform the following functions:

- Identify a need for new policy, procedure, or guideline
- Develop draft document
- Submit draft for review
- Adopt policy, procedure, or guideline
- Publish document
- Implement document's directives
- Evaluate effectiveness.

In addition, the organization should be familiar with standards and the processes for adopting them into law. The following sections describe the functions involved in creating policies, procedures, and guidelines and provide definitions of important terminology for completing the process. Skills associated with developing policy recommendations for meeting agency training goals can be found in **Skill Sheet 18-2**.

Policies

A **policy** is a set of rules that organizations develop, adopt, and use as a basis or foundation for decision-making. Policies help organizations address issues or problems. Training policies serve to guide an organization's function on a day-to-day basis and are often accompanied by procedures. For AHJ policies to be effective, they must have the following characteristics:

- Written using language that is simple, concise, and respectful

Figure 18.4 The practice of maintaining rapid intervention teams/crews during training evolutions is an example of a policy instituted for student safety.

- Adopted through a process that provides reasoned feedback
- Supported by the organization's administration and training manager

In some cases, policies are developed at the local level. In others, organizational training policies are based upon state, provincial, or federal laws or standards. For example, the *U.S. Code of Federal Regulations (CFR)* identifies specific requirements for the training of responders to hazardous materials incidents.

These federal regulations often serve as the basis for locally adopted training requirements **(Figure 18.4)**. Sometimes, these regulations, such as Respiratory Protection Training, *CFR 1910.134*, require development of local policies and procedures. A resource for developing local policies include standards such as NFPA 1201, *Standard for Providing Emergency Services to the Public*, and NFPA 1500™, *Standard on Fire Department Occupational Safety and Health Program*.

Procedures and Guidelines

Level III Instructors ensure that instructors comply with training policies and program goals. To facilitate this, training managers develop AHJ **procedures** and **guidelines** that outline the approved methods in fulfilling obligations as well as what autonomy instructors have to make decisions. Procedures and guidelines are management tools and may either accompany policies or exist independently. Consider the following two definitions and explanations:

A *procedure* identifies the steps that must be taken to fulfill the intent of a policy and is written to support a policy. For example, a policy may state that all live fire-training evolutions must be conducted in compliance with NFPA 1403, *Standard on Live Fire Training Evolutions*. The procedures accompanying this policy consist of steps the instructor takes to ensure that the policy is achieved, including such actions as the following:

- Establish the Incident Command System (ICS) based on the National Incident Management System (NIMS) model.
- Inspect the training structure **(Figure 18.5)**.
- Provide a safety officer for evolutions.
- Require use of appropriate protective equipment.
- Use qualified instructors knowledgeable of fire behavior.

Well-defined procedures offer consistency in implementing a policy. Procedures are indispensable in programs where adherence to policy may directly impact training such as the following:

- Implementing safety precautions during training
- Hiring or evaluating personnel
- Acquiring structures

Figure 18.5 Inspecting training structures and props is an important procedure that should be part of the safety policies in training organizations.

A *guideline* identifies a philosophy. Guidelines may be part of a policy or exist independently that simultaneously provide direction and autonomy for achieving the goal of the guideline or policy. Examples include the following:

- An organization may have a policy for conducting training in inclement or extreme weather conditions.
- Included in the policy are guidelines that provide information that aids decision-making such as when it is appropriate to cancel or reschedule training.
- The information in the guidelines gives instructors the autonomy they need to consider all factors involved before making a decision.

Identifying a Need for a New Policy, Procedure, or Guideline

The steps for determining the need for a new AHJ policy, procedure, or guideline are as follows (**Figure 18.6**):

Step 1: Identify the problem — Determine whether the problem or development requires a new policy, procedure, or guideline to be resolved appropriately. Some situations may be best addressed using a narrow focus and will not require formal organization-level change.

Step 2: Collect the data to evaluate the need — Data may come from personnel interviews, product literature, or activity reports. Determine whether it is quantitative or qualitative.

Step 3: Select the evaluation model — Determine whether an evaluation model is goal-based, process-based, or outcome-based.

Step 4: Establish a timetable for making the needs evaluation — Determine the amount of time required to evaluate the problem. The complexity of the problem and volume of information to be evaluated factor into this estimation.

Step 5: Conduct the evaluation — Follow the recommended steps for the model most appropriate for the situation.

Step 6: Select the best response to the need — Determine the best policy, procedure, or guideline to solve a problem. This determination may include no policy, procedure, or guideline.

Step 7: Select alternative responses — Select a second-best choice to accommodate a situation in which a contingency plan is necessary because of external influences. Personal safety should not be compromised for any personnel in any long-term decision.

Step 8: Establish a revision process or schedule — Create a revision process as part of the policy, procedure, or guideline. Revision may be a general process for all, or one that is specific to each policy, procedure, or guideline.

Step 9: Recommend the policy, procedure, or guideline that best meets the need — Determine whether the recommended policy, procedure, or guideline needs to be formally adopted by the jurisdiction because they may have the effect of law. Formal approval requires they be supported by documentation.

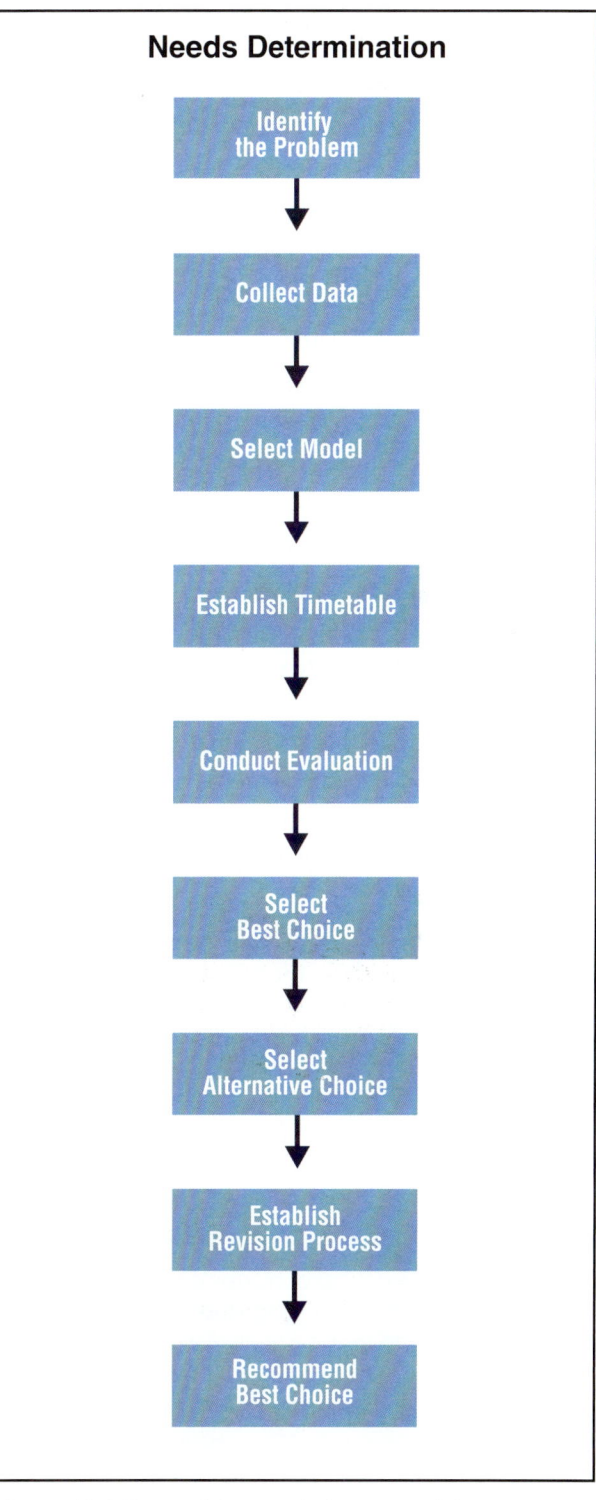

Figure 18.6 Determining needs for new policies can be made less complicated by following a step-by-step process.

Developing a Draft Document

After a need is identified, an individual or committee develops a draft of the document according to AHJ procedures in order to help meet training program goals. The following guidelines will assist in developing a draft document:

- Determine whether a policy, procedure, or guideline is the most appropriate for the issue or incident or all three are needed.
- Select the least restrictive type of document.
- Use similar policies, procedures, or guidelines as references when possible.
- Follow AHJ or national models for format, terminology, and organization.
- Include the date the document becomes effective.
- List the person or unit responsible for managing the related policy, procedure, or guideline and any other policy, procedure, or guideline. Additional requirements for each type of document:
  — Policy must also state to whom or what it applies and the specific rules of application.
  — Procedure must contain the steps to follow and the policy to which it applies.
  — Guideline must include to whom or what it applies and the guideline statement.
- Conduct research on the legality of the new policy, procedure, or guideline because the documented issue or incident may have legal implications. The process includes:
  — Determine whether the issue or incident is addressed in organizational policies.
  — Review NFPA standards that address the issue or incident.
  — Seek similar issue or incident resolutions from other organizations.
- Submit the draft document to the organization's legal counsel for review. It is the training manager's responsibility to ensure that any document is legally and administratively valid before concluding the draft development step.

Submitting a Draft for Review

The draft document should be submitted for AHJ review and comment. Review and comment opportunities are especially important for documents that are controversial or affect multiple groups in an organization. For example, a proposed policy on attendance at training sessions certainly affects students more than instructors or administrators.

When the obligations necessitate a policy, procedure, or guideline to enforce an action, then it is appropriate to adopt one. Use the following review-process steps:

Step 1: Allow personnel to respond with feedback and input on any document that affects them. Those responsible for managing the fire and emergency services responders and other affected should review the draft document.

Step 2: Provide a comment period for the draft document and review comments.

Step 3: Evaluate comments and amend the draft document as necessary.

> **Lack of Support for New Policies, Procedures, and Guidelines**
>
> A critical consideration about feedback on draft documents is that those affected by a new policy, procedure, or guideline may not always support the change. Lack of support does not necessarily indicate that the policy, procedure, or guideline is not necessary. Ultimately, the training manager and the organization's chief executive officer are held accountable to the obligations of providing safe and effective operations in the training program.

Adopting a Policy, Procedure, or Guideline

The policy, procedure, or guideline is ready for adoption once feedback has been evaluated, and if the document has been amended. The appropriate manager or administrator endorses the document. For example, a policy affecting the training program is given to the training manager for endorsement. When a policy affects the organization, the endorsement must come from the chief executive officer. An endorsement signals to organizational personnel that the policy, procedure, or guideline has official sanction.

Publishing a Document

Per the AHJ, once the document has the necessary signatures, the policy, procedure, or guideline is ready for publication. Anyone potentially affected by the document must be informed of the change or addition of duties. Memos often inform personnel, but when the document is implementing a substantial change or addressing a critical issue, the best method for communicating the change is a face-to-face meeting with personnel and supervisors (**Figure 18.7**). A face-to-face meeting provides the following opportunities for personnel to:

- Ask questions
- Gain clarification
- Ensure understanding

Figure 18.7 Meet in person to implement policies when the changes will be significant or complicated.

Regardless of the method used, everyone with an interest in the policy, procedure, or guideline should be informed before it takes effect. Without this communication, instructors and affected personnel may not be able to comply with the changes.

Implementing a Document's Contents

Improper implementation of AHJ policies, procedures, and guidelines is the primary cause for the failure of personnel to accept and adhere to the contents of those documents. The intent of most of these documents is to create a change in behavior in order to achieve training program goals. To ensure that personnel learn, adopt, and practice these changes, they must be educated about the new policy, procedure, or guideline.

Acceptance of change requires that personnel know the reason for the change, understand its benefits, and accept it as an improvement. Other requirements of the implementation step include providing the necessary equipment, support, and training required by the new policy, procedure, or guideline.

Implementation of any policy, procedure, or guideline must be consistent, fair, and documented. Credibility of new requirements as well as the administration that supports them can be destroyed when personnel believe implementation and enforcement are inconsistent.

Evaluating Effectiveness

When the policy, procedure, or guideline is implemented, it must be monitored. Chief officers, managers, and supervisors can observe the new requirement in use and determine its effectiveness based on the established criteria. If observations indicate that the new requirement does not provide the necessary change, the policy, procedure, or guideline process of development should be evaluated.

Interviews with concerned personnel may indicate that additional education, support, or changes would help people accept the requirements. Monitoring of the policy, procedure, or guideline should continue. A periodic review should be performed to determine whether the document requires revision, replacement, or abandonment based on changes in the operating environment or completion of training program goals.

Standards that Influence Training

Standards are key elements in any training program. Organizations adopt standards to provide the basis for performance or operational requirements. The most common standards that fire and emergency services organizations in North America use are from the NFPA. These standards address many issues including professional qualifications, firefighter health and safety programs, and organizational structure. Training managers often make decisions based on standards.

NOTE: Refer to Appendix C of the manual for a list of NFPA standards that apply to instructors.

Other standards used in the fire and emergency services come from government. For example, many of the policies and requirements for hazardous materials programs are found in federal laws. Also, most states and provinces have specific requirements for emergency medical services (EMS) training and certification.

Adoption of a Standard

An example of the adoption of a standard into law occurred in Florida. In response to three fatalities that occurred during two live-fire training incidents, the state adopted NFPA 1403 as a state law. By 2007, all live-fire training had to be conducted at a certified training center or supervised by a certified live-fire instructor at an acquired structure.

Most standards reflect state/provincial or national norms for fire and emergency services. For example, professionals from appropriate fire and emergency services organizations develop NFPA standards through a consensus process. NFPA members have the opportunity to ratify or reject the proposed standards. This process allows standards to change over time to reflect the needs and practices of the fire and emergency services.

Many standards can affect an organization's training program; therefore, it is advisable for training managers to learn about standards that affect their organizations. For example, when an organization participates in a state or provincial hazardous materials first responder certification program, the training manager should review NFPA 1072, *Standard for Hazardous Materials/Weapons of Mass Destruction Emergency Response Personnel Professional Qualifications* (**Figure 18.8**).

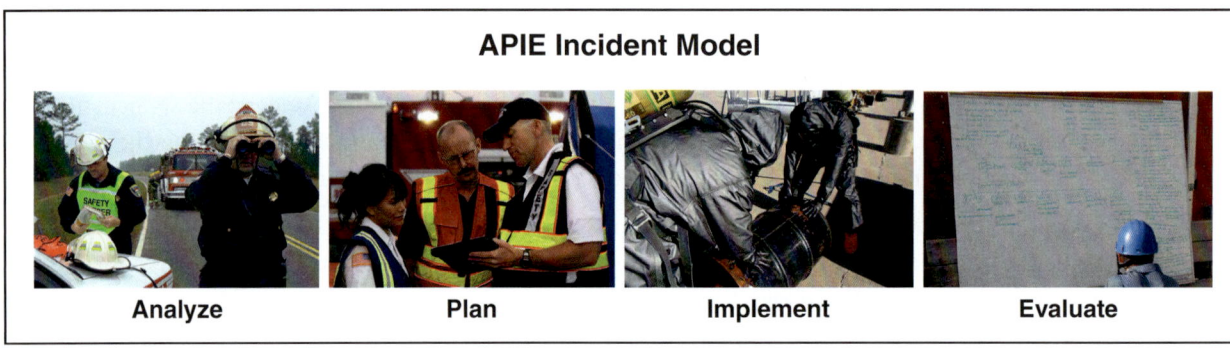

Figure 18.8 An instructor teaching hazardous materials response should be familiar with the simple response model, APIE (Analyze, Plan, Implement, Evaluate), a problem-solving process devised by the National Fire Protection Association.

When there is an NFPA or other standard used for identifying organizational training requirements or operations, the organization should adopt it. Adopting a standard gives it formal authority in the organization and allows the training manager to enforce the standard's requirements.

Of course, adoption also allows the training manager to hold organizational personnel accountable for the requirements. The adoption process for a standard is the same as that used for adopting policies, procedures, or guidelines.

Formulating Budget Needs

Organizations use budgets to ensure that funding is available for their expenses. The type of budget system and the funding source will vary between types of organizations and levels of government. Level III Instructors must be familiar with the budget types used in their jurisdiction, the legal requirements, and their responsibility for formulating budget needs for instruction and training. Legal mandates may exist at the national and state/provincial levels that apply to national, state/provincial, municipal, and rural AHJs. Budget development generally follows an accepted model in most jurisdictions.

To accurately formulate budget needs, the Level III Instructor must identify and document the types of resources that the AHJ will need in order to meet its training goals. **Resource management** takes into account the costs of using the resources that are already at the AHJ's disposal; in contrast, a **resource analysis** will reveal what the AHJ will need to purchase versus what is available. Performing a detailed resource analysis will help the Level III Instructor identify necessary components such as equipment, instructional materials, or other items needed to complete the AHJ's training objectives.

It is important to note that Level III instructors working in fire departments will have a very different budgeting outcome compared with directors of state fire training. In many cases, state fire training will have to estimate their expenses and the amount of income needed for the year. The income can come through fees, contracts, grants, or state appropriations. This means the agency will likely start the year in a deficit and may need the entire year to accrue the planned budget amount by the end of the fiscal year. In municipal fire departments the appropriation is provided, and they draw down the account(s) throughout the year, hoping not to expend more than was appropriated. Most municipal departments may also need to receive permission from their governing body if they are going to receive grants and other assistance outside of the appropriated or approved budget amount provided by the governing body. State laws usually will regulate how this is handled by municipalities.

NOTE: Skills associated with formulating budget needs can be found in **Skill Sheet 18-3**.

Equipment

As discussed in previous chapters, equipment used by the AHJ can include:

- Computers
- Video projectors
- Training props
- Rescue or EMS manikins
- Simulations
- Firefighting equipment such as:
  — Hose
  — Nozzles
  — Ladders
  — Smoke machines
  — Self-contained breathing apparatus for training staff use

Larger departments may fund equipment specifically for use by the training division to prevent having to take equipment off of an apparatus. This allows the department to remain in service if required to respond to regular alarms. The Level III Instructor should possess a thorough knowledge of current resources at the AHJ's disposal. When the resource analysis presents a need for additional equipment, the Level III Instructor should document the need during the budget formulation process.

Sources of Instructional Materials

The Level III Instructor may need to seek instructional materials for the training program or individual course. Depending on the capability of the department's training division, instructors may be able to develop some of these materials in-house. Often the materials may need to come from external sources since the course may eventually lead to state certification; therefore, they will become a necessary component of the budget formulation process. The Level III Instructor can find instructional materials from sources such as:

- Publishers
- Governmental agencies (such as FEMA or the National Fire Academy)
- Nonprofit organizations
- State/provincial programs
- Professional organizations (such as the International Association of Fire Chiefs and the International Association of Fire Fighters)

The Level III Instructor also will have a budget for members of the department to attend out-of-department training or to bring in consultants to the department. The AHJ's budget cycle should help plan how and when this money will be used.

Through this analysis, as costs increase for registration to events, items such as travel and other associated costs increase. Level III Instructors should be able to document the need to ask for an increase in their budget in the future. Many organizations prioritize training; this assists the Level III Instructor to ask for additional funding if state or federal mandates require additional training or if the AHJ requires additional program training. Examples can range from leadership training to firefighter mental health to live-fire training involving flammable liquids for dry chemical and foam use. Many fire departments will not have environmentally approved props to support live-fire training and will need to travel to a facility within their state/province or around the country to acquire it.

Maintenance and Repair of a Training Center

For the jurisdictions that have a training center with external props, such as a tower with gas-fired or Class A props, the repair and maintenance of these items are usually the most expensive parts of the facility. While some AHJs may have capital improvement funds to build these centers, many departments do not budget an appropriate amount of funds to maintain the facilities.

> ### Expenditures for Budgeting
> Some expenditures and general timeframes that a Level III Instructor may need to take into account for long-range budgeting include:
> - **Structural engineering inspection** — Every ten years
> - **Standpipe system inspection** — Every two years
> - **Roof inspection** — Every five years
> - **Electrical inspection** — Annually
> - **Gas line inspection** — Annually
> - **Door replacement** — Every five years
> - **Painting** — Every two to three years
> - **Burn prop preventive maintenance** — Annually
> - **Burn prop replacement** — Every fifteen years
>
> **NOTE:** Timeframes for inspections may vary depending on the item or task. Follow all manufacturer and engineer recommendations for inspection timelines.

Some expenditures may require that the Level III Instructor work with administration during the beginning of the budget cycle. In some cases, municipalities may consider these types of expenses as expanded level packages. In other municipalities, they may be part of an equipment replacement fund or included in the AHJ's 5-year Capital Improvement Plan. Whatever process is used, Level III Instructors must realize that if they make the request after the department has made its initial request for the upcoming year's budget to the governing body, it may take upwards of eighteen months before the funding is made available.

Equipment Purchasing Policies

Level III Instructors may be responsible for making the large equipment purchases, known as *capital purchases*, for training established in budgets. As described in Chapter 14, capital purchases are used for very expensive, one-time-purchase resources – for example, an apparatus or a portable training structure. Because of their expense, purchase of capital resources are scrutinized for usefulness to the AHJ and its overall training and curriculum needs. Funding for these purchases may be provided with a mandated use, for example, a bond issue in a community to fund a training facility or a grant for purchasing a simulator. The money must be spent as designated, and the expense must be well documented.

When making capital purchases based upon the AHJ's needs analysis, a bid process ensures the organization is performing due diligence to purchase a quality resource at an appropriate price. The bid process begins with performing research about the capital resource in question and includes the following actions:

- **Survey other jurisdictions** — Contact other jurisdictions who use similar resources.
- **Review standards and regulations** — Any capital resource must be in compliance with standards and regulations applicable to the purchase.
- **Review industry trends** — Any capital purchase should match industry trends to ensure those responders training with the resource will encounter similar resources in the field.
- **Compare various products** — The bid process involves examining multiple products to make an informed decision. Comparing products before the bid process gives the instructor options for which item at which price range would best suit an organization's needs.
- **Determine equipment compatibility** — Any capital purchase must be useable. Equipment that is incompatible with training equipment becomes a waste of money.

When the jurisdiction permits or requires it, a **request for proposal (RFP)** should be developed specifically for capital purchases. After doing so, the bidding process can be conducted. The sections that follow describe these two steps.

NOTE: Some departments may have a bidding process for purchases that are in the operational budget as well as for purchases in the capital budget. If this is the case, then instructors should familiarize themselves with procedures for these bidding processes.

Developing a Request for Proposal (RFP)

An RFP defines the needs of the organization and allows manufacturers or their authorized distributors to decide whether they can meet purchasing specifications. An RFP must have the following components:

- Specific schedule outline
- Bid dates and delivery dates
- Provisions for supplying equipment for scheduled evaluations
- Training dates for the benefit of maintenance technicians and instructors

An RFP also allows the jurisdiction to limit or specify which companies can bid based on the response to the RFP and participation in pre-bid meetings. The RFP process reduces the number of bidders to companies that are capable of meeting the purchasing specifications. Companies are eliminated from consideration in the following situations:

- Failure to meet delivery deadlines
- Unable to provide the required performance bonds
- Lack of financial support to complete the contract
- Documented history of contract violations

Before writing the RFP, an instructor should consult legal counsel and the AHJ purchasing laws to determine what kinds of controls can legally be placed on bids or bidders. The selection of bidders must not be subjective or arbitrary.

RFPs may also include language about evaluating equipment to be purchased before the bid process can continue. Evaluating a piece of equipment usually involves inspecting the item through direct contact. The RFP or other organizational document should include a checklist of criteria to which the particular piece of equipment should conform. For example, respiratory protection evaluation criteria may include, but are not limited to, the following factors:

- Maneuverability
- Flexibility
- Effect on vision
- Ease of donning
- Ease of doffing
- Effect on workload
- Comfort
- Durability
- Ease of operation
- Compatibility with operational procedures

In addition to evaluating equipment, Level III Instructors should also review any available product data. Some factors to consider include:

- **Features** — List the various features and accessories available with the equipment.
- **Durability** — Answer the following questions: How sturdy is the equipment? Are plastic parts easily broken? Will the equipment stand up to rough treatment?
- **Lifecycle cost** — Include the initial purchase price, which may have to be estimated based on the list price, and the cost of annual maintenance, parts, and support paid in increments over the life expectancy of the equipment to determine lifecycle cost.
- **Maintenance requirements** — Determine maintenance requirements by considering the manufacturer's suggested maintenance schedule, the level of technician certification and training, and whether maintenance can be done in-house or by a contract vendor approved by the manufacturer **(Figure 18.9)**.

Figure 18.9 Maintenance of some equipment may be done in-house by a qualified person.

- **Infrastructure** — Answer the following questions: What infrastructure supports the department's equipment? What changes or investments are required to redesign the equipment maintenance facility, modify systems, and retrofit apparatus mounting hardware?

Creating Purchasing Specifications

The practice of creating purchasing specifications follows the RFP process so that the purchased equipment supports the AHJ's curriculum and training goals. The purchasing department prepares the wording of the legal requirements that define the legal obligations necessary to meet the specifications. These features are required in all purchasing specifications. The instructor does not develop these particular sections of the specifications but should be aware of them and their effects on prospective bidders. These requirements may include the following:

- Vendor attendance at pre-bid meetings
- Warranties
- Liability or performance bonds
- Specified delivery times
- Payment schedules
- Financial statements

The development of the specifications document is the responsibility of the Level III Instructor **(Figure 18.10, p. 386)**. The language must be clear and concise. Each detail of the design requirement must be included, and nothing should be assumed. Most manufacturers provide sample specifications forms as a guide. Some of the topics that should be included in purchasing specifications are as follows:

- NIOSH/Mine Safety and Health Administration (MSHA) standard requirements or ANSI certification for the intended use where applicable
- NFPA compliance when applicable
- Number of units to be purchased
- Design requirements
- Delivery date
- Warranty
- Accessories
- Training for maintenance technicians
- Training for operational personnel
- Start-up parts inventory
- Acceptance testing
- Technical support
- Penalties for late or no delivery

The Level III Instructor must be aware of purchasing ordinances or laws in the event that specifications are too restrictive or legally prohibited. For example, when a specific feature that meets valid operational requirements is only available from a single manufacturer, an option for bidding for an equal alternative or a method to take exception to the specifications must be included. Similarly, a **restrictive bid (sole-sourced)** includes many specifications that only one manufacturer can meet. Therefore, the purchasing ordinances or laws of the AHJ may prohibit this type of bid. When a specific item of equipment is determined to meet the organization's needs to the exclusion of all others, thereby preventing an open-bid process, a variance or exemption from the approved purchasing process may be required from the jurisdiction's purchasing and legal departments.

\multicolumn{6}{c}{**State Fire Service Training Academy Specification Worksheet**}
ITEM #	**QTY**	**OPTION & CODE #**	**DETAILED DESCRIPTION**	**UNIT PRICE**	**AMOUNT**
1	1	J14	100009341 – 2011 Chevrolet Silverado 3500HD 2WD Reg Cab, 137.5" WB, 59.8" CA WT (CC31003), DUAL REAR WHEEL	$20,329.00	$20,329.00
2	1	L20	GVW 13200 DRW	$0.00	$0.00
3	1	L24	TRANSMISSION, ALLISON 1000 6-SPEED AUTOMATIC, electronically controlled w/overdrive, electronic engine grade braking and tow/haul mode. Includes (KNP) external transmission oil cooler		
4	1	L25	AUXILLARY transmission oil cooler	$0.00	$0.00
5	1	L28	SOLID PAINT, FIRE ENGINE RED IN COLOR	$0.00	$0.00
6	1	L29	MIRRORS, OUTSIDE MANUAL, BLACK, manual-folding	$0.00	$0.00
7	1	L31	POWER DOOR LOCK WITH REMOTE KEYLESS ENTRY	$0.00	$0.00
8	1	L32	POWER WINDOWS	$0.00	$0.00
9	1	L35	AUDIO SYSTEM, AM/FM/CD STEREO	$0.00	$0.00
10	1	L36	TILT STEERING WHEEL	$0.00	$0.00
11	1	L38	SEATS, FRONT 40/20/40 SPLIT-BENCH, 3-passenger, driver and front passenger manual reclining with outboard head restraints and center fold-down armrest w/storage.	$0.00	$0.00
12	1	P69	ENGINE, DURAMAX 6.6L V8 TURBO DIESEL V8, B20-DIESEL COMPATIBLE, (335 hp [249.8 kW]@ 3100 rpm, 685 lb-ft of torque [924.8 N-m]@ 1600 rpm), (Requires [MW7] Allison 1000 6-speed automatic transmission and [ZW9] pickup box. Includes [K40] exhaust brake, [TUV] heavy-duty dual, 730 cold-cranking-amp battery and [K05] engine block heater.)	$7,135.00	$7,135.00
13	1	P72	REAR AXLE, 3.73 RATIO W/LIMITED SLIP REAR AXLE	$0.00	$0.00
14	1	P86	DAYTIME RUNNING LIGHTS	$0.00	$0.00
15	1	P95	ELECTRIC BRAKE CONTROLLER	$177.00	$177.00
16	1	P96	ADD FRONT TOW HOOKS	$0.00	$0.00
17	1	P98	CRUISE CONTROL	$228.00	$228.00
18	1	J108	HEAVY DUTY FLOOR MATS	$76.00	$76.00
			TOTAL	**$27,945.00**	**$27,945.00**

Figure 18.10 Features required from purchased equipment should be explained in detail so they can be checked against a manufacturer's documentation.

Once the jurisdiction's finance or purchasing officer approves, the purchasing department issues requests to qualified bidders and sets a date for the opening of the bids. The bids may only be returned to and handled by the purchasing department. Once received, qualified bids are given to the instructor who is responsible for evaluation. When all of the data are collected and reviewed, the Level III Instructor selects or recommends the equipment that best meets training division needs.

NOTE: Skills associated with creating agency specifications for purchasing training equipment can be found in **Skill Sheet 18-4**.

Bid Evaluation and Awarding of the Contract

The instructor may or may not be involved in the evaluation and awarding of bids for equipment purchases. If an instructor is part of the process, the person involved in the evaluation and awarding of bids should help evaluate and score them based on the original purchasing specifications. A matrix or spreadsheet can be created with the specific requirements listed down the side and individual bidders listed across the top. Values can be assigned to each requirement and inserted into the corresponding box, depending on whether the bidder exceeded, met, or failed to meet the specification. Scoring must be equitable and well documented. In most jurisdictions in the U.S., this information and process may be subject to the Freedom of Information Act (or applicable state statutes and local ordinances) and outside review.

Following evaluation, a recommendation is made as to which bidder should be awarded a purchase contract. The legal department writes a contract, and the AHJ awards the bid to the winning supplier. The contract binds the supplier to meet the specifications and the jurisdiction to pay for the goods or services.

Administration of the contract is the responsibility of the purchasing department. This department acts on behalf of the emergency services organization that receives the goods or services. The emergency services organization is then responsible for the following actions regarding the purchased equipment, materials, or apparatus:

Figure 18.11 New items, including apparatus, may be placed into service without fanfare.

- Accept
- Test
- Inventory
- Store
- Maintain
- Place into service **(Figure 18.11)**

What This Means to You

The budgeting and purchasing processes can be complicated. Following your organization's policies and procedures is crucial to your success and reputation as an instructor. You may not be an expert in the financing of the organization, but you are expected to use funds wisely and make thoroughly documented purchases. Work closely with the individuals in the organization who are experts in the budgeting and purchasing process. Relying on their expertise will help ensure that funds are spent as intended.

Human Resource Management

The Level III Instructor may relinquish some teaching responsibility and assume more responsibility for supervising instructors and other staff members. As a result, Level III Instructors may have certain human resource responsibilities, including selecting instructional staff, helping ensure that instructors maintain appropriate qualifications, and advertising positions, all of which will help the AHJ achieve instructional goals.

Instructional Staff Selection

Selecting competent instructional staff is a key management function of Level III Instructors. In addition, determining whether roles are short-term or long-term aids in the selection process. When the instructor is only going to teach one course or topic, then the role may be considered short-term. When the instructor is expected to perform the function for an extended period, such as five years, then the role is long-term and the selection criteria will be different.

Next, the Level III Instructor should determine the instructors' roles in the organization and the qualifications necessary to teach training programs. Establishing their positions in the organizational structure gives importance to the instructor role. Instructors are essential to program planning.

Instructor Qualifications

Level III Instructors function as intermediaries between administration and trainees. Instructors actively apply knowledge and skills on the job at various ranks or positions in the organization. Their experiences are critical components of their qualifications.

Few instructors are qualified to teach all topics in a modern fire and emergency services curriculum. They tend to specialize and become proficient in teaching knowledge and skills in their areas of expertise. The Level III Instructor should base instructor selection upon the topics and the designated instructor roles in the organization and training program.

In addition to topic expertise, instructors must have credibility with students (**Figure 18.12**). Personnel see that instructors have credibility when they display technical proficiency, evidence of formal training, and instructional experience. Other factors affecting credibility include rank, reputation, and respect within an organization.

Instructors must be effective communicators. Topic expertise and credibility are only effective when instructors transmit knowledge and demonstrate skills that program participants understand and apply.

NOTE: See Chapter 1, The Instructor as a Professional, for additional information on characteristics of effective instructors.

Figure 18.12 Instructors develop credibility with students by displaying topic expertise and effective communication.

Along with the technical qualifications, an instructor must also be qualified and certified. Qualifications can be based on legally adopted or nationally recognized certification standards. For example, fire instructors should meet the Level I Instructor requirements of NFPA 1041. When there are no qualified personnel, Level III Instructors need to provide instructor-training programs to qualify and develop in-house instructors. Training, program resources, and information are generally available from local, state, or provincial training organizations.

NOTE: Skills associated with selecting and evaluating instructional staff applicants can be found in **Skill Sheet 18-5**.

Position Advertising

After instructor roles and qualifications are determined, the Level III Instructor must advertise or market the position to prospective candidates in accordance with AHJ policies. An organization may have policies or other requirements, such as required rank or service time, that limit the number of applicants. However, these requirements must also be advertised according to AHJ policy and employment laws so that interested personnel may choose an appropriate course of action. When the position involves a change in working conditions, such as increase in pay, change in hours of work, or other benefits, personnel policies or union/management agreements may require that the process follow a set of steps. These steps are developed in accordance with the overall AHJ training goals and must follow local, state/provincial, and federal laws.

Personnel Evaluations

The personnel evaluation process begins with a meeting between the Level III Instructor and the new instructor or staff member. At this point, the AHJ's expectations for the job performance of the employee have been established. AHJ policies typically require that both parties outline and agree upon performance standards, as well as schedules for probationary and annual evaluations.

Some fire and emergency services organizations have a probationary period for new instructors or staff members. The job performance requirements that were established initially provide the basis for performance evaluations during this probationary period. The Level III Instructor or supervisor must monitor the job performance of new instructors or staff members and provide appropriate feedback.

Feedback may include additional information about expectations or praise for meeting or exceeding expectations. This probationary period is most effective when the supervisor provides sufficient information and access to appropriate resources for improvement. In addition, the new instructor should be encouraged to ask questions to clarify misunderstandings.

Fire and emergency services personnel should also be familiar with the concept of 360-degree feedback. It is similar to the type of size-up that occurs at an emergency incident when the Incident Commander (IC) requests situation reports from all sides of the incident. The need for tactical changes becomes apparent quickly, and the IC then responds accordingly.

When the concept is applied to the human resources program, the process and the results are similar. The information used in the performance evaluation comes from people with professional contact to the person being evaluated. The information that is gathered is based on observations.

Responses must remain confidential to protect the people providing the information. Confidentiality also encourages respondents to speak freely and offer constructive criticism.

Feedback should also support a plan for the improvement of behaviors perceived to be below standard. Evaluations and the interpretation of results should be administered and performed by a professional trained in this technique.

NOTE: Level III Instructors may not be responsible for formal, performance-based evaluations if they are not the official head of their division or do not have direct reports assigned to them.

NOTE: Skills associated with creating a performance-based evaluation plan for instructional staff can be found in **Skill Sheet 18-6**.

Chapter Review

Answer the following questions to review the information provided in this chapter.

1. What functions does a record-keeping system perform?
2. What is meant by disclosure of information?

3. What is the role of a Level III Instructor in conducting an AHJ-level needs analysis?
4. What is the process of developing organizational training policies, procedures, and guidelines?
5. What influence do standards have on training policies?
6. How do Level III Instructors formulate budget needs?
7. How do Level III Instructors write equipment purchasing specifications so that the equipment is appropriate and will support the curriculum?
8. What human resources responsibilities does the Level III Instructor have?
9. What role does the Level III Instructor have with personnel evaluations?

Discussion Questions

The following questions are intended to generate discussion, expand your understanding of the chapter text, and allow you to think critically about what you have learned. Answers to these questions may vary.

1. What resources affect record-keeping systems?
2. Why do some types of records have privacy requirements attached to them?
3. What is the purpose of an AHJ-level needs analysis?
4. What are some challenges Level III Instructors experience when developing training policies, procedures, and guidelines?
5. What are some pros and cons of using standards in training programs?
6. What is one experience you have had as an instructor with formulating budget needs?
7. What aspects of the bid process can be the most challenging?
8. What experiences have you had with instructor staff that were not qualified for the position they were hired for?
9. What experiences have you had with a positive or negative personnel evaluation?

Key Terms

Disclosure — Legal term referring to the act of giving out information either voluntarily or to meet legal requirements or agency policy requirements.

Family Educational Rights and Privacy Act (FERPA) — Legislation that provides that an individual's school records are confidential and that information contained in those records may not be released without the individual's prior written consent.

Guideline — Statement that identifies a general philosophy; may be included as part of a policy.

Policy — Organizational principle that is developed and adopted as a basis for decision-making.

Procedure — Outline of the steps that must be performed in order to properly follow an organizational policy.

Request for Proposal (RFP) — Public document that advertises an organizational need to manufacturers or individuals who may be able to meet that need.

Resource Analysis — Strategic planning tool used by an organization to determine the need and use of resources required for operations.

Resource Management — Process by which the authority having jurisdiction (AHJ) develops and oversees the resources required for its operations.

Restrictive Bid (Sole-Sourced) — Bid that includes many specifications that only one manufacturer can meet; *also known as* Sole-Sourced Bid.

18-1
Organize a training record system that can be readily accessed. [NFPA 1041, 6.2.2]

Task Steps

Step 1: Review AHJ policies and procedures on training records.

Step 2: Review current training record system.

Step 3: Identify and address type of data to be kept.

Step 4: Identify and address how data is entered and verified for accuracy.

Step 5: Identify and address essential general security provisions.

Step 6: Using consistent assignment of data formatting, develop a standardized form to document training activities **(Figure 18.13)**.

Step 7: Enter documented training records into system **(Figure 18.14)**.

Step 8: Ensure system functions according to AHJ policies and procedures.

Figure 18.13

Figure 18.14

SKILL SHEETS

18-2
Develop policy recommendations for meeting agency training goals.
[NFPA 1041, 6.2.3]

Task Steps

Step 1: Review AHJ policies and procedures.

Step 2: Review standards and process for adopting document into law.

Step 3: Identify need for new policy, procedure, or guideline **(Figure 18.15)**.

Step 4: Develop a draft written policy proposal to support training goal, using correct grammar, in same style and format as AHJ policies and procedures.

Step 5: Submit draft document for review **(Figure 18.16)**.

Step 6: Make any necessary changes identified during the document review.

Step 7: Adopt policy, procedure, or guideline.

Step 8: Publish document.

Step 9: Implement document's directives.

Step 10: Evaluate effectiveness.

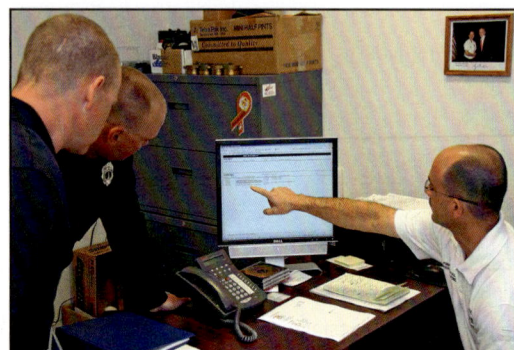

Figure 18.15

Figure 18.16

18-3
Formulate budget needs. [NFPA 1041, 6.2.6]

Task Steps

Step 1: Perform resource analysis based on training goals.

Step 2: Document needs based on resource analysis.

Step 3: Categorize needs based on resource analysis, using categories such as:
- a. Personnel
- b. Equipment
- c. Maintenance and repair
- d. Instructional materials

Step 4: Prioritize needs to achieve training goals **(Figure 18.17)**.

Step 5: Formulate overall budget needs request.

Figure 18.17

SKILL SHEETS

18-4
Create agency specifications for purchasing training equipment.
[NFPA 1041, 6.2.7]

Task Steps

Step 1: Identify equipment purchase need based on curriculum, training goals, and agency guidelines **(Figure 18.18)**.

Step 2: Justify equipment purchase.

Step 3: Review costs of equipment purchase.

Step 4: Follow AHJ policies and procedures/guidelines for equipment purchase.

Step 5: Develop purchasing specifications for equipment.

Step 6: Complete procurement form.

Step 7: Submit procurement form **(Figure 18.19)**.

Figure 18.18

Figure 18.19

18-5
Select and evaluate instructional staff applicants. [NFPA, 1041 6.2.4]

Task Steps

Step 1: Review AHJ policies and procedures for selecting instructional staff.
Step 2: Review/revise current job descriptions for instructional staff.
Step 3: Identify method(s) for selection process.
Step 4: Select evaluation instrument, per AHJ.
Step 5: Evaluate instructional staff applicants **(Figure 18.20)**.
Step 6: Select instructional staff as needed.

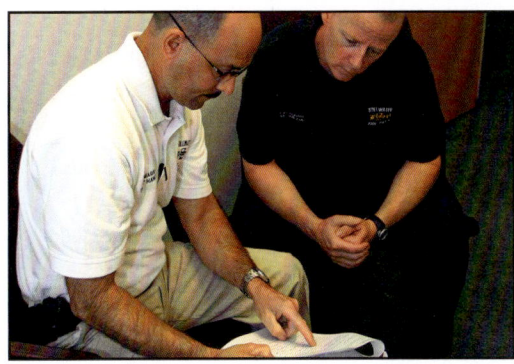

Figure 18.20

SKILL SHEETS

18-6
Create a performance-based evaluation plan for instructional staff.
[NFPA, 1041, 6.2.5]

Task Steps

Step 1: Review AHJ policies and procedures **(Figure 18.21)**.

Step 2: Review job requirements of instructional staff.

Step 3: Identify areas of evaluation.

Step 4: Develop evaluation criteria in accordance with AHJ requirements.

Step 5: Identify evaluation technique(s) to be used, based on AHJ job descriptions.

Step 6: Develop evaluation instrument(s), as needed, for lecture and skills-based sessions.

Step 7: Develop timeline for conducting evaluations **(Figure 18.22)**.

Step 8: Following technical writing style, complete evaluation instrument(s) to include the following areas:

    a. Goal

    b. Scope

    c. Plan

    d. Objectives

    e. Action items

    f. Supporting data

    g. Plan evaluation and revision process

Step 9: Submit completed evaluation instrument(s) according to AHJ policy.

Figure 18.21

Figure 18.22

Appendices

Contents

Appendix A
Chapter and Page Correlation to NFPA 1041 Requirements......................399

Appendix B
Americans with Disabilities Act (ADA) and World Wide Web Consortium (W3C)..401

Appendix C
NFPA Standards Applicable to Training 402

Appendix D
Short Answer Evaluation Rubric...403

Appendix E
Risk Management Formulas.........404

Appendix F
Level I Instructor Evaluation Rubric..406

Appendix A
Chapter and Page Correlation to NFPA 1041 Requirements

Section A: Instructor Level I

NFPA 1041 Competencies	Chapter References	Page References
4.2.1	1, 3	14, 16, 47
4.2.2	1, 3, 4	14, 49-52, 54-57, 63-68
4.2.3	3, 4, 10	58, 63-68, 85, 205-207
4.2.4	10	209-212
4.2.5	9, 10	200, 205-209
4.3.1	3	51-52
4.3.2	2, 3, 4, 6, 7, 8	31-34, 41-43, 48-53, 56-58, 69-71, 85, 108-112, 141-146, 161-164, 168-170
4.3.3	2, 3, 4	41-43, 51-52, 63-71, 85
4.4.2	2, 3, 4, 5, 6, 8	41-43, 53, 55, 57-58, 71-83, 89-101, 103, 131-133, 165-178
4.4.3	2, 3, 4, 6, 7, 8	25-31, 35-41, 54, 57-58, 64, 107-135, 137-138, 141-142, 165-181
4.4.4	2, 3, 6, 7, 8	31-43, 56-57, 127-128, 137-138, 142-157, 165-166
4.4.5	4, 5, 6, 8	71-82, 93-95, 103, 115-116, 126-127, 131-133, 137-138, 166-167
4.5.2	1, 7, 9	17-21, 146-147, 185-192, 198-199
4.5.3	9	185-187, 192-196, 199
4.5.4	9	185-187, 195, 200-201
4.5.5	9	196, 201

Section B: Instructor Level II

NFPA 1041 Competencies	Chapter References	Page References
5.2.2	14	297-302, 315
5.2.3	14	302-307, 316
5.2.4	14	307-310, 316
5.2.5	14	310-313, 317
5.2.6	14, 15	293-297, 321-324, 328-329
5.3.2	11	215-244
5.4.2	11	227-229, 232-234, 244-245
5.4.3	12	249-259, 261
5.5.2	13	265-286, 288-289
5.5.3	15	324-328, 330

Section C: Instructor Level III

NFPA 1041 Competencies	Chapter References	Page References
6.2.2	18	371-374, 380, 391
6.2.3	18	375-379, 392
6.2.4	18	388-389, 395
6.2.5	16, 18	341-342, 389, 396
6.2.6	16, 18	334-336, 381-383, 393
6.2.7	18	383-387, 394

Concluded on next page.

Section C: Instructor Level III *(Concluded)*

NFPA 1041 Competencies	Chapter References	Page References
6.2.8	17	360-362, 367
6.3.2	16, 18	334-342, 346, 374-375
6.3.3	16	334-341, 347, 349
6.3.4	16	338-339, 349
6.3.5	16	338-339, 348
6.3.6	16	338-339, 349
6.5.2	17, 18	360-362, 366, 374
6.5.3	16, 17	341-342, 356-357, 359-360, 364
6.5.4	17	353-359, 365
6.5.5	16	342-344, 350

Appendix B
Americans with Disabilities Act (ADA)

The Americans with Disabilities Act prohibits discrimination against a qualified individual with a disability regarding any of the following:

- Application
- Hiring
- Advancement
- Discharge
- Compensation
- Job training
- Other terms, conditions, and privileges of employment

The act also prohibits certain questions of job applicants, including those related to:

- Medical history
- Workers' compensation or health insurance claims
- Absenteeism due to illness, mental illness, and past treatment for alcoholism

Employers must provide reasonable accommodations for disabled workers, and training organizations must provide them for disabled students. Definitions included in the act are as follows:

- **Disabled person** — One who has a physical or mental impairment that limits one or more life activities, has a record of such impairment, and is regarded as having the impairment.
- **Qualified individual with disability** — Person with a disability who, with or without reasonable accommodations, can perform the essential functions of the position.
- **Reasonable accommodations** — Facilities (such as restrooms, telephones, parking spaces, and drinking fountains) remodeled and made readily accessible to and usable by individuals with disabilities. Includes the following:
  — Acquiring or modifying equipment or devices
  — Adjusting or modifying examinations or training materials or policies appropriately
  — Providing qualified readers or interpreters
  — Adjusting work schedules
  — Providing other reasonable accommodations for disabled individuals

World Wide Web Consortium (W3C)

To assist students with disabilities, the World Wide Web Consortium (W3C) has published accessibility standard guidelines (WCAG 2.1) that make Web content more accessible. These recommendations can enhance learning for a range of disabilities including visual, physical, speech, cognitive/learning and neurological disabilities. Many of the suggestions have been built into desktop applications. For example, Microsoft PowerPoint® offers a presentation inspector tool that allows an instructor to perform a self-check of developed content. This tool will find and suggest changes in content that students with disabilities may find difficult to read or see.

Appendix C
NFPA Standards Applicable to Training

NFPA standards that affect training are as follows:

NFPA 472	Standard for Competence of Responders to Hazardous Materials/Weapons of Mass Destruction Incidents
NFPA 473	Standard for Competencies for EMS Personnel Responding to Hazardous Materials/Weapons of Mass Destruction Incidents
NFPA 600	Standard on Facility Fire Brigades
NFPA 1001	Standard for Fire Fighter Professional Qualifications
NFPA 1002	Standard for Fire Apparatus Driver/Operator Professional Qualifications
NFPA 1003	Standard for Airport Fire Fighter Professional Qualifications
NFPA 1021	Standard for Fire Officer Professional Qualifications
NFPA 1031	Standard for Professional Qualifications for Fire Inspector and Plan Examiner
NFPA 1033	Standard for Professional Qualifications for Fire Investigator
NFPA 1035	Standard on Fire and Life Safety Educator, Public Information Officer, Youth Firesetter Intervention Specialist and Youth Firesetter Program Manager Professional Qualifications
NFPA 1041	Standard for Fire Service Instructor Professional Qualifications
NFPA 1051	Standard for Wildland Firefighting Personnel Professional Qualifications
NFPA 1061	Standard for Public Safety Telecommunications Personnel Professional Qualifications
NFPA 1072	Standard for Hazardous Materials/Weapons of Mass Destruction Emergency Response Personnel Professional Qualifications
NFPA 1081	Standard for Facility Fire Brigade Member Professional Qualifications
NFPA 1201	Standard for Providing Fire and Emergency Services to the Public
NFPA 1401	Recommended Practice for Fire Service Training Reports and Records
NFPA 1403	Standard on Live Fire Training Evolutions
NFPA 1404	Standard for Fire Service Respiratory Protection Training
NFPA 1500™	Standard on Fire Department Occupational Safety, Health, and Wellness Program
NFPA 1521	Standard for Fire Department Safety Officer Professional Qualifications
NFPA 1710	Standard for the Organization and Deployment of Fire Suppression Operations, Emergency Medical Operations, and Special Operations to the Public by Career Fire Departments
NFPA 1720	Standard for the Organization and Deployment of Fire Suppression Operations, Emergency Medical Operations, and Special Operations to the Public by Volunteer Fire Departments

Appendix D
Short Answer Evaluation Rubric

Question 1: Why is it necessary to immobilize victims of a vehicle accident?

Example of correct answer: Once extrication begins, victims have more room to move within the compartment and movement may aggravate their injuries; immobilization helps minimize these injuries by preventing or limiting movement.

	1 - Needs Improvement - 2		3 - Effective	4 - Exceeds Minimum Standards - 5	
Student Response	Did not answer questions. Written response is completely false. Evaluator cannot decifer answer content.	Written response is unorganized and/or illegible to the evaluator. Follows minimal learned material.	Written response covers the minimum amount of material and is correctly conveyed to the evaluator.	Written response is clear, organized, and covers the desired response of the question.	Written response exceeds expectations. Critical thinking skills and attention to detail are highly apparent.

Overall Scores: Student Answer [ ]   **Overall Score:** _____

PASS ☐   FAIL ☐

Appendix D • Short Answer Evaluation Rubric   **403**

Appendix E
Risk Management Formulas

The following formulas may be used to calculate the frequency or incident rate and the severity of incidents.

The Occupational Safety and Health Administration (OSHA) calculates the frequency (incident rate) as follows:

$$N/EH \; 200{,}000 = IR$$

Where:

    N = number of injuries and/or illnesses

    EH = total hours worked by all employees during the calendar year

    200,000 = base for 100 full-time equivalent employees (provides standardization between agencies and companies)

    IR = incident rate

OSHA calculates the severity as follows:

$$LWD/EH \; 200{,}000 = S$$

Where:

    LWD = loss work days

    EH = total hours worked by all employees during the calendar year

    200,000 = base for 100 full-time equivalent employees

    S = severity rate

Another method is to assign values to the frequency and severity in the following formula:

$$R = S \; IR$$

Where:

    R = risk

    S = severity

    IR = incident rate

Assessment of Severity

8.	Extreme	Multiple deaths or widespread destruction may result from hazard.
7.	Very High	Potential death or injury or severe financial loss may result.
6.	High	Permanent disabling injury may result.
5.	Serious	Loss time injury greater than 28 days or considerable financial loss.

4.	Moderate	Loss time injury of 4 to 28 days or moderate financial loss.
3.	Minor	Loss time injury up to 3 days.
2.	Slight	Minor injury resulting in no loss of time or slight financial loss.
1.	Minimal	No loss of time injury or financial loss to organization.

Assessment of Incident Rate

7.	Frequent	Daily to weekly.
6.	Very Likely	Weekly to monthly.
5.	Likely	Monthly to yearly.
4.	Occasional	Annually 1 to 3 years.
3.	Rare	Every 4 to 9 years.
2.	Exceptional	Every 10 to 50 years.
1.	Unlikely	Greater than 50 years.

To apply calculations for organizations that have more than 100 full-time employees, see the Bureau of Labor Statistics' information on how to compute a company's incidence rate for safety management. The BLS offers an online calculator for quick determination of incidence rates and how they compare to organizational averages. Go to: https://www.bls.gov/iif/osheval.htm.

Appendix F
Level 1 Instructor Evaluation Rubric

Classroom Management	1 - Needs Improvement - 2		3 - Effective	4 - Exceeds Minimum Standards - 5	
Student Relations	Oral, written, and nonverbal communication with students is inconsiderate, as characterized by insensitivity, demeaning language and condescension.	Oral, written, and nonverbal communication may not be considerate or respectful.	Oral, written, and nonverbal communication with students is considerate and respectful.	Oral, written, and nonverbal communication with students is considerate and positive, demonstrating genuine respect for individual students and the class as a whole.	Oral, written, and nonverbal communication with students is considerate and positive. There is abundant evidence of mutual respect and trust between teacher and student, as well as among students.
Preparation	Materials and equipment are not ready at the start of the lesson or instructional activity.	Materials and equipment are usually not ready at the start of the lesson or instructional activity.	Ensures materials and equipment are ready at the start of the lesson or instructional activity (most of the time).	Materials and equipment are ready at the start of the lesson or instructional activity.	Materials and equipment are ready at the start of the lesson or instructional activity and learning environment is conducive to the activity.
Student Recognition	There is no evidence that the teacher recognizes student progress or achievement.	There is some evidence that students are recognized for their progress and achievement; however, recognition is sporadic.	Recognizes student progress and achievement at significant intervals and encourages behaviors that would result in student success.	Students are informed regularly regarding their progress and achievement and are provided opportunities to improve and achieve academic success.	Students are informed regularly regarding their progress and achievement and are coached or mentored into higher academic achievements.
Assessment Practices	Assessment is inconsistent and insufficient to determine student's overall progress and is not based on the district's grading policy.	Assessment is inconsistent and is not based on district's grading policy.	Formative and summative assessments are recorded consistently based on district's grading policy and are used to guide instruction.	Formative and summative assessments are recorded consistently based on district's grading policy and are used to develop and evaluate instruction.	Formative and summative assessments are recorded consistently based on district's grading policy and utilized to develop, refine, and evaluate instruction.

Overall Scores:

Student Relations:	
Preparation:	
Student Recognition:	
Assessment Practices:	
Involves All Learners:	
Question Techniques:	
Explains Content:	
Models:	

Overall Score: _____

PASS ☐   FAIL ☐

Concluded on next page.

Instructional Effectiveness	1 - Needs Improvement - 2		3 - Effective	4 - Exceeds Minimum Standards - 5	
Involves All Learners	Students are not mentally engaged in active learning experiences during any significant portion of the class.	A few students dominate the lesson, and only a few students are minimally engaged.	Engages most students in active learning experiences 80 percent of the class time.	An overwhelming majority of students are cognitively engaged and exploring content in active learning experiences.	All students are cognitively engaged. Students initiate or develop their own activities to enhance their learning.
Question Techniques	Does not ask any type of questions or use questioning techniques during the lesson to involve all learners.	All or most questions used are recall questions.	Uses questioning techniques throughout the lesson, scaffolding to at least the mid-level of Bloom's taxonomy. Provides wait time for some student response and does random checking to ensure the involvement of all learners.	Uses consistently high-quality and varied questioning techniques, scaffolding to the higher levels of Bloom's taxonomy. Provides adequate wait time for most students to respond.	Uses consistently high-quality and varied questioning techniques, scaffolding to the higher levels of Bloom's taxonomy. Leads students to formulate many of their own questions. Provides adequate wait time for most students to respond.
Explains Content	There is no attempt to use a variety of activities to support instructional outcomes and no attempt to differentiate tasks to address a variety of student needs/learning styles/multiple intelligences.	Attempts, but does not successfully use a variety of activities (e.g. modeling, visuals, hands-on activities, demonstrations, gestures, body language; and thematic instruction) to support instructional outcomes and meet varied student needs/learning styles/multiple intelligences.	Uses a variety of activities (e.g. modeling, visuals, hands-on activities, demonstrations, gestures, body language and thematic instruction) to support the instructional outcomes and meet varied student needs/learning styles/multiple intelligences	Successfully uses a variety of activities (e.g. modeling, visuals, hands-on activities, demonstrations, gestures, body language, and thematic instruction) to support the instructional outcomes and meet varied student needs/ earning styles/multiple intelligences.	Continually seeks out new strategies to support instructional outcomes and cognitively challenge diverse learners. Willingly shares discoveries and successes with colleagues. Student needs are included in planning for methods of instructional delivery.
Models	Does not demonstrate or model the desired skill or process.	Demonstration or modeling of the desired skill or process is infrequent and unclear to students.	Provides demonstrations and modeling of the desired skill or process that are clear and precise to students.	Demonstrations are clear and precise to students with anticipation and preemptive action to avoid possible students' misunderstanding (Level 4).	Demonstrations will match all characteristics of Level 4. Additionally, teacher's modeling will assist students in achieving the lesson's stated objective. Students will demonstrate the skill or process.

Index

A

Academic misconduct, 195, 197
Access at training grounds, 100
Accident investigations, 259
Accreditation, 164, 182
Accrediting organizations, 164
Acquired building, 182
Acquired structures, 172–176
    advanced planning for use, 173–176
    asbestos issues, 174
    defined, 182
    fuel materials, 175
    inspection for use in training, 173
    notice and documentation, 174–175
    permits and inspections, 174
    structural condition, 174
    training possibilities, 174
    water supply, 175–176
    weather conditions, 174
Acronyms
    APIE, 380
    L-E-A-S-T method of discipline, 152
    SMART, 220
Act of commission, 178
Act of omission, 178
Active learning, 118
ADA. *See* Americans with Disabilities Act (ADA)
Administration, training program, 371–396
    AHJ-level needs analysis, 374–375
    budget needs, 381–383, 393
    disclosure of information, 374
    equipment purchasing, 383–387
    human resource management, 388–389, 395
    personnel evaluations, 389, 396
    policies, procedures, and guidelines, 375–379, 392
    record-keeping, 371–373, 391
    standards influencing training, 380
Administration classification, 188–189, 197
Administrative support for safety at training evolutions, 249–251
Advanced knowledge level, 57
Advertising the position, 389
Affective (attitude) domain of learning
    description, 29, 30–31
    learning objective development, 223
    test items, 280
After action reports, 209
Age and student diversity, 32
Agenda-based process, 121, 135
AHJ. *See* Authority Having Jurisdiction (AHJ)

Aircraft incident training props, 81, 177
Airport training areas, 98
Alternatives, 270, 271, 287
American National Standards Institute (ANSI)
    industry standards, 224
    purchasing specifications, 385
    purpose of, 163
American Psychological Association (APA), 226
American Red Cross for CPR training, 300
American Society of Mechanical Engineers (ASME), 224
American Society of Safety Engineers (ASSE), 164
Americans with Disabilities Act (ADA)
    categories of disabilities, 143
    provisions, 19
    requirements, 224
Analyses for research, 224
Analysis (analyze) level of learning, 221, 222, 282
Analyze (analysis) level of learning, 221, 222, 282
Anatomical manikins, 80
Anatomical models, 74–75
Ancillary components, 229–231
    assignment sheet, 231, 242
    defined, 229, 234
    handouts, 230, 238
    skill sheets, 230, 239
    study sheet, 231, 241
    worksheet, 230, 240
Andragogy, 26, 44, 115
Animations, 112
ANSI. *See* American National Standards Institute (ANSI)
APA (American Psychological Association), 226
APIE incident model, 380
Apparatus operations, indoor facilities for, 93. *See also* Vehicle
Apple Keynote, 84
Application (apply) level of learning, 221, 222, 282
Application, four-step method of instruction, 113, 114
Apply (application) level of learning, 221, 222, 282
Asbestos at acquired structures, 174
ASME (American Society of Mechanical Engineers), 224
ASSE (American Society of Safety Engineers), 164
Assessment. *See* Testing and evaluation
Assignment sheet, 231, 242
Association during instruction, 27, 215
Associative phase, 124, 135
Atmospheric issues at training evolutions, 258
Attendance records, 206, 311
Attendance reports, 207
Attending as listening skill, 40
Attention-seeking students, 154

Attitude (affective) domain of learning
    description, 29, 30–31
    learning objective development, 223
    test items, 280
Attitude, positive, 297
Attitude, unsafe, 252
Audience-centered speakers, 107
Audiovisual teaching aids, 71, 93–94
Auditing of records, 312
Auditorium seating arrangement, 90
Authority, sharing of, 295
Authority Having Jurisdiction (AHJ)
    acquired structures for training evolutions, 173
    adoption of codes and standards, 20
    budgets, 304, 382
    computer-based training, 132
    course evaluations, 325
    defined, 22
    facility and prop inspections, 166
    human resource management, 388–389
    instructor competency, 28
    instructor evaluations, 323
    lesson plan revision, 234
    needs analysis, 374–375
    policies, procedures, and guidelines, 375–379
    public access to records and reports, 313
    purchasing, 302, 306, 308–310
    purpose of, 1
    scheduling training sessions, 211
    test security, 191
    training history for needs analysis, 336
Autism as a disability, 144
Autonomous phase, 124–125, 135
Autonomy, psychomotor skill instruction, 125

B

Baby Boomer characteristics, 33
Backdraft indicators, 250
Behavior
    disruptive, 154–156
    student behavior management, 147–151
    unsafe, 251–253
Behavior (performance) statement, 66
Bias
    grading, 194–195
    lesson plan, 218–219
    test, 189, 269
Bids, 384, 385, 387
Biological needs, 29
Bioterrorism training, 162
Blended E-learning, 136
Blended learning, 131
Blindness as a disability, 145
Block grants, 306, 314

Blog, 133, 135
Bloom's Taxonomy, 268
Blu-Ray presentation aids, 76
Body language in nonverbal communication, 39
Body of reports, 208
Bored students, 153
Brainstorming, 120
Budgets, 302–307
    AHJ policies, 304
    budget requests, 304, 305, 306–307
    capital, 304, 314
    equipment funding, 381
    expenditures, 382
    funding for training, 18
    funding needs determination, 302, 304
    funding sources, 304–306
        budget requests, 304, 305
        charitable contributions, 305
        data collection, 224
        grants, 305–306
        taxes, 304
    instructional materials, 382
    line items, 302
    operational, 304
    resource analysis, 381
    resource management, 381
    skills needed for formulating, 393
    sources of, 224
    for training, 18
    training center maintenance and repair, 382–383
    training resources needs analysis, 316
    unexpected cost of training, 304
Burn building, 100, 102

C

CAFC/ACCP (Canadian Association of Fire Chiefs/ Association Canadienne des Chefs de Pompiers), 164
Canada
    Canadian Association of Fire Chiefs (Association Canadienne des Chefs de Pompiers)(CAFC/ACCP), 164
    Canadian Centre for Occupational Health and Safety (CCOHS), 162
    Canadian Human Rights Act, 19
    Office of the Privacy Commissioner of Canada, 313
    Ontario, Municipal Freedom of Information and Protection of Privacy Act (MFIPPA), 313
    Underwriters Laboratories of Canada (ULC), 163
Capital budget, 304, 314
Capital funds, 308
Capital Improvement Plan, 383
Capital purchases, 383
CAR (Customer Approval Rating), 361, 363

Case study
    defined, 125, 136
    group discussions, 119
    lesson activity development, 228–229
    structured exercises, 125–126
Casualty
    IAFC statistics, 249
    IAFF statistics, 249
    NFPA statistics, 249
    simulations, 80
    USFA statistics, 249
CBL (competency-based learning), 107, 127–128, 136
CBT. *See* Computer-based training (CBT)
CCOHS (Canadian Centre for Occupational Health and Safety), 162
CDC (Centers for Disease Control and Prevention), 162, 171
Centers for Disease Control and Prevention (CDC), 162, 171
Certification applications, 206
Certification records, 311
Certification reports, 207
Certification vs. competence, 28
CFR. See Code of Federal Regulations (CFR)
Characteristics of effective instructors, 11–15
    competencies, 12
    conflict-resolution skills, 14–15
    desire to teach, 12
    empathy, 14
    fairness, 15
    honesty, 15
    ingenuity, creativity, and flexibility, 14
    interpersonal skills, 13
    leadership abilities, 12–13
    motivation, 12
    personal integrity, 15
    preparation and organization, 14
    sincerity, 15
Charitable contributions, 305, 309
Chat sessions, 133
Checklists
    Live Fire Evaluation Sample Checklist, 173
    performance tests, 279–280
    skills evaluations, 186
Chevron seating arrangement, 90
Chicago style, 226
Choices, 270, 271
Circled chair seating arrangement, 90, 91
City miniatures, 74
Civil Rights Act, 19
Class continuity, 59. *See also* Learning environment continuity
Class handouts, 73–74
Class size and interaction, 52

Classroom environment, 89–95
    audiovisual equipment, 71, 93–94
    comfort facilities, 95
    desktops and tables, 91
    emergency exits, 91, 95
    Internet, phone, and cable television access, 95
    lighting, 92
    noise level, 93
    online environments, 96
    power outlet access, 94–95
    safety hazards, 91, 95
    seating arrangements, 89–92
    temperature and ventilation, 92–93
    visual distractions, 95
Classroom instruction
    competency-based learning, 107, 127–128, 136
    four-step method of instruction, 112–114
        application, 113, 114
        evaluation, 113, 114
        preparation, 112–114
        prepared classroom lesson, 137
        prepared practical (psychomotor) lesson, 138
        presentation, 113, 114
    instructional methods, 115–125
        asking questions, 121–124
        class discussions, 118–121
        defined, 115
        demonstration, 116–117
        illustrated lecture, 115–116
        psychomotor skill instruction, 124–125
    presentation techniques, 107–112
        effective speaker characteristics, 107–108
        methods of sequencing, 110
        organizing the presentation, 109
        planning, 108–109
        transitions, 110–112
    structured exercises, 125–127
        case studies, 125–126
        field and laboratory experiences, 127
        role playing, 126
        simulations, 126–127
    teaching strategies, 128–135
        blended or hybrid learning, 131
        computer simulation, 131–133
        flipping the classroom, 131
        individualized instruction, 134–135
        instructor-led training (ILT), 128–131
        self-directed learning, 71, 133–134
        student-centered learning, 131
Classroom issues, 151–156
    disruptive, nonparticipating students, 153–154
        attention-seeking students, 154
        talkative students, 153–154
    disruptive behavior, reasons for, 154–156

instructor-caused behavior, 154–155
student-caused behavior, 155–156
disruptive students, 152
L-E-A-S-T method, 152
nondisruptive, nonparticipating students, 152–153
bored or uninterested students, 153
shy or timid students, 153
students who feel superior, 151–152
Closed questions, 122, 123
CMS (course management system), 133
Coaching, 149–150, 158
Code of ethics
defined, 22
disciplinary action ethical issues, 157
Fire Fighter Code of Ethics, 19
personal integrity, 15
Code of Federal Regulations (CFR)
CFR 1910.134, Respiratory Protection Training, 376
policy development, 376
purpose of, 224
29 *CFR* 1910.120, Hazardous Waste Operations and Emergency Response, 20
29 *CFR* 1926.651, Specific Excavation Requirements, 20
Codes. *See also* Laws
adoption by the Authority Having Jurisdiction, 20
defined, 22
legal precedent, 20
Cognition
defined, 30, 44
test item level of cognition and difficulty, 268
Cognitive (knowledge) domain of learning, 29, 30, 282–283
Cognitive levels of learning, 220–222, 282–283
Cognitive phase, 124, 136
Collaborative relationships, 18
Collapse operations, training props, 80, 81
Comfort facilities in the classroom, 95
Committee meetings, 374–375
Communication, interpersonal
characteristics, 35–36
classroom model, 36
elements, 37–38
interference, 38
interpersonal, 35–40
nonverbal component, 39–40
purposes for, 36–37
verbal component, 38
Competence vs. certification, 28
Competency as instructor characteristic, 12
Competency reports, 207
Competency-based learning (CBL), 107, 127–128, 136
Complacency as unsafe behavior, 252
Comprehension (understand) level of learning
description, 220, 282
examples, 221
expressive words, 222
Comprehension barriers of test items, 284
Comprehensive (summative) test, 187, 197, 282
Computer record-keeping systems, 372
Computer simulations, 79, 131–133
Computer testing, 189
Computer-based training (CBT)
audiovisual aids, 71
defined, 136
simulations, 126, 131–133
Computer-generated slide presentation
advantages of, 75
defined, 84
presentation guidelines, 231–232
recommendations for use, 115–116, 232
Conceptualization, 125
Conclusion of reports, 208
Conditions description, 66
Cone of Learning, 26, 44
Conference, teleconferencing, 133
Conference discussion, 119
Conference seating arrangement, 90, 91
Confidentiality. *See* Privacy issues
Conflict resolution
instructor skills, 14–15
supervisor responsibilities, 295, 297
Contact hours, 207
Contract award, 387
Controlled burning, 98, 102
Conversational tone in presentations, 108
Conversions, metric, 4–7
Copyright Act of 1976, 69, 84
Copyright laws and permissions, 69–71
copyright renewal, 69
fair use, 69–70
infringement, 69
invasion of privacy, 70
model release, 70, 84
right of privacy, factors of lawsuits, 70–71
location(s) and subjects(s), 70
permissions, 70
restrictions, 71
rights of individuals, 71
Cost/benefit analysis of curriculum or training, 338, 345
Counseling for students, 148–149, 158
Course, defined, 345
Course consistency
defined, 57, 59
safety factors, 57–58
training and resource materials, 58
Course design evaluations, 359–360, 364
Course development, 333–350
curriculum design, 338

412 Index

design considerations, 339
four-step development model, 334
four-step model, 334
goals, 347
modifications, 344
objectives, 348
pilot course, 340–341, 345
relationship to the training program, 333
Course evaluations, 324–328, 330
Course management system (CMS), 133
Course performance ratings, 361
Course schedules, 18
CPR training, 300
Create (synthesis) level of learning, 221, 222, 283
Creativity as instructor characteristic, 14
Credibility of the instructor, 48
Criteria, 186, 197
Criterion-reference learning, 136
Criterion-referenced assessments
defined, 197
purpose of, 185, 186–187, 281
validity and reliability, 342
Critical criteria, 230, 234
CSR (Customer Satisfaction Rating), 361
Culture
concept of words, 34, 35
eye contact and, 39
student diversity, 34
student variables concerning, 142
Curriculum, defined, 345
Curriculum development, 333–350
course identification, 338
evaluation, 341–342
four-step model, 334
implementation, 339–341, 349
modifications, 344
outcome identification, 338
pilot course, 340–341, 345
relationship to the training program, 333
sequence courses, 338–339
skills needed, 349
Customer Approval Rating (CAR), 361, 363
Customer Satisfaction Rating (CSR), 361
Cutaway models, 74, 75

D

Data collection for research, 224
Data organization and interpretation for training programs, 375
Deaf-blindness as a disability, 144
Deafness as a disability, 144
Delivery skills for presentations, 108
Demonstration
actual equipment for, 75

instructional methods, 116–117
psychomotor skills, 167–168
safety, 116
skills, 117
Department of Agriculture grants, 305–306
Department of Defense (DoD), 162
Department of Homeland Security (DHS)
grants, 305
purpose of, 162
Department of Transportation (DOT)
grants, 305
instructor familiarization with standards, 17
safe workplace policies, 162
Developmental delay as a disability, 144
DHS. *See* Department of Homeland Security (DHS)
Diagram displays, 73
Diagramming ideas, 110
Direct questions, 122, 123
Director of discussions, 120
Disability
accommodations for students, 143, 145
Americans with Disabilities Act (ADA)
categories of disabilities, 143
provisions, 19
requirements, 224
autism, 144
blindness, 145
categories under IDEA law, 144–145
deaf-blindness, 144
deafness, 144
developmental delay, 144
emotional disturbance, 144
hearing impairment, 144
Individuals with Disabilities Education Act (IDEA), 143–145
instructor challenges, 17
intellectual disability, 144
language impairment, 145
learning, 19, 22, 143
multiple disabilities, 144
orthopedic impairment, 145
other health impairment, 145
Rehabilitation Act, 143
special needs of students, 142–146
specific learning disability, 145
speech impairment, 145
student, challenges of, 17
traumatic brain injury, 145
visual impairment, 145
Disagreements. *See* Conflict resolution
Disciplinary action
data collection, 156
documentation, 156–157
legal and ethical issues, 157

Disclosure of information, 374, 390
Discovery zone, 49
Discussion method of instruction, 118–121
    active learning, 118
    leading discussions, 120
    lesson activity development, 227–228
    small groups, 119
    techniques, 120–121
    whole groups, 118–119
Disruptive, nonparticipating students, 153–154
Disruptive behavior, 154–156
Distance learning
    computer-based training, 131–133
    defined, 136
    video presentations, 78
Distractors, 271, 287
Disuse of learning, 27, 215
Diversity of students, 31–35
    adult learner characteristics, 31–32
    age, 32
    challenges of, 17
    cultural and ethnic background, 34
    examples, 31
    gender, 34
    generational characteristics, 33
    individual student needs, 141–142
    instructor ramifications, 31
    learning styles and learner characteristics, 34–35
    self-regulated learning, 35
Document camera, 78
Documentation
    acquired structure, training in, 174–175
    draft document, 378
    facility and prop inspections and repairs, 167
    publishing, 379
    training, 18
DoD (Department of Defense), 162
Domains of learning. *See* Learning domain
Donald Kirkpatrick's Four Levels of Evaluation, 266
Donations, 305, 309
DOT. *See* Department of Transportation (DOT)
Dot Comer characteristics, 33
Draft document, 378
Dumpster fire props, 177
Duplicated materials, 73–74
DVD presentation aids, 76

E
Easel pads, 72
Education. *See also* Training
    defined, 11, 22
    entities provided with, 11
EEO (Equal Employment Opportunity), 342

EEOC (Equal Employment Opportunity Commission), 189
Effect of learning by adults, 27, 215
Egress at training grounds, 100
E-learning, 136
Emergency Management Institute (EMI), 162
EMI (Emergency Management Institute), 162
Emotional disturbance as a disability, 144
Empathy as instructor characteristic, 14
Employment tests, 189
Empowering instructors, 295–297
    celebrating instructor accomplishments, 295
    conflict resolution, 295, 297
    positive examples/attitudes, 297
    recognizing quality performance, 295, 296
    sharing authority, 295
Enthusiasm for teaching, 12
Environment
    classroom. *See* Classroom environment
    instructor knowledge of, 18
    learning environment, 89–103
        classroom, 89–95
        online, 96
        set up, 103
        training ground, 96–101
    learning environment as learning obstacle, 41
    learning environment continuity, 54–57
        defined, 59
        instructional resource variations, 56
        instructor changes, 54–55
        knowledge level differences, 57
        learner characteristics, differences in, 56
        weather variations, 55
    testing and environment, 190
    training evolutions, 257–258
        atmosphere, 258
        soil, 258
        water, 257
    training ground. *See* Training ground environment
Environmental Protection Agency (EPA), 162, 258
EPA (Environmental Protection Agency), 162, 258
Equal Employment Opportunity Commission (EEOC), 189
Equal Employment Opportunity (EEO), 342
Equipment. *See also* Materials and equipment, instructional
    actual equipment for demonstrations, 75
    purchasing policies, 383–387
        awarding of the contract, 387
        bid evaluation, 387
        purchasing specifications, 385–387, 394
        request for proposal (RFP), 383–385
Essay test items, 275–277

Esteem needs, 29
Ethernet ports, 95
Ethics
  defined, 22
  disciplinary action ethical issues, 157
  Fire Fighter Code of Ethics, 19
  personal integrity, 15
Ethics, *Fire Fighter Code of Ethics*, 19
Ethnic background, student diversity, 34
Evaluate (evaluation) level of learning, 221, 222, 282
Evaluation. *See also* Testing and evaluation
  course, 324–328, 330
  curriculum, 341–342
  defined, 186, 197
  Four Levels of Evaluation, 266
  four-step method of instruction, 113, 114
  instructor skills needed, 350
  lesson plan, 233–234, 243
  as listening skill, 41
  personnel, 389, 396
  supervisory instructor, 321–330
    course evaluations, 324–328, 330
    evaluation tools, 321–322
    findings from evaluations, 328
    processes, 322–324, 329
  testing instruments, 342–344
    test validity and reliability, 342
    test-item analysis, 343–344
    test-result analysis, 342–343
  training program. *See* Training program evaluation
Everyone Goes Home initiative, 68
Evolutions. *See* Training evolutions
Examples
  comprehension (understand) level of learning, 221
  diversity, 31
  hazard exposure training, 170
  lesson outline, 65, 67
  positive, 297
  remote training ground, 97–98
  research, 224
  training props, 80–81
Exercise of skills learned, 27, 215
Exercises. *See* Structured exercises
Exits in classrooms, 95
Experience as instructor competency, 12
Experience of adult learners, 26–27, 28
Expert opinions, 224
Exterior fire training, 171, 176–177
Eye contact, 39

F
Facial expression in nonverbal communication, 39
Failure, fear of, 150
Fair use, 69–70, 84

Fairness as instructor characteristic, 15
Family Educational Rights and Privacy Act (FERPA)
  confidentiality, 207
  defined, 390
  disclosure of information, 374
  provisions, 19
  test score record privacy, 313
Fan seating arrangement, 90
Fatality investigation and prevention programs, 180–181
FDSOA (Fire Department Safety Office Association), 164
Federal Emergency Management Agency (FEMA), 162
Federal government agencies for training, 161–162
Feedback
  evaluation, 196, 201
  positive reinforcement, 30, 131
  psychomotor domain of learning, 30
  purpose of, 125
FEMA (Federal Emergency Management Agency), 162
FERPA. *See* Family Educational Rights and Privacy Act (FERPA)
Field experiences, 127
File sharing, 133, 136
Finances. *See* Budgets
Fire and emergency services profession, instructor obligations to, 16
Fire Department Safety Office Association (FDSOA), 164
Fire Fighter Code of Ethics, 19
Fire Fighter Fatality Investigation and Prevention Program, 180
Firefighter Close Calls, 68
Fitness, health, and wellness programs for safety, 164, 251
Flammable/combustible liquid live-fire training, 177
Flammable/combustible liquid pit, 101, 102
Flexibility as instructor characteristic, 14
Flipping the classroom, 131
Foam application training, 177
Foreseeability, 179–180, 182
Formative (progress) test, 187, 197, 282
Formative assessments, 353
Formative evaluation, 354, 363
Foundations of learning. *See* Learning foundations
Four Levels of Evaluation, 266
Four-Level Training Evaluation Model, 356
Four-step development model for course or curriculum
  designing course or curriculum, 338–339
    course design, 339
    course objectives, 348
    curriculum design, 338–339
    needs analysis, 347
    overview, 334
  evaluating course or curriculum, 341–342
    modifications, 344
    overview, 334
    skills needed, 342

test validity and reliability, 342
test-item analysis, 343–344
test-result analysis, 342–343
identifying training needs
cost/benefit analysis, 338
gap analysis, 337
job performance requirements, 336–337
needs analysis, 335–336
overview, 334–338
implementation of course or curriculum, 334, 339–341, 349
Four-step method of instruction, 112–114
application, 113, 114
defined, 136
evaluation, 113, 114
preparation, 112–114
prepared classroom lesson, 137
prepared practical (psychomotor) lesson, 138
presentation, 113, 114
Fuels at acquired structures, 175
Funding. *See* Budgets

G

Gap analysis, 337, 345, 374
Gatekeeper of discussions, 120
Gender and student diversity, 34, 35
Generation Z characteristics, 33
Generational characteristics, 33
Gen-X characteristics, 33
Gen-Y characteristics, 33
Gestures in nonverbal communication, 39–40
Goals of training, 294
Goals-based evaluation, 355, 363
Grading tests, 192
Grants, 305–306, 308, 314
Ground ladder, training evolutions for lifting and placing, 169
Group discussions, 118–121, 227–228
Guide to Building Fire Service Training Centers (NFPA 1402), 164
Guided group discussion, 119
Guideline
adopting, 379
AHJ development of, 376
approach for number of test items, 284
defined, 377, 390
evaluating effectiveness, 379
implementation, 379
lack of support for new guidelines, 378
needs determination, 377
Web Content Accessibility Guidelines (WCAG), 76

H

Handouts, 230, 238
Hands-on experience, 28
Hands-on training, 30
Harassment, 19, 21, 22
Hazard, defined, 254
Hazard and risk analysis, 253–254, 259
Hazard exposure training, 170–178
acquired structures, 172–176
examples of scenarios, 170
exterior and wildland fires, 176–177
increased training, 177–178
live-fire training, 171–177
NIST study on fuel pallet size, 172
purpose-built structures, 176
safety during training, 171
swift water rescue, 178
Hazard identification, 167
Hazardous materials, NFPA 1072 standards for response personnel, 380
Hazardous Waste Operations and Emergency Response (29 *CFR* 1910.120), 20
Heading of reports, 208
Health Insurance Portability Accountability Act (HIPAA), 20
Hearing impairment as a disability, 144
Heating, ventilation, and air conditioning (HVAC) in the classroom, 92–93
Hierarchy of needs, 29
High learning abilities, 146
HIPAA (Health Insurance Portability Accountability Act), 20
Hollow square seating arrangement, 90, 91
Honesty as instructor characteristic, 15
Horseshoe seating arrangement, 90, 91
Hoseline, training evolutions for unloading, 169
Human resource management, 388–389
instructional staff selection, 388
instructor qualifications, 388, 395
position advertising, 389
Humor used in presentations, 108
HVAC in the classroom, 92–93
Hybrid learning, 131
Hydrocarbons released at live-fire exercises, 258

I

IAFC. *See* International Association of Fire Chiefs (IAFC)
IAFF. *See* International Association of Fire Fighters (IAFF)
IAP. *See* Incident action plan (IAP)
ICS (Incident Command System), 376
IDEA (Individuals with Disabilities Education Act), 143–145

IFSAC (International Fire Service Accreditation Congress), 164
Illustrated lecture, 115–116
Illustration displays, 73
ILT. *See* Instructor-led training (ILT)
Immunity, 182
IMS. *See* Incident Management System (IMS)
Incident action plan (IAP)
    defined, 259
    training ground safety analysis, 97
    training supervision, 254–255
Incident Command System (ICS), 376
Incident Management System (IMS)
    duties and functions, 254
    NFPA 1561 standards, 254
    operational training, 255, 256
    skills training, 255, 256
    training evolution evaluation, 255–257
    training plan or incident action plan, 254–255
Incident records, 311
Incident/injury records, 206
In-class assignments, 74
Independent learning, 136
Individualized instruction, 71, 134–135, 136
Individuals with Disabilities Education Act (IDEA), 143–145
Industry standards, 224
Ingenuity as instructor characteristic, 14
Injury
    IAFC statistics, 249
    IAFF statistics, 249
    NFPA statistics, 249
    records, 206, 311
    traumatic brain injury, 145
    USFA statistics, 249
Inspections
    acquired structures, 173, 174
    facilities and props, 166–167
    training ground safety, 58
Instructional design evaluations, 359–360, 365
Instructional materials. *See also* Materials and equipment, instructional
    defined, 84
    planning for use of, 63
Instructional methods, 115–125
    asking questions, 121–124
    class discussions, 118–121
    defined, 115
    demonstration, 116–117
    illustrated lecture, 115–116
    psychomotor skill instruction, 124–125
    selection, 227
Instructional planning, 47–59
    course consistency, 57–58
        safety factors, 57–58
        training and resource materials, 58
    learning environment continuity, 54–57
        defined, 59
        instructional resource variations, 56
        instructor changes, 54–55
        knowledge level differences, 57
        learner characteristics, differences in, 56
        weather variations, 55
    planning to teach, 47–51
        organization, 47
        session logistics, 49–51
        session preparation, 48–49
    presentation, 108–109
    training aids, 51–53
        class size and interaction, 52
        defined, 51, 59
        learning objectives, 52
        learning pace, 53
        lesson content, 52
        practice factors, 53
        selection, 51–53
Instructional/contact hours, 207
Instructor
    changes in, 54–55
    characteristics, 11–15
    credibility, 48
    defined, 11, 22
    evaluations, 321–330
    multiple instructors, 129
    obligations to self, 16
    observing other instructors, 49
    priorities, 17
    qualifications, 388, 395
    as safety role model, 165–166
    skill level for safe training, 166
    staff selection, 388
    strengths evaluations, 328
    team teaching, 129
    weakness evaluations, 328
Instructor-caused disruptive behavior, 154–155
Instructor-led training (ILT), 128–131
    features, 128–129
    generating and maintaining student interest, 129–130
    multiple instructors, 129
    positive/negative reinforcement, 131
    reinforcing learning, 130
Intellectual disability, 144
Intensity of learning, 28, 216
Interactive display systems, 77–78
Interactive television (ITV), 133, 134
International Association of Fire Chiefs (IAFC)
    casualty and injury statistics, 249
    safety information and model programs, 164

source of instructional materials, 382
International Association of Fire Fighters (IAFF)
   casualty and injury statistics, 249
   instructional materials, 382
   safety information and model programs, 164
International Fire Service Accreditation Congress (IFSAC), 164
International Society of Fire Service Instructors (ISFSI), 164
Internet. *See also* Websites; *specific Computer entries*
   access for online videos, 79
   access in the classroom, 95
   class materials, 70
   as information source for research, 225–226
Interpersonal communication, 35–40
   characteristics, 35–36
   classroom model of communication, 36
   elements, 37–38
   interference, 38
   nonverbal component, 39–40
   purposes for, 36–37
   verbal component, 38
Interpersonal skills as instructor characteristic, 13
Interpretive exercise, 277
Introduction of reports, 208
Invasion of privacy, 70, 84
Investigations
   accident, 259
   training fatalities, 180–181
ISFSI (International Society of Fire Service Instructors), 164
ITV (interactive television), 133, 134

J

Job performance requirements (JPRs)
   defined, 84
   learning objectives vs., 68
   NFPA 1041 standards, 1
   training needs identification, 336–337
JPR. *See* Job performance requirements (JPRs)

K

Key points, 230, 234
Keystoning, 77
Kirkpatrick, Donald, 266
Kirkpatrick's Four-Level Training Evaluation Model, 266, 356
Knowledge (cognitive) domain of learning, 29, 30, 282–283
Knowledge (remember) level of learning
   description, 220, 282
   examples, 221
   expressive words, 222
Knowledge, lack of, as unsafe behavior, 252–253

Knowledge level of students, 57
Knowledge of subject as instructor competency, 12, 16
Knowles, Malcolm, 26–27
Knowles' assumptions of adult learners, 26–27

L

Laboratory experiences, 127
Language barriers of test items, 284
Language impairment as a disability, 145
Large-screen images, 76–77
Laws. *See also* Codes; Legal requirements; Regulations; Standards
   Americans with Disabilities Act (ADA)
      categories of disabilities, 143
      provisions, 19
      requirements, 224
   applicable to instructors, 19–20
   Canadian Human Rights Act, 19
   Civil Rights Act, 19
   copyright. *See* Copyright laws and permissions
   Copyright Act of 1976, 69, 84
   Family Educational Rights and Privacy Act (FERPA)
      confidentiality, 207
      defined, 390
      disclosure of information, 374
      provisions, 19
      test score record privacy, 313
   Health Insurance Portability Accountability Act (HIPAA), 20
   Individuals with Disabilities Education Act (IDEA), 143–145
   Municipal Freedom of Information and Protection of Privacy Act (MFIPPA), 313
   Open Records Act, 313
   ordinances, 20, 22
   Privacy Act, 19
   Title VII of the Civil Rights Act, 19
   Workforce Investment Act of 1998, 142
Laws of Learning, 27–28, 215
Leadership
   ability of instructors, 12–13
   instructor qualities, 13
Learner characteristics
   class continuity, 56
   defined, 44
   student diversity, 34–35
Learning disability, 19, 22, 143
Learning domain, 29–31
   affective (attitude)
      description, 29, 30–31
      learning objective development, 223
      test items, 280
   cognitive (knowledge), 29, 30, 282–283
   defined, 44
   psychomotor (skills), 29, 30, 55

418 Index

Learning environment, 89–103
    classroom, 89–95
    online, 96
    set up, 103
    training ground, 96–101
Learning environment continuity, 54–57
    defined, 59
    instructional resource variations, 56
    instructor changes, 54–55
    knowledge level differences, 57
    learner characteristics, differences in, 56
    weather variations, 55
Learning foundations, 25–29
    Knowles' assumptions of adult learners, 26–27
    Maslow's hierarchy of needs, 29
    sensory-stimulus theory, 25–26
Learning levels
    cognitive, 220–222, 282–283
    psychomotor, 222–223
Learning management system (LMS), 132, 133, 136
Learning objectives
    action verbs, 66
    conditions description, 66
    defined, 66, 84
    job performance requirements vs., 68
    overview, 66–68
    performance (behavior) statement, 66
    standards criteria, 66
Learning orientation of adults, 27
Learning outcome
    defined, 66, 84
    lesson plan creation, 216
    outcome, defined, 187
    outcomes-based evaluation, 355, 363
Learning pace, 53
Learning plateau, 42–43, 44
Learning principles, 25–44
    factors affecting learning, 41–43
        areas of student frustration, 42
        learning obstacles, 41–42
        learning plateaus, 42–43
    foundations of learning, 25–29
        Knowles' assumptions of adult learners, 26–27
        Maslow's hierarchy of needs, 29
        sensory-stimulus theory, 25–26
    interpersonal communication, 35–40
        characteristics, 35–36
        classroom model of communication, 36
        elements, 37–38
        interference, 38
        nonverbal component, 39–40
        purposes for, 36–37
        verbal component, 38
    learning domains, 29–31
        affective (attitude), 29, 30–31, 223, 280
        cognitive (knowledge), 29, 30, 282–283
        defined, 44
        psychomotor (skills), 29, 30, 55
    listening skills, 40–41
    motivation, 43
    student diversity, 31–35
        adult learner characteristics, 31–32
        age, 32
        cultural and ethnic background, 34
        gender, 34
        generational characteristics, 33
        learning styles and learner characteristics, 34–35
        self-regulated learning, 35
Learning styles, 34–35, 44
Lease/purchase, 309
Leases, 309
L-E-A-S-T method of discipline, 152
Lecture review duplicated materials, 73
Legal liability. *See* Liability
Legal precedent, 20, 22
Legal requirements. *See also* Laws
    research, 224
    training records, 312–313
        Open Records Act, 313
        privacy, 312–313
        public access, 313
        retention length, 312
Lesson content, 52
Lesson outline, 64–68
    benefits of, 66
    creating, 217
    development, 227, 237
    examples, 65, 67
    overview, 66–68
    resources, 68
Lesson plan
    adapting a prepared plan, 65, 85
    benefits of, 63–64
    components, 64, 218
    creation, 216–232
        ancillary components, 229–231
        bias elimination, 218–219
        conducting research, 223–226
        instructional method selection, 227
        learning objectives, 219–223, 236
        lesson activity, 227–229
        by Level II instructor candidates, 245
        outlines, 217, 227, 237
        steps, 235
        technology tools, 231–232
    development, 215

        development process, 217
        evaluation, 233–234, 243–244
        first use, 232–233
        revision, 234, 244
    Level I Instructor
        characteristics, 11–15
        classroom instruction, 107–138
            competency-based learning, 127–128
            four-step method of instruction, 112–114, 137–138
            instructional methods, 115–125
            presentation techniques, 107–112
            structured exercises, 125–127
            teaching strategies, 128–135
        instructional materials and equipment, 63–85
            lesson plans, 63–68
            teaching aids, 71–83
        instructional planning, 47–59
            course consistency, 57–58
            learning environment continuity, 54–57
            planning to teach, 47–51
            training aids, 51–53
        learning environment, 89–103
            classroom, 89–95
            online, 96
            set up, 103
            training ground, 96–101
        learning principles, 25–44
            domains of learning, 29–31
            factors affecting learning, 41–43
            interpersonal communication, 35–40
            learning foundations, 25–29
            listening skills, 40–41
            motivation, 43
            student diversity, 31–35
        records, reports, and scheduling, 205–212
        report writing, 208
        responsibilities, 1–2, 16–19
        skills-based training, 161–182
            evolution control, 168
            hazard exposure training, 170–178
            instructor as safety role model, 165–166
            legal liability, 178–181
            psychomotor skills demonstrations, 167–168
            resources, 161–164
            safe training, planning for, 166–167
            simple training evolutions, 168–170
        student interaction, 141–158
            classroom issues, 151–156
            disciplinary action, 156–157
            individual student needs, 141–146
            student behavior management, 147–151
            students' rights, 146–147
        testing and evaluation, 185–201
            feedback, 196

        grade reporting, 195, 200
        student assessment, 185–187
        test administration, 190–192
        test bias, 189
        test classifications, 187–189
        test scoring, 192–195
        test security, 195–196
Level II Instructor
    budgeting, 302, 304–307
    lesson plan development, 215–245
        evaluation and revision, 233–234
        plan creation, 216–232
        plan first use, 232–233
    purchasing, 307–310
    report writing, 208
    responsibilities, 2
    scheduling, 297–302, 303
    supervisory duties, 293–297
    supervisory instructor evaluations, 321–330
    test item construction, 265–289
        Donald Kirkpatrick's Four Levels of Evaluation, 266
        formatting and item arrangement, 267–268
        instructions and time requirements, 268–269
        level of cognition and difficulty, 268
        student evaluation instruments, 269–280, 288
        test planning, 281–285
        test scoring, 285–286
        testing bias, 269
    training evolution supervision, 249–261
        accident investigation, 259
        environmental issues at training evolutions, 257–258
        IMS to supervise training, 254–257
        safety challenge, 249–254
    training record maintenance, 207
    training records management, 310–313
Level III Instructor
    course and curriculum development, 333–350
        designing, 338–339, 347, 348
        evaluating testing instruments, 342–344, 350
        evaluating the curriculum, 341–342
        four-step model, 334
        identifying training needs, 334–338, 346
        implementation, 339–341, 349
    prerequisites, 333–334
    report writing, 208
    responsibilities, 2
    training program administration, 371–396
        AHJ-level needs analysis, 374–375
        budget needs, 381–383, 393
        disclosure of information, 374
        equipment purchasing, 383–387
        human resource management, 388–389, 395
        personnel evaluations, 389, 396
        policies, procedures, and guidelines, 375–379, 392

record-keeping, 371–373, 391
standards influencing training, 380
training program evaluation, 353–367
methodologies, 353–354
plans, 356–360, 364, 365
results and recommendations, 360–362
types and categories, 354–355
Liability
act of commission, 178
act of omission, 178
defined, 178, 182
foreseeability, 179–180
negligence, 178
reduction of, 178, 180
risk during training, 180–181
vicarious liability, 179, 182
Libraries as information source, 225
License reports, 207
Life experiences, 31–32
Lighting
classroom, 92
training ground, 100
Listening skills, 40–41
Literacy of students, 142
Live burn exercises, 102
Live Fire Evaluation Sample Checklist, 173
Live streaming, 133
Live-fire exercises, 98, 102
Live-fire training
acquired structures, 172–176
defined, 102
exterior and wildland fires, 171, 176–177
hazard exposure training, 171–177
NFPA 1403 standards. *See* NFPA 1403, *Standard on Live Fire Training Evolutions*
NIST study on fuel pallet size, 172
purpose-built structures, 176
remote sites, 98–99
safety during, 171
LMS (learning management system), 132, 133, 136
Local safety and health agencies, and safe workplace policies, 162–163
Logistics
defined, 49, 59
duties, 50
materials and equipment, 50, 51
preparation steps, 50

M

Mager, Robert F., 66
Mager Model of learning objective development, 220
Maintenance
teaching aids, 81–82
training center, 382–383

training records, 207, 317
Management directives, 18
Manikins as training aids, 51, 52, 80
Marker boards, 72–73
Maslow, Abraham, 29
Maslow's hierarchy of needs, 29
Mastery learning, 127, 136
Mastery of skills, 125
Matching test items, 273–274
Materials and equipment, instructional, 63–85. *See also* Teaching aids
actual equipment for teaching, 75
budget needs, 381–382
equipment purchasing policies, 383–387
awarding of the contract, 387
bid evaluation, 387
purchasing specifications, 385–387, 394
request for proposal (RFP), 383–385
lesson plans, 63–68
components, 64, 65
outlines and learning objectives, 64–68, 85
logistics for instruction, 50, 51
teaching aids, 71–83
audiovisual, 71, 93–94
benefits of, 82–83
cleaning, care, and maintenance of, 81–82
effective use of, 71, 72
nonprojected aids, 72–75
projected aids, 75–79
simulators, 79–80
training props, 80–81, 82–83
training and resource materials, 58
Mathematical approach for number of test items, 284
Measurement conversions, 4–7
Media transitions, 112
Medical record privacy, 20
Mentoring, 151
Methodologies, training program evaluation, 353–354
formative assessment, 353
mixed methods research, 354
preference scales, 354
qualitative, 353–354
quantitative, 354
summative assessment, 353
types and categories, 354–355
formative evaluation, 354
goals-based evaluation, 355
outcomes-based evaluation, 355
process-based evaluation, 355
summative evaluation, 355
Metric conversions, 4–7
MFIPPA (Municipal Freedom of Information and Protection of Privacy Act), 313
Microsoft PowerPoint, 84

Millennial characteristics, 33
Mine Safety and Health Administration (MSHA), 385
Miniatures as teaching aids, 74, 75
Mixed methods research, 354, 363
Mnemonic device, 41, 44
Model
    anatomical, 74–75
    APIE incident model, 380
    appropriate and safe behavior by instructors, 16
    classroom model of communication, 36
    cutaway, 74, 75
    field and laboratory experiences, 127
    Four-Level Training Evaluation Model, 356
    four-step model. *See* Four-step development model for course or curriculum
    instructor as safety role model, 165–166
    Mager Model of learning objective development, 220
    model release (permission), 70, 84
    nonprojected teaching aids, 74–75
    research, 224
    SMART model, 220
    teaching aids, 74–75
Model release (permission), 70, 84
Motivation
    adult learners, 27
    instructor characteristic, 12
    instructor techniques, 43, 150
    psychological barriers to, 150
    self-motivation of students, 150
    student diversity, 32
Motor vehicle. *See* Vehicle
Moulage kits, 80, 84
MSHA (Mine Safety and Health Administration), 385
Multimedia projectors, 76–77
Multiple disabilities, 144
Multiple-choice test item, 270–272
Municipal Freedom of Information and Protection of Privacy Act (MFIPPA), 313

N

NAFTD (North American Fire Training Directors), 164
National Board on Fire Service Professional Qualifications (ProBoard), 164
National Fallen Firefighters Foundation, 68
National Fire Academy (NFA)
    adult education, 25
    copyright permission, 69
    Fire Fighter Code of Ethics, 19
    lesson outline resources, 68
    safe workplace policies, 162
    training program evaluation, 356–357
    training records, 311
National Fire Protection Association (NFPA). *See also specific NFPA standard*
    casualty and injury statistics, 249
    industry standards for research, 224
    issues addressed by standards, 380
    purpose of, 163
    as source of standards, 17
National Incident Management System (NIMS)
    incident action plans, 254
    NIMS-ICS, 170
    skills training, 255
    training procedures and guidelines, 376
National Institute for Occupational Safety and Health (NIOSH)
    lesson outline resources, 68
    live-fire training in acquired structures, 171
    purchasing specifications, 385
    safe workplace policies, 161
    training fatality investigations, 180–181
National Institute of Standards and Technology (NIST)
    lesson outline resources, 68
    safe workplace policies, 162
    study on fuel pallet size in training evolutions, 172
National Society of Fire Executive Officers, *Fire Fighter Code of Ethics*, 19
National Volunteer Fire Council (NVFC), 164, 249
National Wildland Coordinating Group (NWCG), 177
Needs analysis
    budgeting, 316
    course or curriculum development, 335–336, 345–347
    prerequisites, 374
    training program, 374–375
Needs determination
    policy, procedure, or guideline, 377
    purchasing, 309
    scheduling training sessions, 298, 300
Negative reinforcement, 131
Negative suggestion effect of true-false questions, 272
Negligence
    defined, 22, 182
    failure to take reasonable precautions, 179
    liability for, 21, 178
New York, State of v. Baird, 181
Nexter characteristics, 33
NFA. *See* National Fire Academy (NFA)
NFPA. *See* National Fire Protection Association (NFPA)
NFPA 1001, *Standard for Fire Fighter Professional Qualifications*, 20, 181
NFPA 1021, *Standard for Fire Officer Professional Qualifications*, 1
NFPA 1041, *Standard for Fire Service Instructor Professional Qualifications*
    course and curriculum development, 333
    hazard exposure training, 170
    human resource management, 388
    job performance requirement components, 68

lesson plan development, 215
Level III Instructor prerequisites, 333–334, 339
live-fire training supervision, 98
provisions, 1, 164
record-keeping, 373
report writing, 208
student evaluation instruments, 265
testing and evaluation, 185
NFPA 1072, *Standard for Hazardous Materials/Weapons of Mass Destruction Emergency Response Personnel Professional Qualifications,* 380
NFPA 1142, water supply requirements, 175
NFPA 1201, *Standard for Providing Emergency Services to the Public,* 376
NFPA 1401, *Recommended Practice for Fire Service Training Reports and Records,* 310, 372
NFPA 1402, *Guide to Building Fire Service Training Centers,* 164
NFPA 1403, *Standard on Live Fire Training Evolutions*
acquired structures, 98, 172
exterior and wildland fires, 176
fuel materials, 175
instructor familiarity with provisions, 20
instructor requirements for safety, 251
Live Fire Evaluation Sample Checklist, 173
policy development, 376
provisions, 164
purpose-built structures, 176
safety for selected structures, 171
safety in training, 181
student-to-instructor ratio, 168
training ground, 96
training records, 310
NFPA 1410, *Standard on Training for Initial Emergency Scene Operations,* 164
NFPA 1500™, *Standard on Fire Department Occupational Safety, Health, and Wellness Program*
instructor familiarity with provisions, 20
live-fire training, 98
policy development, 376
provisions, 164
risk-management plan, 250
NFPA 1561, *Standard on Emergency Services Incident Management System and Command Safety,* 254
NFPA 1583, *Standard on Health-Related Fitness Programs for Fire Department Members,* 164
NFPA 1584, *Standard on the Rehabilitation Process for Members During Emergency Operations and Training Exercises,* 164
NIMS. *See* National Incident Management System (NIMS)
NIMS-ICS, 170
NIOSH. *See* National Institute for Occupational Safety and Health (NIOSH)

NIST. *See* National Institute of Standards and Technology (NIST)
Noise
in learning environments, 93
training ground, 100
Nominal group process, 120–121, 136
Nondisruptive, nonparticipating students, 152–153
Nonprofit organization grants, 306
Nonprojected teaching aids, 72–75
class handouts, 73–74
duplicated materials, 73–74
easel pads, 72
illustration or diagram displays, 73
marker boards, 72–73
models, 74–75
recordings, 75
Nonverbal communication, 39–40
Nonverbal transitions, 112
Normalization of deviance, 252, 260
Norm-referenced assessments
defined, 197
purpose of, 185, 186, 281
North American Fire Training Directors (NAFTD), 164
Note-taking guide, 74
Notices for acquired structure training, 174–175
NVFC (National Volunteer Fire Council), 164, 249
NWCG (National Wildland Coordinating Group), 177

O
Objective test item, 270
Objectives
defined, 187
learning, 219–223
affective domain, 223
cognitive levels of learning, 220–222
identifying, 282
Mager Model, 220
psychomotor levels of learning, 222–223
SMART model, 220
steps, 236
training aid selection, 52
useful words to express, 222
writing procedure, 236
learning objectives basics
action verbs, 66
conditions description, 66
defined, 66, 84
job performance requirements vs., 68
overview, 66–68
performance (behavior) statement, 66
standards criteria, 66
psychomotor, 223
SMART model, 220
training, 294

Obligations of instructors, 16
Observation
    instructor performance evaluation, 323
    of other instructors, 17
    to reinforce skills, 167
    training program evaluation, 357
Obsolete terms and their replacements, 35
Occupational Safety and Health Administration (OSHA)
    applicability of regulations, 20
    defined, 22, 182
    regulations, 20
    safe workplace policies, 162
    as source of standards, 17
    training ground safety, 96
    training records, 311
Occupational safety and health program standards, 20
Office of the Privacy Commissioner of Canada, 313
Online environments, 96
Online instruction, 136
Ontario, Municipal Freedom of Information and Protection of Privacy Act (MFIPPA), 313
Open questions, 122, 123
Open Records Act, 313
Operating funds, 308
Operational budget, 304, 314
Operational steps, 230, 234
Operational training, 255, 256
Oral tests
    administration classification, 188
    designing and conducting, 277
    illustrated, 270
    purpose of, 277
    scoring, 194, 199
Ordinance, 20, 22
Organization
    for class sessions, 14
    effective instructor characteristics, 108
    oral presentations, 109
    planning to teach, 47
Organizations
    apathy toward, 17
    evaluations, 358
    instructor obligations to, 16
    professional and accrediting organizations, 164
    promotion of, 18
    standards-writing, 163–164
Orthopedic impairment as a disability, 145
OSHA. *See* Occupational Safety and Health Administration (OSHA)
Other health impairment as a disability, 145
Outcome, defined, 187
Outcomes-based evaluation, 355, 363
Outdoor learning environment, 96. *See also* Training ground environment

Overhead questions, 122, 123

P

Pace of learning, 53
Parking lots for training grounds, 97, 98
Pedagogy, 115
Peer assistance, 150–151
Performance
    defined, 187
    evaluation, 295, 296, 322–324
    measurement, 358–359
    recognition of, 295, 296
    tests
        administration, 191–192, 198
        administration classification, 189
        checklists for evaluation, 279–280
        illustrated, 270
        purpose of, 277
        scoring, 194, 199
        skills evaluation, 278, 289
Performance (behavior) statement, 66
Performance-based competency, 127
Performance-based learning, 136
Permissions, 70, 84
Permits for acquired structures, 174
Personal appearance of instructors as nonverbal communication, 40
Personal integrity as instructor characteristic, 15
Personal protective equipment (PPE), for high-threat incidents, 162
Personal style used in presentations, 108
Personnel evaluations, 389, 396
Personnel files, security in records, 313
Physical limitations as unsafe behavior, 253
Physiological manikins, 80
Physiological needs, 29
Picture identification test item, 275
Pilot course, 340–341, 345
Planning. *See* Instructional planning
Planning, test, 281–285
    appropriate test items, 283–285
        avoiding giving clues to test answers, 284
        language and comprehension barriers, 284
        number of test items, 283–284
        test item difficulty, 283
        test usability, 284–285
        validity and reliability, 285
    learning objectives, 282
    purpose and classification, 281–282
Plateau, learning, 42–43, 44
Poise as nonverbal communication, 40
Policy
    adopting, 379
    agency training goals, 392

budgets, 304
defined, 375, 390
evaluating effectiveness, 379
implementation, 379
lack of support for new policies, 378
needs determination, 377
purchasing, 383–387
review, 147–148
safe workplace, 162–163
training program administration, 375–376
Portable fire extinguisher, training evolutions for using, 169
Position advertising, 389
Positive attitude of instructors during presentations, 108
Positive reinforcement (feedback), 30, 131
Postincident critique or analysis, 255–256, 260
Post-Millennial characteristics, 33
Posture in nonverbal communication, 40
Power outlet access in the classroom, 94–95
PowerPoint, 76, 84
Practice factors, 53
Practice of skills, 125
Precourse materials, 73
Preference scales, 354
Preparation for class
four-step method of instruction, 112–114
instructor effectiveness, 14
psychomotor skill instruction, 125
Prescriptive (pretest) test, 187, 197, 282
Presentation for class, four-step method of instruction, 113, 114
Presentation techniques, 107–112
distracting behaviors to avoid, 109
effective speaker characteristics, 107–108
methods of sequencing, 110
organizing the presentation, 109
planning, 108–109
transitions, 110–112
Pretest (prescriptive) test, 187, 197, 282
Primacy of learning, 28, 216
Priority levels of training, 298
Privacy Act, 19
Privacy issues
copyright law, invasion of privacy, 70
medical records, 20
training records, 312–313, 317
ProBoard (National Board on Fire Service Professional Qualifications), 164
Procedure
adopting, 379
AHJ development of, 376
defined, 376, 390
evaluating effectiveness, 379
implementation, 379

lack of support for new procedures, 378
needs determination, 377
safety in training, 376
Process-based evaluation, 355, 363
Profession of instructors, 16
Professional development, 294–295
Professional organizations, 164
Professional qualifications
National Board on Fire Service Professional Qualifications (ProBoard), 164
NFPA 1001, *Standard for Fire Fighter Professional Qualifications*, 20, 181
NFPA 1021, *Standard for Fire Officer Professional Qualifications*, 1
NFPA 1041. *See* NFPA 1041, *Standard for Fire Service Instructor Professional Qualifications*
NFPA 1072, *Standard for Hazardous Materials/ Weapons of Mass Destruction Emergency Response Personnel Professional Qualifications*, 380
Programs
administration. *See* Training program administration
defined, 345
evaluation. *See* Training program evaluation
Fire Fighter Fatality Investigation and Prevention Program, 180
fitness, health and wellness, 251
four-step development model, 334
health-related fitness programs (NFPA 1583), 164
standards on fire department programs. *See* NFPA 1500™, *Standard on Fire Department Occupational Safety, Health, and Wellness Program*
Progress (formative) test, 187, 197, 282
Projected teaching aids, 75–79
changing technology, 76
computer-generated slide presentations
advantages of, 75
defined, 84
presentation guidelines, 231–232
recommendations for use, 115–116, 232
considerations for use, 76
disadvantages, 76
interactive display systems, 77–78
keystoning, 77
multimedia projectors/large-screen images, 76–77
video conferencing, 79
video presentations, 78–79
visual presenters, 78
Web Content Accessibility Guidelines (WCAG), 76
Projector bulbs, 82
Props. *See* Training props
Prostheses, 80
Provincial safety and health agencies, and safe workplace policies, 162–163
Proximity of instructors, 40

Psychological barriers to learning, 150
Psychomotor (skills) domain of learning, 29, 30, 55
Psychomotor, defined, 187
Psychomotor levels of learning, 222–223
Psychomotor skills
    demonstrations, 167–168
    domain of learning, 29, 30, 55
    instruction, 124–125
    testing, 278
Public announcement (PA) system, 78
Public domain, 69, 84
Publishing
    documents, 379
    scheduling, 301–302, 303
Purchase orders, 309–310
Purchasing, 307–310
    equipment specifications, 394
    funding sources, 308–309
    needs determination, 309
    purchase orders, 309–310
    request for proposal (RFP), 383–385
        company choices, 384
        components, 383
        equipment evaluation, 384
        product evaluation, 384–385
        purchasing specifications, 385–387, 394
    training resources, 307–308
    vendor contacts, 309
Purpose classification, 187, 197
Purpose-built structures, 176, 182

Q

Qualification reports, 207
Qualifications. *See* Professional qualifications
Qualitative methodology, 353–354, 363
Quantitative methodology, 354, 363
Questions
    answering student's questions, 124
    asking, 121–122
    closed, 277
    open, 277
    responding to student's answers, 123–124
    types, 122–123

R

Rapid intervention team (RIT), 172
Ratio of students to instructors, 168
Readiness to learn, 27, 215
Reasonable accommodation, 19, 22
Recency of learning, 28, 216
Recommended Practice for Fire Service Training Reports and Records (NFPA 1401), 310, 372
Recommended practices for research, 224
Recordings as teaching aids, 75

Records
    attendance records, 311
    certification records, 311
    defined, 205, 211
    format, 206
    incident/injury records, 311
    maintenance and security of, 317
    managing training records, 310–313
        auditing procedures, 312
        legal requirements, 312–313
        record management systems, 311–312
        training information, 310–311
        types of training records, 311
    NFPA 1401 standards, 310
    record-keeping, 371–373
        automated systems, 371–372
        electronic databases, 372
        information included, 371
        instructor prerequisites, 373
        resources, 372
        roll-call form, 373
        skills needed, 391
    test records, 311
    training records. *See* Training records
    training schedules, 311
Redirected questions, 122, 123
Reference material citations, 226
Refreshments during training, 95
Regulations. *See also* Laws
    defined, 20, 22
    standards vs., 21
    29 *CFR* 1910.120, Hazardous Waste Operations and Emergency Response, 20
    29 *CFR* 1926.651, Specific Excavation Requirements, 20
Reinforcing learning, 130
Relay questions, 122, 123
Reliability of tests, 285, 342, 345
Remedial knowledge level, 57
Remember (knowledge) level of learning
    description, 220, 282
    examples, 221
    expressive words, 222
Remembering as listening skill, 41
Remote training ground, 96–100
    access/egress, 100
    environment considerations, 99
    examples, 97–98
    lighting, 100
    live-fire training, 98–99
    noise, 100
    vehicle traffic, 99–100
    weather conditions, 99
Replicas, 75
Reports

after action reports, 209
components, 210
defined, 205, 211
format, 206
grade, 195, 200
NFPA 1401 standards, 310, 372
public access to, 313
security, 312–313, 317
training, 207
writing, 208–209
Request for proposal (RFP), 383–385
company choices, 384
components, 383
defined, 390
equipment evaluation, 384
product evaluation, 384–385
purchasing specifications, 385–387, 394
Research
data collection, 224
information sources, 224–226
mixed methods research, 354, 363
purpose of, 223–224
reference material citations, 226
steps, 223
uses for, 223
Resource analysis, 381, 390
Resource management, 381, 390
Resources. *See also* Budgets
federal government agencies, 161–162
instructional resource variations, 56
professional and accrediting organizations, 164
request, 206
standards-writing organizations, 163–164
state/provincial and local agencies, 162–163
training
instructor challenges, 18
needs analysis, 316
scheduling, 297
Respiratory Protection Training (*CFR* 1910.134), 376
Responding as listening skill, 41
Responsibilities
challenges, 17–18
Level I Instructor, 1–2, 16–19
Level II Instructor, 2
Level III Instructor, 2
obligations, 16
Restrictive bid (sole-sourced), 385, 390
RFP. *See* Request for proposal (RFP)
Rhetorical questions, 123
Rights of students, 146–147
Risk, defined, 254
Risk-management plan, 250, 260
RIT (rapid intervention team), 172
Role model for safety, instructor as, 165–166

Role playing, 126, 229
Roll-call form, 373

S
Safety
classroom environment, 95
course consistency, 57–58
demonstrations, 116
fitness, health, and wellness programs, 251
inspecting training grounds, 58
need for, 29
planning for safe training, 166–167
training environments, 18
training evolutions, 249–254
hazard and risk analysis, 253–254
organizational and administrative support, 249–251
unsafe behavior, 251–253
training risks, 180–181
unsafe behavior, 251–253
Safety officer to oversee training, 165, 166
SCBA (self-contained breathing apparatus), 172
Scenario, 125. *See also* Case study
Scheduling
availability of instructors and facilities, 300
considerations for use, 209, 211
course schedules, 18
creation of, 301
factors affecting scheduling, 297–298, 299
needs determination, 298, 300
procedures, 212
publishing, 301–302, 303
records, 206, 311
requirements determination, 300
revisions, 302
steps for creation of schedules, 297, 315
training coordination, 300–301
training priority levels, 298
training resources, 297
training sessions, 209–211
Scoring rubric, 188, 193, 197
Scoring tests, 192–195
grading bias, 194–195
grading tests, 194
oral tests, 194, 199
performance tests, 194, 199
scoring rubric, 188
security, 313
selection of method, 285–286
written tests, 192–194, 199
Seating arrangement, 52, 89–92
Security
assessment materials, 207
distance learning, 132
records and reports, 312–313, 317

Index **427**

tests and answer sheets, 191
Selection tests, 189
Self-actualization, 29
Self-concept, 26
Self-confidence, 32
Self-contained breathing apparatus (SCBA), 172
Self-directed learning, 71, 133–134, 136
Self-regulated learning, 35
Self-study materials, 73
Senses for learning, 26
Sensory memory, 26, 44
Sensory-stimulus theory, 25–26
Sequencing of presentations, 110
Sexual harassment, 19, 22
Short-answer/completion test item, 275
Shy students, 153
Similar situations for research, 224
Simple training evolutions, 168–170
Simulations, training, 126–127
Simulators, 79–80
    anatomical/physiological manikins, 80
    casualty simulations, 80
    computer simulations, 79
    virtual reality, 79–80
Sincerity as instructor characteristic, 15
16 Firefighter Life Safety Initiatives, 68
Skill sheets, 230, 239
Skills (psychomotor) domain of learning, 29, 30, 55
Skills, defined, 187
Skills demonstration, 117
Skills test. *See* Performance, *subhead* tests
Skills training, 255, 256
Skills-based training, 161–182
    evolution control, 168
    hazard exposure training, 170–178
        acquired structures, 172–176
        examples, 170
        exterior and wildland fires, 176–177
        increased training, 177–178
        live-fire training, 171–177
        purpose-built structures, 176
        safety during, 171–172
    instructor as safety role model, 165–166
    legal liability, 178–181
    psychomotor skills demonstrations, 167–168
    resources, 161–164
        federal government agencies, 161–162
        professional and accrediting organizations, 164
        standards-writing organizations, 163–164
        state/provincial and local agencies, 162–163
    safe training, planning for, 166–167
    simple training evolutions, 168–170
SMART model for writing learning objectives, 220

Smoke building, 102
Smokehouse, 100, 102
Social needs, 29
Social networking, 133, 136
Social Security number, security in records, 312
Soil environmental issues at training evolutions, 258
Special needs students, 142–146
Specific Excavation Requirements (29 *CFR* 1926.651), 20
Specific learning disability, 145
Speech impairment as a disability, 145
Standard for Fire Fighter Professional Qualifications (NFPA 1001), 20, 181
Standard for Fire Officer Professional Qualifications (NFPA 1021), 1
Standard for Fire Service Instructor Professional Qualifications. See NFPA 1041, *Standard for Fire Service Instructor Professional Qualifications*
Standard for Hazardous Materials/Weapons of Mass Destruction Emergency Response Personnel Professional Qualifications (NFPA 1072), 380
Standard for Providing Emergency Services to the Public (NFPA 1201), 376
Standard on Emergency Services Incident Management System and Command Safety (NFPA 1561), 254
Standard on Fire Department Occupational Safety and Health Program. See NFPA 1500™, *Standard on Fire Department Occupational Safety, Health, and Wellness Program*
Standard on Health-Related Fitness Programs for Fire Department Members (NFPA 1583), 164
Standard on Live Fire Training Evolutions. See NFPA 1403, *Standard on Live Fire Training Evolutions*
Standard on the Rehabilitation Process for Members During Emergency Operations and Training Exercises (NFPA 1584), 164
Standard on Training for Initial Emergency Scene Operations (NFPA 1410), 164
Standards. *See also* Laws
    adopting, 380
    adoption by the Authority Having Jurisdiction, 20
    criteria, 66
    defined, 22, 380
    influence on training, 380
    instructor familiarization with, 17
    legal precedent, 20
    regulations vs., 21
    standards-writing organizations, 163–164
State safety and health agencies, and safe workplace policies, 162–163
Statistics for research, 224
Stem, 270, 287
Strategies for teaching. *See* Teaching strategies
Structured exercises, 125–127
    case studies, 125–126

field and laboratory experiences, 127
role playing, 126
simulations, 126–127
Student
   disruptive students. *See* Classroom issues
   diversity, 31–35
      adult learner characteristics, 31–32
      age, 32
      cultural and ethnic background, 34
      examples, 31
      gender, 34
      generational characteristics, 33
      instructor challenges, 17
      instructor ramifications, 31
      learning styles and learner characteristics, 34–35
      self-regulated learning, 35
   engagement in the learning process, 116
   generating and maintaining interest, 129–130
   instructor obligations to, 16
   knowledge level of, 57
   power access in the classroom, 94–95
   priorities, 17
Student evaluation instruments, 269–280
   essay, 275–277
   interpretive exercises, 277
   matching, 273–274
   multiple-choice, 270–272
   objective test item, 270
   oral tests, 277
   performance (skills) tests, 277–280
   short-answer/completion, 275
   skills for developing test items, 288
   subjective test item, 270
   true-false, 272–273
   written tests, 270–277
Student interaction, 141–158
   classroom issues, 151–156
      disruptive, nonparticipating students, 153–154
      disruptive behavior, reasons for, 154–156
      L-E-A-S-T method, 152
      nondisruptive, nonparticipating students, 152–153
   disciplinary action, 156–157
   individual student needs, 141–146
      diversity, 141–142
      high learning abilities, 146
      low literacy levels, 142
      special needs, 142–146
   student behavior management, 147–151
      coaching, 149–150
      counseling, 148–149
      mentoring, 151
      motivating and encouraging students, 150
      peer assistance, 150–151
      policy review, 147–148

   students' rights, 146–147
Student-caused disruptive behavior, 155–156
Student-centered learning, 131
Study guides, 270
Study sheet, 231, 241
Subjective test item, 270
Suggested practices for research, 224
Summary of reports, 208
Summative (comprehensive) test, 187, 197, 282
Summative assessments, 353
Summative evaluation, 355, 363
Supervision of other instructors, 293–297
   empowering instructors, 295–297
      celebrating instructor accomplishments, 295
      conflict resolution, 295, 297
      positive examples/attitudes, 297
      recognizing quality performance, 295, 296
   goals and objectives, 294
   knowledge skills needed, 293–294
   professional development, 294–295
Supervision of training evolutions, 249–261
   accident investigation, 259
   environmental issues, 257–258
      atmosphere, 258
      soil, 258
      water, 257
   IMS to supervise training, 254–257
      duties and functions, 254
      evaluation of training evolutions, 255–257
      training plan or incident action plan, 254–255
   multiple instructors, 261
   safety challenge, 249–254
      hazard and risk analysis, 253–254
      organizational and administrative support, 249–251
      unsafe behavior, 251–253
   supervisory instructor evaluations, 321–330
      course evaluations, 324–328, 330
      findings from evaluations, 328
      processes, 322–324, 329
      tools, 321–322
Supervisor surveys, 357–358
Supplementary texts as lesson outline resources, 68
Survey, 336, 357–358
Synthesis (create) level of learning, 221, 222, 283

T

Tabletop miniatures, 74
Tabletop simulation, 127
Take-home assignments, 74
Talkative students, 153–154
Teaching aids, 71–81
   audiovisual, 71, 93–94
   effective use of, 71, 72
   nonprojected aids, 72–75

class handouts, 73–74
duplicated materials, 73–74
easel pads, 72
illustration or diagram displays, 73
marker boards, 72–73
models, 74–75
recordings, 75
Teaching desire as instructor characteristic, 12
Teaching strategies, 128–135
blended or hybrid learning, 131
computer simulation, 131–133
flipping the classroom, 131
individualized instruction, 134–135
instructor-led training (ILT), 128–131
features, 128–129
generating and maintaining student interest, 129–130
multiple instructors, 129
positive/negative reinforcement, 131
reinforcing learning, 130
self-directed learning, 71, 133–134
student-centered learning, 131
Team teaching, 129
Technology tools, 231–232
Teleconferencing, 133, 136
Telephone access in the classroom, 95
Television
cable access in the classroom, 95
interactive television (ITV), 133, 134
as teaching aid, 78–79
Temperature of the classroom, 92–93
Test item
affective domain, 280
analysis, 343–344
construction. *See* Test item construction
defined, 287
learning objectives, 265
Test item construction, 265–289
analysis, 343–344
student evaluation instruments, 269–280
essay, 275–277
interpretive exercises, 277
matching, 273–274
multiple-choice, 270–272
objective test item, 270
oral tests, 277
performance (skills) tests, 277–280
short-answer/completion, 275
skills for developing test items, 288
subjective test item, 270
true-false, 272–273
written tests, 270–277
test considerations, 265–269
Donald Kirkpatrick's Four Levels of Evaluation, 266
formatting and item arrangement, 267–268
instructions and time requirements, 268–269
level of cognition and difficulty, 268
test bias, 269
test planning, 281–285
appropriate test items, 283–285
avoiding giving clues to test answers, 284
cognitive levels of learning, 282–283
language and comprehension barriers, 284
learning objectives, 282
number of test items, 283–284
purpose and classification, 281–282
test item difficulty, 283
test usability, 284–285
validity and reliability, 285
test scoring method selection, 285–286
Testing and evaluation, 185–201. *See also* Evaluation
bell curve, 186
criterion-referenced assessments
defined, 197
purpose of, 185, 186–187, 281
validity and reliability, 342
feedback, 196, 201
four-step method of instruction, 113, 114
grade reporting, 195, 200
norm-referenced assessments
defined, 197
purpose of, 185, 186, 281
records, 206, 311
reports, 207
scoring, 192–195
grading bias, 194–195
grading tests, 194
oral tests, 194
performance tests, 194, 199
scoring rubric, 188
written tests, 192–194
security, 207
skills checklist, 186
student assessment, 185–187
test administration, 190–192
oral tests, 198
performance tests, 191–192, 198
written tests, 190–191, 198
test bias, 189
test classifications, 187–189
administration classification, 188–189
formative tests, 187, 197, 282
oral tests, 188
performance tests, 189
prescriptive tests, 187, 197, 282
purpose classification, 187
summative tests, 187, 197, 282
written tests, 188–189

test scoring, 192–195
   oral tests, 199
   performance tests, 199
   scoring rubric, 188
   security, 313
   selection of method, 285–286
   written tests, 199
test security, 195–196
Testing instruments
   bias, 269
   defined, 287
   reliability, 342, 345
   validity, 342, 345
Test-result analysis, 342–343
Theater seating arrangement, 90
Thorndike, Edward L., 27–28
Timekeeper of discussions, 120
Timid students, 153
Title slide, 231
Title VII of the Civil Rights Act, 19
Tools and equipment, actual equipment for demonstrations, 75. *See also* Equipment; Materials and equipment, instructional
Touch as nonverbal communication, 40
Traditional seating arrangement, 90
Training
   defined, 11, 22
   education, 11, 22
   entities provided with, 11
   environmental issues, 257–258
   funding and resources, 18
   hands-on, 30
   IMS evaluation, 255–257
   NFPA 1403 standard for live fire training evolutions. *See* NFPA 1403, *Standard on Live Fire Training Evolutions*
   NIST study on fuel pallet size, 172
   priority levels, 298
   safety, 249–254
      hazard and risk analysis, 253–254
      organizational and administrative support, 249–251
      unsafe behavior, 251–253
   simple, 168–170
   supervision. *See* Supervision of training evolutions
   weather variations and, 55
Training aids
   class size and interaction, 52
   cost of, 52
   defined, 51, 59
   learning objectives, 52
   learning pace, 53
   lesson content, 52
   manikins, 51, 52
   practice factors, 53
   selection, 51–53
Training centers, 164, 382–383
Training evolutions
   control of, 168
   simple training evolutions, 168–170
   standards. *See* NFPA 1403, *Standard on Live Fire Training Evolutions*
Training ground environment, 96–101
   facilities included, 96
   incident action plan safety analysis, 96, 97
   permanent facilities, 100–101
   remote sites, 96–100
      access/egress, 100
      environment considerations, 99
      examples, 97–98
      lighting, 100
      live-fire training, 98–99
      noise, 100
      vehicle traffic, 99–100
      weather conditions, 99
Training needs, identifying
   cost/benefit analysis, 338
   gap analysis, 337
   job performance requirements, 336–337
   needs analysis, 335–336, 346
   overview, 334
Training program administration, 371–396
   AHJ-level needs analysis, 374–375
   budget needs, 381–383, 393
   disclosure of information, 374
   equipment purchasing, 383–387
   human resource management, 388–389, 395
   personnel evaluations, 389, 396
   policies, procedures, and guidelines, 375–379, 392
   record-keeping, 371–373, 391
   standards influencing training, 380
Training program evaluation, 353–367
   methodologies, 353–354
   plans, 356–360
      assessing achievements or outcomes, 357
      course and instructional design evaluations, 359–360
      creating, 364, 365
      creation of, 364–365
      NFA evaluation plan, 356–357
      observation of instructors, 357
      organizational evaluations, 358
      performance measurement, 358–359
      supervisor surveys, 357–358
   results and recommendations, 360–362, 366, 367
   types and categories, 354–355
      formative evaluation, 354
      goals-based evaluation, 355
      outcomes-based evaluation, 355

process-based evaluation, 355
summative evaluation, 355
Training props
aircraft, 177
benefits of use, 82–83
dumpsters, 177
examples, 80–81
exterior and wildland fires, 176–177
inspecting and repairing, 166–167
subjects taught with, 80
Training records
attendance records, 311
certification records, 311
incident/injury records, 311
legal requirements, 312–313
Open Records Act, 313
privacy, 312–313
public access, 313
retention length, 312
maintenance and security of, 317
management of, 310–313
auditing procedures, 312
legal requirements, 312–313
record management systems, 311–312
training information, 310–311
types of training records, 311
needs analysis, 336
NFPA 1401 standards, 310
safety, 249–254
test records, 311
training schedules, 311
types, 205, 206–207
Training reports, 207
Training resources
instructor challenges, 18
needs analysis, 316
scheduling, 297
Transitions, 110–112
animations, 112
media, 112
nonverbal, 112
purpose of, 111
verbal, 111–112
Traumatic brain injury as a disability, 145
Trench operations
regulations, 20
training props, 80, 81
Trends, 224
True-false test item, 272–273
29 *CFR* 1910.120, Hazardous Waste Operations and Emergency Response, 20
29 *CFR* 1926.651, Specific Excavation Requirements, 20

U

U shape seating arrangement, 90, 91
UL (Underwriters Laboratories Inc.), 163
ULC (Underwriters Laboratories of Canada), 163
Understand (comprehension) level of learning
description, 220, 282
examples, 221
expressive words, 222
Understanding as listening skill, 41
Underwriters Laboratories Inc. (UL), 163
Underwriters Laboratories of Canada (ULC), 163
Uninterested students, 153
United States Department of Transportation (DOT). *See* Department of Transportation (DOT)
United States Fire Administration (USFA)
casualty and injury statistics, 249
grants, 305
higher education for adults, 25
purpose of, 162
safety challenge of training evolutions, 249–250
Urban/wildland interface, 182
USB drive presentation aids, 76
User-generated video websites, lesson outline resources, 68
USFA. *See* United States Fire Administration (USFA)

V

Validity of tests, 285, 342, 345
Vehicle
apparatus operations, indoor facilities for, 93
driving course, 101, 102
incident props, 80, 81
traffic at training grounds, 99–100
Vendors, 309
Ventilation in the classroom, 92–93
Verbal communication, 38
Verbal transitions, 111–112
Verbalization, 125
Vicarious liability, 179, 182
Video cassette recorders (VCRs), 76
Video conferencing, 79
Video presentations, 78–79
Video recording of presentation to increase instructional effectiveness, 109
Virtual reality simulations, 79–80
Visual distractions in the classroom, 95
Visual impairment as a disability, 145
Visual presenters, 78
Visualization, 125
Vocal characteristics
effective instructor characteristics, 108
nonverbal communications, 39
Vocal inferences in nonverbal communications, 39
Volunteer organization training scheduling, 300

W

Water
- environmental issues at training evolutions, 257
- NFPA 1142 requirements, 175
- supply at acquired structures, 175–176
- swift water rescue, 178

WCAG (Web Content Accessibility Guidelines), 76
Weapons of mass destruction (WMD), 162, 380
Weather
- acquired structures and, 174
- scheduling training sessions, 298, 300
- training activities and, 55
- training ground considerations, 99

Web conferencing, 133, 136
Web Content Accessibility Guidelines (WCAG), 76
Websites. *See also* Internet; *specific computer entries*
- fire and emergency services, 68
- user-generated video websites, 68
- Web Content Accessibility Guidelines (WCAG), 76

Wiki, 133, 136
Wildland fires
- defined, 182
- live-fire training, 176–177

Wildland/urban interface, 177, 182
Wireless networks (Wi-Fi), 95
WMD (weapons of mass destruction), 162, 380
Wording used in presentations, 108
Workbooks, 270
Workforce Investment Act of 1998, 142
Worksheets, 230, 240
Written tests, 270–277
- administration, 190–191, 198
- administration classification, 188–189
- essay, 275–277
- illustrated, 270
- interpretive exercises, 277
- matching, 273–274
- multiple-choice, 270–272
- objective test item, 270
- scoring, 192–194, 199
- short-answer/completion, 275
- subjective test item, 270
- true-false, 272–273

Index by Nancy Kopper